Polymer-Based Additive Manufacturing

Declan M. Devine
Editor

Polymer-Based Additive Manufacturing

Biomedical Applications

Springer

Editor
Declan M. Devine
Athlone Institute of Technology
Athlone, Co. Westmeath, Ireland

ISBN 978-3-030-24531-3 ISBN 978-3-030-24532-0 (eBook)
https://doi.org/10.1007/978-3-030-24532-0

This Springer imprint is published by the registered company Springer Nature Switzerland AG
The registered company address is: Gewerbestrasse 11, 6330 Cham, Switzerland

To my parents, Peggy and Corny (RIP)
To my Princesses Jessica, Livia, Eloise and Madelyn

Preface

Additive manufacturing technologies, also referred to as 3D printing among other names, were traditionally used as a prototyping tool for form and function during the product development cycle. However, recently, there has been a significant surge in the use of 3D printing technologies for the production of a variety of end products not least so in the biomedical field. This is due to many reasons including patent lapsing of the original designs, which has led to an increase in the number of companies developing machines which in turn has led to rapid improvements in machine resolution along with the reduction in machine costs. 3D printing machines now cost a few hundred dollars (or Euros), and this has proved transformative as this technology is accessible to the general public as well as major corporations. This transformation promises to change manufacturing, from mass production to personally customised products, and business models of central output with extensive supply chains to locally manufactured and supplied products that can be manufactured in the home, office or hospital locations.

3D printing uses a wide range of materials including clay, metals, polymer and composite materials to build complex parts which may not be possible using traditional fabrication techniques in a layer-by-layer fashion. However, this book will concentrate on the use of polymer-based additive manufacturing processes as these processes were the first to be developed and as such are the best understood processes with a range of feedstocks available for the production of the final parts. This versatility, combined with a low machine, processing and feedstock costs, has enabled polymer-based additive manufacturing to enter a range of biomedical applications such as orthodontic braces and hearing aids. As research continues in this field, coupled with the growing need for complex treatments for modern ailments, the use of additive manufacturing to produce patient-specific implants in terms of shape and composition is set to continue to grow.

This book aims to address all aspects of product development for polymer-based additive manufacturing. Chapter 1 of this book discusses the history and main polymer-based additive manufacturing techniques and special material considerations which need to be considered when selecting a particular process/ material combination. While additive manufacturing enables unparalleled design freedom, Chap. 2 discusses factors in the process of designing for additive manufacturing. Chapter 3 examines the modelling of additive manufactured polymers including shape memory polymers. The subsequent chapters discuss practical applications of additive manufacturing with relevance to the biomedical industry. Chapter 4 discusses the application of 3D printing of tooling for the production of prototype tools for use in mass production techniques commonly used in the manufacture of medical devices. Chapter 5 discusses current applications of additive manufacturing in biomedical applications, while Chap. 6 continues this *with reference to* orthopaedic applications. *Chapters* 8 and 9 *are focused* on bioprinting hardware and *bioinks* and applications in tissue engineering. To complete the production cycle, Chap. 10 concentrates on post-processing considerations. Chapter 11 focuses on regularity considerations for devices produced using additive manufacturing, while Chap. 12 attempts to forecast the future directions of 3D printing in the biomedical field. While forecasting is an inaccurate science, it is almost *certain* that 3D printing will change the face of healthcare and medical device manufacturing in ways not yet imagined.

Athlone, Co. Westmeath, Ireland Declan M. Devine

Contents

Contributors

Farah Alwani Azaman Athlone Institute of Technology, Co. Westmeath, Ireland

Margaret E. Brennan Fournet Athlone Institute of Technology, Co. Westmeath, Ireland

Ross Burdis Trinity Centre for Bioengineering, Trinity Biomedical Sciences Institute, Trinity College Dublin, Dublin, Ireland

Department of Mechanical and Manufacturing Engineering, School of Engineering, Trinity College Dublin, Dublin, Ireland

Laura Ruiz Cantu Faculty of Engineering, University of Nottingham, Nottingham, UK

Yuan Yuan Chen Athlone Institute of Technology, Co. Westmeath, Ireland

Maurice Dalton Materials Research Institute, Athlone Institute of Technology, Athlone, Ireland

Declan M. Devine Materials Research Institute, Athlone Institute of Technology, Athlone, Co. Westmeath, Ireland

Aleksandra Foerster Faculty of Engineering, University of Nottingham, Nottingham, UK

Evert A. Fuenmayor Materials Research Institute, Athlone Institute of Technology, Athlone, Co. Westmeath, Ireland

Noel M. Gately Materials Research Institute, Athlone Institute of Technology, Athlone, Ireland

Suzan Gunbay Materials Research Institute, Athlone Institute of Technology, Athlone, Co. Westmeath, Ireland

Katie Hardiman Irish Manufacturing Research, Mullingar, Ireland

Conor Hayes Materials Research Institute, Athlone Institute of Technology, Athlone, Ireland

Andrew V. Healy Materials Research Institute, Athlone Institute of Technology, Athlone, Co. Westmeath, Ireland

Eoin P. Hinchy Confirm Centre Smart Manufacturing, School of Engineering, University of Limerick, Limerick, Ireland

Michael Hopkins Jr Materials Research Institute, Athlone Institute of Technology, Athlone, Ireland

Guang Hu Materials Research Institute, Athlone Institute of Technology, Athlone, Co. Westmeath, Ireland

Daniel J. Kelly Trinity Centre for Bioengineering, Trinity Biomedical Sciences Institute, Trinity College Dublin, Dublin, Ireland

Department of Mechanical and Manufacturing Engineering, School of Engineering, Trinity College Dublin, Dublin, Ireland

Department of Anatomy, Royal College of Surgeons in Ireland, Dublin, Ireland

Advanced Materials and Bioengineering Research Centre (AMBER), Royal College of Surgeons in Ireland and Trinity College Dublin, Dublin, Ireland

John G. Lyons Athlone Institute of Technology, Athlone, Co. Westmeath, Ireland

Ian Major Materials Research Institute, Athlone Institute of Technology, Athlone, Co. Westmeath, Ireland

Yiqi Mao School of Mechanical and Vehicle Engineering, Hunan University, Changsha, China

John McDonald Teleflex Medical Euorpe Ltd, Athlone, Co. Westmeath, Ireland

Martin Neff Arburg GmbH + Co KG, Loßburg, Germany

Jessica Nulty Trinity Centre for Bioengineering, Trinity Biomedical Sciences Institute, Trinity College Dublin, Dublin, Ireland

Department of Mechanical and Manufacturing Engineering, School of Engineering, Trinity College Dublin, Dublin, Ireland

Steven Pollack Carbon, Inc., Redwood City, CA, USA

Rossana Schipani Trinity Centre for Bioengineering, Trinity Biomedical Sciences Institute, Trinity College Dublin, Dublin, Ireland

Department of Mechanical and Manufacturing Engineering, School of Engineering, Trinity College Dublin, Dublin, Ireland

Xin Shu Department of Mechanical Engineering, State University of New York at Binghamton, Binghamton, NY, USA

School of Mechanical and Vehicle Engineering, Hunan University, Changsha, China

Christopher Tuck Faculty of Engineering, University of Nottingham, Nottingham, UK

Chaitra Venkatesh Athlone Institute of Technology, Co. Westmeath, Ireland

Ricky Wildman Faculty of Engineering, University of Nottingham, Nottingham, UK

Pu Zhang Department of Mechanical Engineering, State University of New York at Binghamton, Binghamton, NY, USA

Chapter 1
Polymer-Based Additive Manufacturing: Historical Developments, Process Types and Material Considerations

Steven Pollack, Chaitra Venkatesh, Martin Neff, Andrew V. Healy, Guang Hu, Evert A. Fuenmayor, John G. Lyons, Ian Major, and Declan M. Devine

Abstract 3D printing is a manufacturing technique where parts are built in a layer-by-layer fashion. The earliest embodiments were based on generating successive layers of solidified materials at the air liquid interface through exposure with actinic radiation and the lowering of the growing object into a vat of photoactive material. One of the earliest concepts was by Otto Munz in 1956 where he describes "photo-glyph recording". However, many today consider Charles Hull the father of 3D printing as it was his patent that coined the term "stereolithography" and was the first to commercialize a 3D printing apparatus.

Since then, there has being a multitude of technologies developed which fall into the category of 3D printing. While photochemical resin curing remains one of the most widely utilised 3D printing techniques, other techniques based on the extrusion of polymers, cutting sheets or films and thermally or chemically binding powdered materials have all reached the commercial markets. This chapter aims to give an outline of the most common polymer-based 3D printing techniques available today and describes some of the material and processing constraints which should be considered when selecting these systems for 3D printing of biomedical parts.

Keywords 3D printing · Polymers · Stereolithography · Fused filament fabrication · Selective laser sintering · PolyJet · InkJet · Freeformer · Laminated object manufacturing

---

S. Pollack
Carbon, Inc., Redwood City, CA, USA

C. Venkatesh · A. V. Healy · G. Hu · E. A. Fuenmayor · J. G. Lyons
I. Major · D. M. Devine (✉)
Athlone Institute of Technology, Co. Westmeath, Ireland
e-mail: ddevine@ait.ie

M. Neff
Arburg GmbH + Co KG, Loßburg, Germany

D. M. Devine (ed.), *Polymer-Based Additive Manufacturing*,
https://doi.org/10.1007/978-3-030-24532-0_1

1.1 Introduction

Three-dimensional (3D) printing, also referred to as additive manufacturing (AM), rapid prototyping (RP), or solid freeform fabrication, is a process which involves the manufacturing of an object or structure by deposition or binding of materials layer-by-layer (Sachs et al. 1990; Conner et al. 2014; Gross et al. 2014; Prasad and Smyth 2016). A recent publication by Norman et al. (2016) indicates that the American Society of Mechanical Engineers (ASTM) favour the term additive manufacturing; however, these terms are routinely used interchangeably as is the case in this body of work.

The origins of 3D printing date back to 1986 when Charles Hull developed, patented and commercialised the first 3D printer. The technique was called stereolithography apparatus-1 (SLA-1) and was capable of manufacturing a 3D-printed object without the need for moulds or other tooling (Prasad and Smyth 2016). Further to this work, Hull also developed the STL (Standard Tessellation Language) file format. Today the STL file format is seen as the gold standard for communication of data between the computer-aided design (CAD) software and 3D printers, and it is in this that all the information corresponding to each surface of the 3D model is stored in the form of triangulated sections (Gross et al. 2014)

There has been a multitude of 3D printing methods developed since the patenting of Hull's SLA-1. These methods can be separated depending on the way of layer deposition and the material feedstock. However, regardless of the 3D printing technique employed, they all share similarities in the method which they create an object, as the model must be first designed and generated through the use of computer-aided design (CAD) software, e.g. SolidWorks, AutoCAD and Creo Parametric, which is then converted into a STL file and transferred to the specific additive manufacturing device for fabrication. This STL file is cut into "slices" with each of them containing the information required for each layer of the part. A platform is created to support any overhanging structures, and the part is built one slice at a time.

Additive manufacturing is a simple, automated process which is fast and inexpensive (Hiemenz 2011). The fundamental layer-by-layer manufacturing mechanisms of additive manufacturing provides a wide range of advantages over conventional manufacturing processes. These include design freedom and superior levels of design complexity with fewer or single-step manufacturing required, reduced or elimination of post-production finishing, reduced or elimination of tooling and fixtures, shorter cycle times for both designs and processing, multiple material manufacturing, savings in energy and start-up cost (Gao et al. 2015) and reduction or elimination of post-production assembly.

1.2 Stereolithography (SLA)

The concept of creating a three-dimensional object from a liquid precursor through photochemistry has a long history (Wohlers 2005). The earliest embodiments were based on generating successive layers of solidified materials at the air liquid interface through exposure with actinic radiation and the lowering of the growing object into a vat of photoactive material. One of the earliest concept was by Otto Munz in 1956 where he describes "photo-glyph recording" (Munz 1956). In 1981, Hideo Kodama of Nagoya Municipal Industrial Research Institute described the use of UV exposure controlled by a mask pattern or via a scanning fibre transmitter (Kodama 1981) to create a 3D object. In 1984, Jean Claude André, Alain Le Mehauté and Olivier de Witte filed their patent application (Andre et al. 1984) for this process (their patent was ultimately abandoned). Around the same time, Charles Hull filed his patent (Hull and Arcadia 1986), coining the term "stereolithography", and was the first to commercialize a 3D printing apparatus. A variety of photochemically reactive polymers have been used including free radical, cationic and thiol-ene chemistry (Ligon et al. 2017).

The original commercial SLA consisted of a photopolymer reservoir, or vat, which is underneath a movable platform. At each stage of the "build", the platform, or in latter stages, the most recently created layer, is lowered just below the surface of the resin in order to create a thin layer, and the laser beam passes back and forth over the liquid solidifying it onto the prior layer as it passes (Fig. 1.1). More recently, some manufacturers have utilized digital light processors (Hornbeck 1997) to create an image of the layer in UV light to cure the entire layer at once. This leverages the technology for high-resolution projectors used in the movie industry and developed by Texas Instruments.[1] Others such as UNIZ and Photocentric have

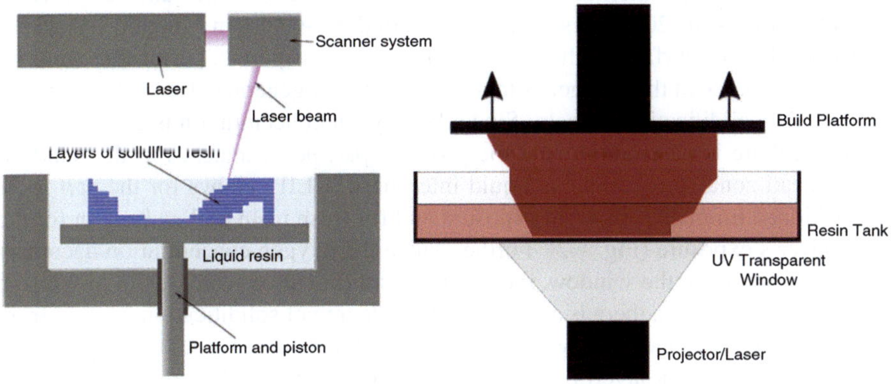

Fig. 1.1 (Left) Conventional SLA utilizes a vat of resin which is lowered as each layer is cured. (Right) Inverted SLA utilizes a UV transparent window through which the UV light penetrates from underneath, with the platform and the build plate moving upwards after each pass

[1] Retrieved January 22, 2019, from http://www.ti.com/dlp-chip/overview.html.

developed liquid crystal arrays which act as "shutters" create the image of the cross section (sometimes referred to as LCD-SLA).

Frequently, a wiper is drawn across the topmost layer to ensure a uniform liquid film for the next layer. Upon completion of the print, the part is pushed up out of the unreacted resin in the vat, and any sacrificial supports are removed, and the parts are post-cured (a final, overall dose of UV) in order to remove any unreacted resin from the printing process. One consequence of this geometric arrangement, is that the depth of the resin vat must be at least as deep as the part being built is tall and the vat must be filled. Other manufacturers have developed their own SLA systems including EnvisionTEC and RPS. Voit (Voit et al. 2016) has described a concept utilizing a dense "Z fluid" upon which the active photopolymer floats so as to minimize the amount of photopolymer used.

Another approach has been to have the resin reservoir with a UV transparent window with the UV light penetrating from underneath the platform and the build plate moving upwards after each pass (sometimes referred to as inverted SLA) (Ligon et al. 2017). This greatly reduces the quantity of photoactive resin used during the build. However, a consequence of this geometry is that the resin will cure and span the full gap between the build plate and the window at the bottom of the reservoir. This requires a physical delamination of the growing part from the window at each step. In addition to slowing down the process, there are undesirable mechanical forces on the growing part and the reservoir window. The former can cause decreased accuracy, and the latter may require frequent replacement of the reservoir/window. Nevertheless, companies like Formlabs and EnvisionTEC have been very successful in marketing their systems for a variety of applications including dental and medical device applications.

In all the above-mentioned systems, there is full solidification of the growing part, either at the liquid/air interface or at the liquid/window interface, leading to inherently layered internal structures. Recently, DeSimone and co-workers (Tumbleston et al. 2015) described an inverted SLA system wherein oxygen is introduced at the surface of the window through a transparent, permeable window. The consequence of this oxygen is to deactivate photo-generated free radicals which initiate the solidification process. Since the oxygen concentration is greatest at the window, there is no reaction, and the growing part never attaches to the window. This "dead zone" or continuous liquid interphase (CLIP) allows for the part to be repositioned immediate after an exposure and for fresh resin to flow back in for the next layer's exposure (Fig. 1.2). Further, since the oxygen concentration decreases with distance from the window, the extent of solidification increases in proportion to that distance. Thus, there is a gradient in the degree of solidification moving from the growth front to the already solidified object. Each subsequent layers then interdigitate with the prior layer(s), creating a more monolithic object with substantially less mechanical anisotropy. In theory, other reactive systems could be similarly deactivated by introduction of a suitable agent at the window/resin interface (DeSimone et al. 2015). The company Carbon, Inc. was formed in 2013 based on this technology and the concept of dual cure resins (Rolland et al. 2016).

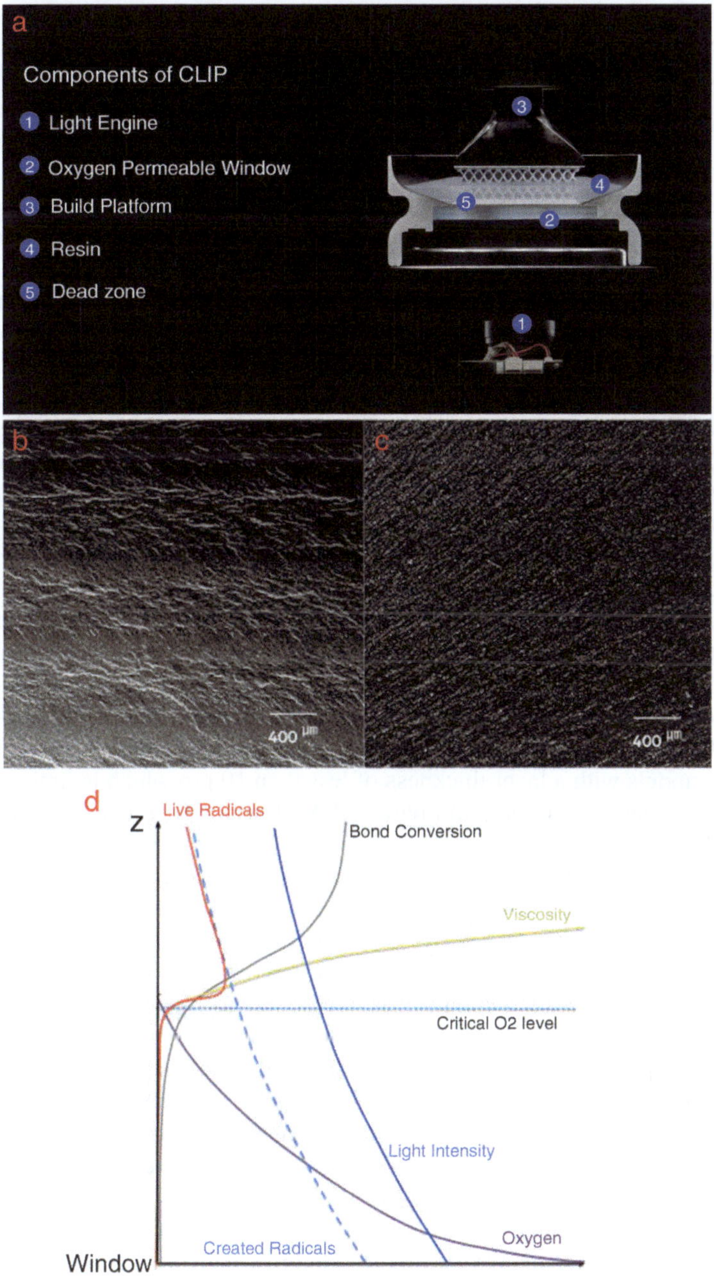

Fig. 1.2 (**a**) Continuous liquid interfase production (CLIP) where (1) light engine, (2) oxygen permeable window, (3) build platform, (4) resin, (5) dead zone. (**b**) Conventional SLA fracture surface. (**c**) CLIP-based print. (**d**) Lower axis generalized concentration; vertical axis is distance from the window. Light intensity/radical generation decreases with distance due to molar absorptivity. Oxygen concentration decreases due to diffusivity and reaction with primary radicals. Below the critical oxygen concentration (top of the "dead zone"), polymerization is possible and increases with distance from the window. Graphic courtesy of Matthew Panzer, Carbon, Inc

As the SLA is the oldest technology, it is also the most advanced and is capable of higher spatial resolutions compared to other 3D printing methods. In conventional and inverted SLA, there is a trade-off between *xy* resolution and build cross section. In laser-based systems, resolution is dictated by the diameter of the focused laser (beam waist) and the angular resolution of the beam steering mirrors (galvanometer). Across various manufacturers and form factors, resolution can range from a 75 to 250 μm spot size. The image is "rastered" across the build plane to solidify the desired areas. Alternatively, the entire image can be projected using a large planar UV source and a digital light processors (DLP): an array of micro-mirrors which can illuminate (or suppress) each "pixel" simultaneously. For the commercially available digital light processors (DLP), these mirrors have a pitch between 8 and 14 μm. The resolution at the image plane (where the layer is solidified) is controlled by the distance of the DLP chip to the window and the imaging optics. Resolution is inversely proportional to build size. Commercial DLP-based SLA systems have resolutions ranging from 35 to 250 μm. Non-commercial systems have been described with pixel sizes down to 10 μm, but with proportionally smaller build cross sections.

Resolution in the "Z" or build direction is limited by the mechanical stability/reproducibility of the movement of the build platform and control of the photopolymerization process. Unfocussed light can still initiate photopolymerization and cause unwanted solidification, so light-absorbing dyes and/or reactive groups in the resin serve to limit polymerization. Build direction resolution can range from 100 μm down to a few microns. Build speed decreases with increase in Z resolution. For highly specialized applications, micro-stereolithography is now capable of building models with a layer thickness of less than 10 μm which is limited only by the width of the concentrated ultraviolet (UV) laser or DLP resolution (Melchels et al. 2010).

Part size limitations are related to the particular technology. For conventional SLA, the physical limitations are the volume of the build vat and the optical distortions of the focused laser as it is steered to larger angles, causing the beam waist to take on a more elliptical geometry, which can make resolution vary as a function of *xy* position. Prodways Technologies has overcome this by utilizing a DLP coupled to an *xy* stage to move the projected image across the top of the vat. The entire "layer" is projected in sections before moving to the next layer. For inverted SLA, the mechanical properties of the window become the limiting issue. Hydrodynamic (suction) forces between the growing part and the window can cause the window to distort, slowing down the print process to allow it to mechanically recover after each print step. This problem increases as a function of window cross section. Also, in inverted SLA, as the window size increases, the overall stiffness of the printer needs to increase to be able to handle the increasing forces generated between the window and the build plate. Finally, for inverted SLA, maintaining constant resolution with increasing cross-sectional area requires either higher resolution DLPs and/or multiple "tiled" DLPs. In principle, LC shutter-based systems are only limited by the technologies for fabricating larger arrays.

One of the limiting factors in the more widespread use of SLA-based technologies for medical device fabrication is the limited range of material properties and biocompatibility. Photopolymerized materials are limited in the range of physical properties they can exhibit, as they are generally non-crystalline and derived from low-to-moderate molecular weight building blocks. As an approach to this, Carbon has developed "dual-cure resins" (Rolland et al. 2016). These use photochemistry to generate the initial 3D object having low "green" strength. Then, a latent heat-activated thermosetting reaction is initiated by baking the part for 4–12 h, dependent on the thermosetting chemistry and final part physical properties. Currently, polyurethane, epoxy and cyanate ester chemistries have been incorporated in Carbon dual-cure resins, leading to material properties equivalent to those from thermoplastic elastomers to high-temperature engineering resins.

In all, the polymers used must have suitable biocompatibility (Amato and Ezzell Jr 2015) and be able to withstand sterilization. This means at a minimum, they should be able to be tested successfully against the ISO 10993-5 (cytotoxicity) and ISO 10993-10 (irritation and sensitization) (*ISO 10993—Biological Evaluation of Medical Devices*, 2001–2018), to allow for extended skin contact and short-term (<24 h) mucosal contact. For other applications, tests such as systemic toxicity or genotoxicity of leachables will need to be assessed. While this data is not required by most regulatory agencies for the resins themselves, as they are generally interested in the behaviour of the finished device, this affords the end user a degree of security that the final device can safety utilize these materials and pass regulatory scrutiny.

These high resolutions and newer materials enable SLA to be used in biomedical applications (Clair 1996; Melchels et al. 2010). As such SLA printing has being used for the production of many biomedical parts. Some of the earliest applications of SLA have been in the dental space. Dentists and other oral care practitioners were some of first adopters of 3D SLA printing using the technology for the generation of dental models. While not formally medical devices, they are integral to developing solutions for dental restorations. Align Technology uses SLA-based 3D printing to generate patient-matched templates for the thermoforming fabrication of their Invisalign® clear aligners (McLure 2017). More recently, dentists and oral surgeons are using specialized software (e.g. 3Shape) and SLA-based 3D printing in the direct fabrication of drill guides to more precisely place dental implants. Lastly, Dentca developed the first FDA-approved resin for the 3D-printed fabrication of denture and denture teeth. Other denture resins have since been cleared by the FDA and other regulatory bodies, and a number of manufacturers are now using this approach.

Another early adopter of 3D SLA is the hearing aid industry. Most major hearing aid manufactures now fabricate the "in-the-ear" components using 3D printing, transforming a scanned image of a model or direct scan of the outer ear and ear canal. Companies like Starkey state the 98% of their hearing aids are created using 3D printing (Sharma 2013). Hard components can be fabricated directly using resins from several manufacturers (e.g. Dreve, Pro3Dure, Vertex) or, using sacrificial "egg shell" or "cocoon" moulds, soft components from thermally cured silicone resins. A related therapy is the Cerezen™ intra-aural device produced by Renew

Healthcare for the treatment of temporomandibular disorder (TMD) which places a device in the ear to restrict excessive motion of the joint ('Cerezen device provides treatment for temporomandibular joint disorders day and night' 2016).

Medical models are another important area for utilization of SLA. Having models fabricated from clear resins allow for the clinician and patient to visual the relative three-dimensional positions of organs and lesions. Using design software like Mimics (Materialise) and simple dyes, a colourless transparent resin can have specific organs and structure coloured to mark tumours, vessels and other points of interest.[2]

Further uptake by the medical device industry will be driven by greater familiarity with the materials and their acceptance by regulatory agencies. FDA's recent technology guidance for the use of additive manufacture bodes well for this future[3] (FDA's Role in 3D Printing n.d.). However, SLA printing is continually finding new applications in the biomedical field, as recent case study by students and faculty at UC Berkeley utilizing Carbon's high-temperature CE 221 cyanate ester resin for creating patient-matched MRI surface coils for restriction of patient motion and enhanced signal-to-noise ratio (Corea 2016; Zamarayeva et al. 2018).

Looking forward, there are a number of academic laboratories beginning to develop bioresorbable 3D-printed devices. A number of groups have demonstrated tissue scaffolds for cartilage regeneration using 3D-printed poly(propylene fumarate) (Luo et al. 2016; Parry et al. 2017). In the cardiovascular field, Ameer and Chen (van Lith et al. 2016) have constructed an inverted SLA printer using Carbon "CLIP" technology and fabricated cardiovascular stents from a novel citrate polyester. Combining novel polymers with these new high-resolution SLA print technologies will lead to new therapies and better outcomes for patients.

1.3 Fused Filament Fabrication

Fused filament fabrication (FFF), also known under the trademark termed fused deposition modelling (FDM), is an extrusion-based process which is one of the most widely utilised 3D printing technologies today (Fig. 1.3). It was developed and patented by Scott Crump of Stratasys Inc. in 1992 (Masood and Song 2004). FFF is a widely implemented method for 3D printing of solid objects (Skowyra et al. 2015). With the invention of FFF, it became possible to not only print functional prototypes but also to print concept models and end-user products (McLouth et al. 2017)

The FFF process fabricates 3D models by extruding materials through a heated print head that moves in *X* and *Y* directions onto a stage layer by layer (Gross et al. 2014). It uses thermoplastic materials which are produced in the form of a filament

[2] https://www.materialise.com/en/blog/how-to-3d-print-multicolored-models-formlabs-sla.

[3] https://www.fda.gov/MedicalDevices/ProductsandMedicalProcedures/3DPrintingofMedicalDevices/ucm500548.htm.

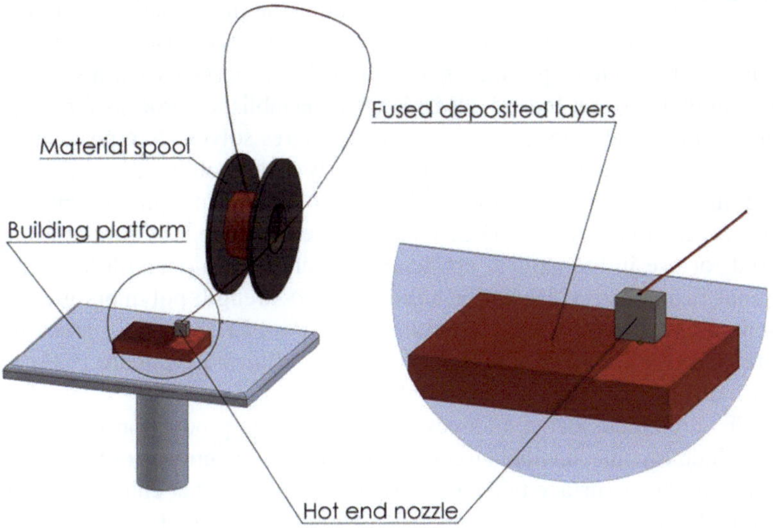

Fig. 1.3 Schematic of a FFF process. The filament is passed from a roll through a heated nozzle using a stepper motor where it is heated to a temperature just above its melting temperature before placement in ultra-thin layers onto a substrate

and then are extruded from movable head on the FFF machine and deposited in ultra-thin layers onto a substrate, translating the dimensions of digital data into an actual printed parts (Wang et al. 2017). The material is heated slightly above its melting point; this is to ensure it will quickly solidify after being deposited and welds to the former layer or printing bed. Modern domestic FFF printers consist of a heated nozzle which is fed filament by a stepper motor. A chamber before the nozzle melts the material and the filament pushed into this cavity forces out the necessary amount of molten material. Common materials used for FDM include polycarbonate (PC), acrylonitrile butadiene styrene (ABS), polyphenylsulfone (PPSF), PC-ABS blends, medical grade polycarbonate, wax, metals and even ceramics (Kruth et al. 1998). It is far from a perfect process as it leaves visible seam lines between layers; overhanging parts need supports to be printed below; build times are long; and the resolution is low compared to other techniques. The high relative viscosity of the thermoplastics limits the use of small diameter nozzles for FFF 3D printing with maximum resolution of 40 μm (Dermanaki Farahani and Dubé 2017). This may become a considerable challenge with using fillers/active agent with nozzle clogging occurring due to reduction in flow properties or degradation of one or more components within the filament (Ivanova et al. 2013).

However, parts do not require chemical post-processing which is highly desirable for biomedical applications, and the material used are cost-effective, and the equipment needed is considered to be simple-to-use and low-cost as compared to other 3D printing techniques (Sood et al. 2010; Bellini and Guceri 2003).

As FFF is an extrusion-based process, it can be highly versatile as in theory any thermoplastic polymer, thermoplastic composite or polymer-based drug delivery platform can be used to produce parts using this process (Kokkinis et al. 2015; Fuenmayor et al. 2018). Indeed, HME is a well-established process for the production of pharmaceutical dosage forms, which ensures solvent-free production, low-cost scale-up and enhanced solubility for poorly water-soluble drugs (Tiwari et al. 2016). It may also be used for the production of parts from high-strength polymers such as polyether ether ketone (PEEK) and polyetherimide. The former is currently approved for use in biomedical applications such as craniomaxillofacial implants and spinal fusion cages. However, although high-strength polymers are desirable for use in biomedical application, polymers such as PEEK require processing temperatures of up to 400 °C which is above the capabilities of standard FFF machines thereby increasing the cost of the overall system.

In FFF some technical challenges may prevent the production of usable parts. These include low mechanical strength compared to injection moulded parts, long build time and poor surface finish (Zaharin et al. 2018). Other challenges surrounding FFF include poor layer adhesion and layer delamination of the fabricated materials. In addition mechanical properties can also result from the FFF process generating void spaces between deposition lines, which adversely effects properties such as stiffness and strength (Ferreira et al. 2017). These can be addressed by the post-processing methods such as microwave treatment, vapour smoothing, metal plating, cold welding, gap filling, polishing, dipping, the use of resin to fill the voids, optimisation of the process parameters and structural optimisation (Dizon et al. 2018; Yao et al. 2017). Li and researchers demonstrated the improvement in mechanical properties of FFF-printed parts following ultrasonic treatment. These results were attributed to ultrasonic vibration causing re-entanglement and deeper interdiffusion, reducing gaps and increasing interfacial area and strength at the raster interface (Li et al. 2018).

The infill density and the printing pattern significantly influence the material properties of the printed samples. The effect of these parameters were studied on the tensile, dynamic mechanical and thermoelectric properties of FFF-printed conductive acrylonitrile butadiene styrene/zinc oxide (CABS/ZnO) in comparison with (ABS/ZnO) (Yun et al. 2018). CABS is a mixture of acrylonitrile butadiene styrene (ABS) and carbon black residue. There was a positive impact on the electrical conductivity and thermal conductivity. However, no change was observed in tensile strength of the nanocomposite due to infill density. Samples printed in line pattern had better tensile properties with high stiffness and storage modulus, while electrical conductivity also improved with line samples when compared to rectilinear samples (Yun et al. 2018). Additionally, Dul et al. demonstrated that the reinforcement effect of the graphene nanoplates in the ABS matrix was optimum at vertical and horizontal orientation when compared to perpendicular orientation (Dul et al. 2016). Mohamed and researchers developed mathematical models for FFF process parameter optimisation using I-optimal design. Parameters such as layer thickness, air gap and build orientation show a significant effect on percentage change. This method is proposed to be used for part quality in computer-generated optimal design in additive manufacturing (Mohamed et al. 2016).

FFF may be employed in various medical applications to produce 3D parts with complex shapes because of its highly controllable channel size and porosity (Zaharin et al. 2018). A common indication where this could prove advantageous is in the 3D printing of stents. Much work has being carried out in this field with PLA or variations the polymer mostly investigated (Chen et al. 2016). While Coronary stents remain the holy grail of 3D-printed stents, easier success may be found elsewhere where pressures are lower and failure have less catastrophic effects. An example of this was demonstrated by Mills et al. who successfully fabricated nasal stents by FFF in porous or mesh form made of multiple polymers in layers (Mills et al. 2017).

One of the most growing area of recent research and development in 3D printing is nanotechnology. Nanotechnology can be seen in various disciplines of science and technology including the areas of polymer-based biomaterials, nanoparticle drug delivery, polymer blends, nanofibers and nanocomposites (Bhatt and Anbarasu 2017).

Nanocomposites are emerging potentially in various industries due to excellent mechanical, electrical and thermal properties. The nanomaterial can be dispersed into the host matrices by 3D printing to produce nanocomposite signifies numerous advantages due to its properties and improves the homogeneity of the product (Ngo et al. 2018). The high viscosity of the thermoplastics can be further increased by the addition of nanofillers which affects the printability of the nanocomposite filament. This can be addressed by selecting suitable mixing strategies while printing for good dispersion of the nanomaterial into the polymer matrix. Various other methods such as solution mixing by sonication and high shear mixing by extrusion can be employed to reinforce the nanomaterial into the polymer matrix before 3D printing (Dermanaki Farahani and Dubé 2017).

3D-printed nanocomposites can be created using filament which incorporate nanoparticles. Shrinking the scale size to the nano-range can change the properties of the materials. A variety of nanomaterials such as carbon nanotubes, graphene and metal nanoparticles are used in 3D printing for various applications (Ivanova et al. 2013). Several researches have been conducted to study the physical properties of the final printed part when they are blended with nanomaterials. Recent studies show that in FFF the mechanical properties of the filament can be increased by addition of the nanofillers with researchers such as Healy et al. reporting that halloysite significantly increased Young's modulus 11% and 25% when the loading was two and six percent, respectively (Healy et al. 2018).

Nevertheless, it has also being reported that the incorporation of nanofillers can also have adverse effects on the properties of the printed parts. Dorigato et al. dispersed multiwalled carbon nanotubes into an ABS matrix and conducted a comparative study of the properties of compression-moulded and 3D-printed materials. While the stiffness and yield properties increased by the addition nanofillers, the 3D-printed samples resulted in loss of ductility and electrical conductivity due to the dependency on the printing direction (Dorigato et al. 2017). Additionally, proper care is necessary during fabrication of the nanocomposite for optimised dispersion of the particles within the matrix as nanoparticles such as carbon nanotubes (CNTs) tend to agglomerate due to their large aspect ratio, chemical structure and small scale (Kim et al. 2018). Functionalisation of the nanoparticles is one of the methods to overcome this issue (Chen et al. 2018). However, processing through screw extrusion using high shear rates may be enough to distribute the nanomaterial without chemical treatment.

Recycling FFF parts has also shown promise. Tian et al. studies recycling of 3D printing of carbon fibre reinforced PLA composites produced by FFF. It was found that the carbon fibre reinforcement enhanced the mechanical properties of PLA (Tian et al. 2016). Additionally, in a subsequent study, the recycling and remanufacturing of the carbon fibre which reinforced PLA composites from the previously 3D-printed carbon fibre filament was assessed. Although aging of the PLA matrix was observed, the recycled composites attained comparatively higher mechanical properties with respect to originally 3D-printed composite (Tian et al. 2017).

1.4 Selective Laser Sintering (SLS)

SLS was discovered by Carl Deckard and Joe Beaman in 1980s (Gross et al. 2014). It is a 3D printing technique that uses a high-power power resource such as laser to sinter powered materials (polymer, resin or metal) to build up object (Wang et al. 2017). In contrast to other additive manufacturing technologies, laser sintering allows the processing of almost any material that consolidates upon heating, including polymers, metals and even ceramics. High requirements of heat sources during the melting process becomes the biggest barrier to widespread use of this method. However, this challenge is lessened when using polymer-based systems as these can be sintered at temperatures which are lower than those of metals or ceramics. Nonetheless, parts produced by laser sintering can reach material properties that are close to those obtained by other manufacturing processes, such as injection moulding (Verbelen et al. 2016).

A SLS machine usually consists of a powder reserve and a CO_2 to sinter the powder. As a first step, the powder bed is filled and preheated to a temperature between the melting and crystallization temperature for semi-crystalline polymers and above its glass transition temperature for amorphous (Wudy et al. 2014; Shirazi et al. 2015). A layer of powder is spread across the build chamber, and then the high energy laser can selectively melt and fuse the thin layer of material together. The surrounding powder particles remain loose in the build chamber. As such SLS is not restricted by the need for support materials as overhanging structures can be supported by the unsinkable powder (Wudy et al. 2014; Chia and Wu 2015). Once completed, the printing part moves down to a predefined depth usually 0.1 mm, and the reserve part moves up to enable a new layer of the power material spread across the build chamber by a blade or a counter rotating roller. This process repeats until the last layer has been printed (Hwa et al. 2017). After production the part is slowly cooled.

The quality of printed part is effected by a multitude of factors from powder composition and morphology, laser energy input, scan spacing and speed and processing (or powder bed) temperature (Dadbakhsh et al. 2016; Macdonald et al. 2017). In terms of powder composition, much research has being conducted into the effect of powder size, shape and homogeneity as thousands of polymeric materials could theoretically be frozen below its glass transition temperature and milled either cryogenically or using standard milling processes to create powders for use in SLS. However, of these materials there are only a few grades of polymers that are commercially available for use in the SLS process. This includes amorphous, semi-crystalline, reinforced or filled polymers. Among these materials, semi-crystalline polyamide is the dominant material used (Dadbakhsh et al. 2016). Verbelen et al. studied a range of polyamide (PA) grades, namely, PA12, PA11 and PA6 to reveal characteristics of these materials that explain it current popularity for laser sintering in an attempt to develop guidelines that can be used for new materials. It was reported that a combination of powder availability, low initial zero-shear viscosity and the low tendency to warp due to a large degree of super-cooling coupled with low crystallization shrinkage were deemed to be important factors. Post-condensation behaviour also appeared to play a major role in establishing these properties. However, this behaviour was deemed detrimental for the recyclability of the powder and the consistency of part properties. Moreover, it was concluded that for SLS to become a widespread manufacturing technique, a larger variety of materials that can readily be processed is necessary (Verbelen et al. 2016). To this end the team at KU Leuven investigated cryogenic milling of thermoplastic polyurethane (TPU). The study investigated both coarse powder with rough surfaces and a fine powder which incorporated extremely fine flow modifiers. This work showed that the fine powders were easier to process, and while course and fine powders were comparable in shear-punch testing, the tensile properties of parts produced using the coarse powder were one third of the fine powders. As the coarse powders were three times those of the fine powders, it is clear that powder size and homogeneity are critical (Dadbakhsh et al. 2016).

Vasquez et al. also examined the characteristics of PA 12 to aid in the selection of materials suitable for SLS. In this work information on the thermal characteristics of the polymer measured by differential scanning calorimetry (DSC) and hot stage microscopy coupled with viscosity data was used to predict the viability of SLS printing of a thermoplastic elastomer (TPE) and a TPU. Thermogravimetric analysis (TGA) was utilized to determine the overall thermal stability region for the polymers. SLS printing of the selected polymer confirmed the viability of this approach for screening the suitability of materials for use in SLS (Vasquez et al. 2014).

1.5 Freeformer

The main drawback for additive manufacturing in the biomedical sector is that most polymers which have being approved by regulatory authorities for use in the manufacture of medical devices cannot be used without at the least the melt production of a filament. This increases cost, complexity of manufacture and lead times. Additionally, extruding of the polymer into filaments may cause a degree of thermal degradation especially in thermally unstable polymers. To overcome this challenge, the ARBURG Plastic Freeforming (APF) process was developed. This process allows the use of standard polymer granules which are commonly used in conventional injection moulding to be utilised in the production of 3D-printed parts directly, without any other modification or pre-processing steps.

The APF concept of using qualified standard granulate enables the production of parts at lower costs as there is a wide range of low-cost materials and dyes available to choose from (Fig. 1.4). In principle, the APF is suitable for all melt processible resins. Indeed there are opportunities to produce bespoke polymer formulations and composites which can be used for the production of 3D-printed parts through standard compounding and granulation processes. With this process there is no critical dimension as is observed with FFF where diameter variations can cause challenges

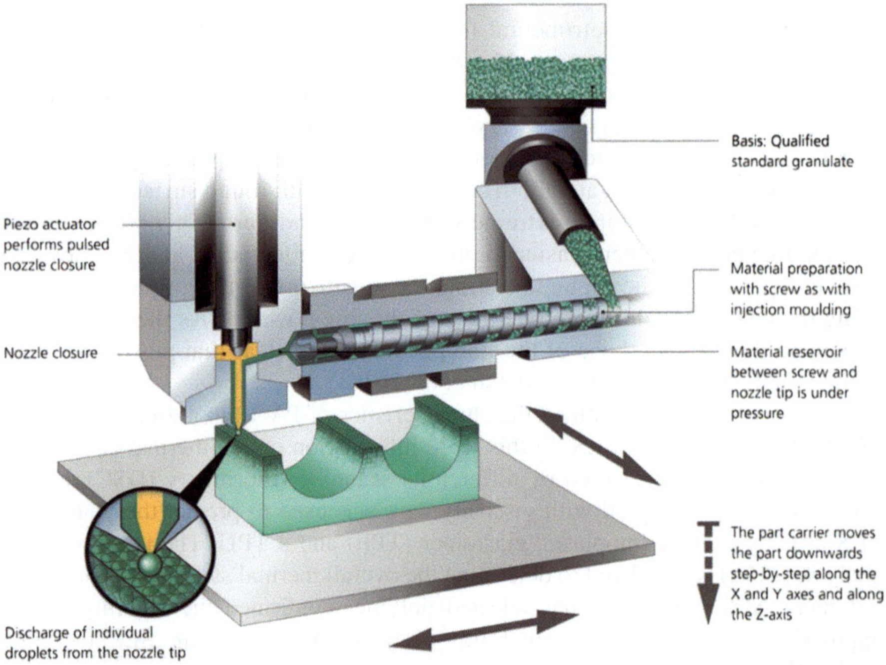

Fig. 1.4 The APF process principle in detail with (inset) a close-up of the surface of a part produced using the APF process: homogeneous, dense layer build-up

in the production of parts. However, to ensure reproducible parts are produced, material qualification, and individual development work, may be necessary for any new formulation or blend. For this the most important parameters are the processing temperature and temperature resistance. This is supplemented by the layer-by-layer geometrical slicing and automatic preparation of the 3D CAD data according to quality and material-specific criteria to create a system-specific numerical control program. Aspects such as edge, filling and strength strategies as well as the build chamber temperature can also be considered. This results in pre-optimised process data for the individual material types.

Developed together with the APF process, the "freeformer" is designed as an open production system. Slice and process parameters are freely programmable and can thus be individually optimised at any time. The material database for the APF process is growing continuously and includes process data for a large number of thermoplastics. These include not only the additive standard materials ABS, PC and PA12 (amorphous) but also partially crystalline PP, as well as special plastics such as the high-temperature material PEI, elastic TPU, medical PLA or biopolymers. Taking this as the starting point, modified original materials can also be used quickly and easily. In terms of biomedical implants or devices, this allows the use of polymeric materials which have previously had regularity approval for use in medical devices. The freeformer integrates material preparation with a special plasticising screw and discharge unit. Plasticising is followed by freeforming without the use of a mould where a nozzle closure controlled by high-frequency piezo technology enables extremely rapid opening and closing of the nozzle. Under a material pressure of up to 500 bar, up to 250 plastic droplets are applied per second. Three moving part carriers are positioned by means of servo motors so that each droplet is placed at the precise point. Solidification of the part is through polymer cooling where the droplets automatically fuse together as they cool in the temperature-controlled build chamber in a layer-by-layer manner.

The surface quality achievable with the APF process features a particularly even and close structure—in all directions (Fig. 1.4). Surface quality, mechanical properties and density are determined by droplet size and process regulation (filling strategy and level). The smaller the plastic droplets and the more closely they are placed next to one another, the finer the structure and the more robust the parts. On the other hand, larger diameters and distances enable shorter build times. Depending on the available nozzle diameters (0.15 mm, 0.2 mm or 0.25 mm), the plastic droplets measure around 0.2–0.3 mm. The layer thickness varies between 0.14 and 0.34 mm, and the minimum wall thickness is 0.6–1.0 mm.

The manufactured products can be used not just as reliable prototypes but also as fully functional individual parts and small-volume batches. This opens up more efficient processes and new design freedoms when developing and producing products. Prototypes can be subjected to stress tests, for example, and the necessary product adjustments can be identified and implemented at an early stage. In the area of pre-series production, small unit volumes can be achieved using the APF method. The integration of functions by combining part production and assembly in a single-process step is a further means of achieving cost efficiency. With this additive process, complete modules can be produced in a fully assembled state. With regard to lightweight construction and weight reduction, bionically optimised parts can be designed and produced without having to factor in the technical production constraints that apply to injection moulding. In conjunction with the corresponding process regulation, the following component properties can be achieved compared to injection-moulded parts, depending on the material:

- Tensile strength is about 90–98% depending on the resin type (tensile test according to DIN EN ISO 527-02).
- Part density up to 95%.
- Surface quality up to Rz = 7.84 µm/Ra = 1.69 µm.

In addition the APF process can be combined with injection moulding in conjunction with Industry 4.0 technologies; this makes it economically viable to manufacture single-unit customized batches in a flexible, automated, digitally networked, cyberphysical production system. The individualisation of mass-produced products in terms of 3D geometry and functionality means that plastic parts can be enhanced in such a way as to generate added value for manufacturers. Serial production for specific jobs will commence "on demand", thanks to IT networking. An injection moulding machine and a freeformer can be linked by means of an automation component for this purpose. A digital manufacturing (DM) code applied to the part during the production process turns it into an information carrier and enables it to control its own production steps; this could also be used for part tracking for regularity purposes. The automation of additive manufacturing also reduces unit costs and improves process control. In addition to the automatic opening and closing of the hood and the loading and unloading of the build chamber with loaded parts, automated assembly with base plates or cleaning cycles can also be realised. Loading takes place with in-line measurement and automatic height correction. In addition, data is exchanged and archived at a higher level via an industrial interface. The freeformer is also equipped with several discharge units as standard. This enables functional parts to be produced using several different polymers with combinations of hard and soft polymers possible. As with injection moulding, self-dyeing is possible in master batches. In the case of complex part geometries, the second component can alternatively be used to construct support structures.

In terms of industries and areas of application, the possibilities of the APF process with its wide range of materials are virtually boundless. In the medical technology sector, for example, it can be used for the additive manufacturing of individually adapted implants and orthotics as well as models used in preparation for surgery. Another example of biomedical implants which may benefit from production using the freeformer is the manufacture of customized resorbable polylactide (PLA)-based implants, conventionally used as screws or pins for securing bones and tissue. These parts are traditionally produced by injection moulding; however, production using the freeformer would enable these implants to be tailored precisely to a particular patient. Another opportunity would be in the production of surgical guides through the use of medical imaging to produce 3D CAD data.

1.6 InkJet Techniques

InkJet printing was initially described by Lord Rayleigh in 1878. It was patented by Siemens for 2D printing in 1951; however, it was 2001 before InkJet printing was used for 3D printing structures (Gross et al. 2014).

InkJet technology can be dived into two separate but related 3D printing techniques. The first technology is also referred to as binder jetting, drop-on-demand or powder-liquid 3D printing. This system was developed at Massachusetts Institute of Technology in 1993 (Wang et al. 2017). The system utilises photocurable polymers laid down by an inkJet head to selectively bind powders which are laid down in a similar manner to SLS 3D printing (Shirazi et al. 2015). This process continues with the build bed being lowered and fresh powder spread across the surface until the final part is built. Unbound powder is removed to get the final part (Wang et al. 2017). The key advantage of this technique is the flexibility of the system and processability at room temperatures. As with SLS, theoretically any polymer material in a powder state can be printed using this technology. Surface porosity can be high using this technique due to unbound particles which may depending on the application be advantageous (Gross et al. 2014). However, its applicability to biomedical applications may be limited as the binder may incorporate contaminates which are biologically toxic and resolution is limited (Gross et al. 2014; Wang et al. 2017).

Powderless InkJet technologies such as the Stratasys, PolyJet system which can print 16 μm photopolymer layers. In this system liquid photopolymer is heated to about 73 °C, jetted onto a surface and instantly cured with UV light. Most commercial printers contain at least two print heads similar to the ones in conventional printing, one for the support material and one for the model material. Additional print heads can be added for higher material throughput or to print multiple materials, which is one major advantage of the process (Mueller et al. 2015). This system is ideally suited for the production of parts which require two or more distinct properties as combinations of resins can easily be build up within the part due to the number of heads employed. Thin layers of cured material pile up while the build tray moves down until the part is finished. The Stratasys Objet500 Connex3 is one of the most advanced commercially available PolyJet printers; it is representative for the current state-of-the-art of inkJet 3D printing (Mueller et al. 2015). However, as the PolyJet system is relatively new (Wang et al. 2017), the number of resins available is somewhat limited, and to the authors knowledge, none are approved for use in implant materials.

1.7 Laminated Object Manufacturing

Laminated object manufacturing was developed by Helisys (now Cubic Technologies). This system generates a 3D part through stacking layers of polymer sheets. Excess material is removed with a razor or laser which traces the desired shape. Layers are combined using an adhesive or through thermal welding (Gross et al. 2014). Thermal bonding is more desirable than adhesive bonding for biomedical applications as the adhesives may be biologically toxic as was discussed in relation to InkJet printing.

This process is an important automated process used for fabrication of large composite structures in aeronautical industry (Zhang et al. 2015), and in this industry is often referred to as automated tape placement (ATP). Most commonly the ATP system utilises a continuous fibre thermoplastic composite tape to generate high-strength composite parts. The tape is laid down in layers with pressure applied with a heated roller during bonding to ensure good contact between layers (Zhang et al. 2015). Engineering materials can be utilised using this process in combination with a heated bed as if the local temperature of the bed or roller is not controlled well enough, the part may delaminate due to insufficient bonding between layers, or the part may suffer structural damage if the temperatures do not enable good adhesion (Gross et al. 2014).

1.8 Summary

As illustrated in this chapter, there are a number of additive manufacturing techniques which all have strengths the weaknesses. The type and tolerance of these machines is constantly increasing. This is due in part to the patent lapsing on the earlier techniques which has allowed new entrants into the market. This has coincided with an increase in the development of a lot of new resins for use in 3D printing as many of the new suppliers have moved to an open source resin approach.

While there are many different techniques, this chapter has concentrated on the most established processes which are more likely to be used in the production of biomedical parts.

Liquid resin-based systems like SLA and PolyJet are capable of the best printing resolutions. However, the resins which are approved for use in biomedical implants are limited. These systems are generally based on thermosetting polymer resins which following curing do not melt on application of heat. Nevertheless, these processes have being used for the production of a variety of products, and as the material technology is developed, this is set to increase.

Conversely, thermoplastic resin-based systems can be used in conjunction with a vast array of regularity approved thermoplastic resins. With FFF, the challenge of delamination between weld lines is a concern; however, these effects can be lessened with post-production processing. The Arburg freeformer system appears promising in this regard; however, again the range of materials which can be processed using this method needs to be expanded.

SLS and binder jetting are both based on powdered polymers, but they differ in how the powders are bound. With SLS, thermal energy is used to sinter the polymer particles together eaving a potentially porous construct whereas binder l jetting utilises adhesives which may in themselves be toxic. However, if this can be overcome, both have potential for use in areas where surface roughness is desirable such as in orthopaedic implants. Similar to the previous methods, laminated object manufacturing uses sheets of polymers which enables the production of parts from a wide variety of polymer resins. However, these can use thermal or chemical bonding techniques similar to the previous techniques.

With this vast array of processes capable of producing complex parts with complete design freedom from a growing number of materials, it is clear that the application of 3D printing in the biomedical field will continue to grow and find new applications which will improve the lives of countless people now and well into the future.

Acknowledgements The authors wish to thank Dr. Joe DeSimone and Mr. Ken Sibley, Esq., both of Carbon, Inc., for some of the historical references.

This work was supported in part by financial support from Athlone Institute of Technology under the Presidents Seed Fund, Enterprise Ireland, funding under the Technology Gateway program, grant number TG-2017-0114, and Science Foundation Ireland (SFI) under Grant Number 16/RC/3918, co-funded by the European Regional Development Fund.

References

Amato, S., & Ezzell, R., Jr. (2015). *Regulatory affairs for biomaterials and medical devices* (p. 203). Cambridge: Elsevier. https://doi.org/10.1016/C2013-0-16313-1.

Andre, J.-C., Le Mehaute, A., & De Witte, O. (1984). Dispositif pour réaliser un modèle de pièce industrialle (Device for fabricating an industrial part model). French Patent App. 2567668, 1984

Bellini, A., & Guceri, S. (2003). Mechanical characterization of parts fabricated using fused deposition modeling. *Rapid Prototyping Journal, 9*(4), 252–264. https://doi.org/10.1108/13552540310489631.

Bhatt, A., & Anbarasu, A. (2017). Nanoscale biomaterials for 3D printing. *IOSR Journal of Pharmacy and Biological Sciences, 12*(03), 80–86. https://doi.org/10.9790/3008-1203068086.

Cerezen device provides treatment for temporomandibular joint disorders day and night. (2016). *British Dental Journal, 220*, 666.

Chen, Y., et al. (2016). A review of multifarious applications of poly (lactic acid). *Polymer-Plastics Technology and Engineering, 55*, 1057–1075. https://doi.org/10.1080/03602559.2015.1132465.

Chen, Y., et al. (2018). Surface-modified halloysite nanotubes reinforced poly(lactic acid) for use in biodegradable coronary stents. *Journal of Applied Polymer Science, 135*(30), 46521. https://doi.org/10.1002/app.46521.

Chia, H. N., & Wu, B. M. (2015). Recent advances in 3D printing of biomaterials. *Journal of Biological Engineering, 9*(1), 4. https://doi.org/10.1186/s13036-015-0001-4.

Clair, J-J. (1996). Stereolithography and the biomedical engineering, Journal of Materials Processing Technology, 57 393-396, https://doi.org/10.1016/0924-0136(95)02072-1

Conner, B. P., et al. (2014). Making sense of 3-D printing: Creating a map of additive manufacturing products and services. *Additive Manufacturing, 1*, 64–76. https://doi.org/10.1016/j.addma.2014.08.005.

Corea, Joseph Russel, "Screen Printed MRI Receive Coils", Doctoral Thesis, Department of Electrical Engineering and Computer Science. 2016 (118 pages).

Dadbakhsh, S., et al. (2016). Effect of powder size and shape on the SLS processability and mechanical properties of a TPU elastomer. *Physics Procedia, 83*, 971–980. https://doi.org/10.1016/j.phpro.2016.08.102.

Dermanaki Farahani, R., & Dubé, M. (2017). Printing polymer nanocomposites and composites in three dimensions. *Advanced Engineering Materials, 20*, 1–9. https://doi.org/10.1002/adem.201700539.

DeSimone, Joseph M. Ermoshkin, Alexander, Ermoshkin, Nikita, Samulski, Edward T., "Continuous Liquid Phase Interphase Printing", US Patent 9,205, 601 B2, Dec. 8, 2015.

Dizon, J. R. C., et al. (2018). Mechanical characterization of 3D-printed polymers. *Additive Manufacturing, 20*, 44–67. https://doi.org/10.1016/j.addma.2017.12.002.

Dorigato, A., et al. (2017). Electrically conductive nanocomposites for fused deposition modelling. *Synthetic Metals, 226*, 7–14. https://doi.org/10.1016/j.synthmet.2017.01.009.

Dul, S., Fambri, L., & Pegoretti, A. (2016). Fused deposition modelling with ABS-graphene nanocomposites. *Composites Part A: Applied Science and Manufacturing, 85*, 181–191. https://doi.org/10.1016/j.compositesa.2016.03.013.

FDA Role in 3D Printing, https://www.fda.gov/consumers/consumer-updates/3rs-3d-printing-fdas-role.

Ferreira, R. T. L., et al. (2017). Experimental characterization and micrography of 3D printed PLA and PLA reinforced with short carbon fibers. *Composites Part B: Engineering, 124*, 88–100. https://doi.org/10.1016/j.compositesb.2017.05.013.

Fuenmayor, E., et al. (2018). Material considerations for fused-filament fabrication of solid dosage forms. *Pharmaceutics, 10*(2), 44. https://doi.org/10.3390/pharmaceutics10020044.

Gao, W., et al. (2015). The status, challenges, and future of additive manufacturing in engineering. *Computer-Aided Design, 69*, 65–89. https://doi.org/10.1016/j.cad.2015.04.001.

Gross, B. C., et al. (2014). Evaluation of 3D printing and its potential impact on biotechnology and the chemical sciences. *Analytical Chemistry, 86*(7), 3240–3253. https://doi.org/10.1021/ac403397r.

Healy, A. V., et al. (2018). Degradable nanocomposites for fused filament fabrication applications. https://doi.org/10.3390/jmmp2020029.

Hiemenz, J. (2011). 3D printing jigs, fixtures and other manufacturing tools. Stratasys, Inc.

Hornbeck, L. J. (1997). Digital light processing for high-brightness high-resolution applications. In M. H. Wu (Ed.), International Society for Optics and Photonics, Bellingham, pp. 27–40. https://doi.org/10.1117/12.273880.

Hull, C. W., & Arcadia, C. (1986). Apparatus for production of three-dimensional objects by stereolithography. Apparatus for Production of Three-Dimensional Objects by Stereolithography. https://doi.org/10.1145/634067.634234.

Hwa, L. C., et al. (2017). Recent advances in 3D printing of porous ceramics: A review. *Current Opinion in Solid State and Materials Science*, 323–347. https://doi.org/10.1016/j.cossms.2017.08.002.

ISO 10993 —Biological Evaluation of Medical Devices (2001–2018.).

Ivanova, O., Williams, C., & Campbell, T. (2013). Additive manufacturing (AM) and nanotechnology: Promises and challenges. *Rapid Prototyping Journal, 19*(5), 353–364. https://doi.org/10.1108/RPJ-12-2011-0127.

Kim, H., et al. (2018). Enhanced dielectric properties of three phase dielectric MWCNTs/BaTiO 3/PVDF nanocomposites for energy storage using fused deposition modeling 3D printing. *Ceramics International*. https://doi.org/10.1016/j.ceramint.2018.02.107.

Kodama, H. (1981). Automatic method for fabricating a three-dimensional plastic model with photo-hardening polymer. *Review of Scientific Instruments, 52*(11), 1770–1773. https://doi.org/10.1063/1.1136492.

Kokkinis, D., Schaffner, M., & Studart, A. R. (2015). Multimaterial magnetically assisted 3D printing of composite materials. *Nature Communications, 6*(1), 8643. https://doi.org/10.1038/ncomms9643.

Kruth, J.-P., Leu, M. C., & Nakagawa, T. (1998). Progress in additive manufacturing and rapid prototyping. *CIRP Annals-Manufacturing Technology, 47*(2), 525–540. https://doi.org/10.1016/S0007-8506(07)63240-5.

Li, G., et al. (2018). Effect of ultrasonic vibration on mechanical properties of 3D printing non-crystalline and semi-crystalline polymers. *Materials, 11*(5). https://doi.org/10.3390/ma11050826.

Ligon, S. C., et al. (2017). Polymers for 3D printing and customized additive manufacturing. *Chemical Reviews, 117*(15), 10212–10290. https://doi.org/10.1021/acs.chemrev.7b00074.

Luo, Y., et al. (2016). Synthesis and biological evaluation of well-defined poly(propylene fumarate) oligomers and their use in 3D printed scaffolds. *Biomacromolecules, 17*(2), 690–697. https://doi.org/10.1021/acs.biomac.6b00014.

Macdonald, N. P., et al. (2017). Comparing microfluidic performance of three-dimensional (3D) printing platforms. *Analytical Chemistry, 89*(7), 3858–3866. https://doi.org/10.1021/acs.analchem.7b00136.

Masood, S. H., & Song, W. Q. (2004). Development of new metal/polymer materials for rapid tooling using fused deposition modelling. *Materials and Design, 25*(7), 587–594. https://doi.org/10.1016/j.matdes.2004.02.009.

McLouth, T. D., et al. (2017). The impact of print orientation and raster pattern on fracture toughness in additively manufactured ABS. *Additive Manufacturing, 18*, 103–109. https://doi.org/10.1016/j.addma.2017.09.003.

McLure, T. (2017). 3D printing moves align technology toward $1.3 billion in sales. *Forbes.*

Melchels, F. P. W., Feijen, J., & Grijpma, D. W. (2010). A review on stereolithography and its applications in biomedical engineering. *Biomaterials, 31*(24), 6121–6130. https://doi.org/10.1016/j.biomaterials.2010.04.050.

Mills, D., et al. (2017). The use of 3D printing in the fabrication of Nasal Stents. *Inventions, 3*(1), 1. https://doi.org/10.3390/inventions3010001.

Mohamed, O. A., Masood, S. H., & Bhowmik, J. L. (2016). Optimization of fused deposition modeling process parameters for dimensional accuracy using I-optimality criterion. *Measurement: Journal of the International Measurement Confederation, 81*, 174–196. https://doi.org/10.1016/j.measurement.2015.12.011.

Mueller, J., Shea, K., & Daraio, C. (2015). Mechanical properties of parts fabricated with inkjet 3D printing through efficient experimental design. *Materials and Design, 86*, 902–912. https://doi.org/10.1016/j.matdes.2015.07.129.

Munz, O. J. (1956). Photo-glyph recording. US Patent 2,775,758, December 25.

Ngo, T. D., et al. (2018). Additive manufacturing (3D printing): A review of materials, methods, applications and challenges. *Composites Part B: Engineering*. https://doi.org/10.1016/j.compositesb.2018.02.012.

Norman, J., et al. (2016). A new chapter in pharmaceutical manufacturing: 3D-printed drug products. *Advanced Drug Delivery Reviews*. https://doi.org/10.1016/j.addr.2016.03.001.

Parry, J. A., et al. (2017). Three-dimension-printed porous poly(propylene fumarate) scaffolds with delayed rhBMP-2 release for anterior cruciate ligament graft fixation. *Tissue Engineering Part A, 23*(7–8), 359–365. https://doi.org/10.1089/ten.tea.2016.0343.

Prasad, L. K., & Smyth, H. (2016). 3D Printing technologies for drug delivery: A review. *Drug Development and Industrial Pharmacy, 42*(7), 1019–1031. https://doi.org/10.3109/03639045.2015.1120743.

Rolland, J., Chen, K., & Poelma, J. (2016). *Methods of producing polyurethane three-dimensional objects from materials having multiple mechanisms of hardening*. US Patent.

Sachs, E., Cima, M., & Cornie, J. (1990). Three-dimensional printing: Rapid tooling and prototypes directly from a CAD model. *CIRP Annals—Manufacturing Technology, 39*(1), 201–204. https://doi.org/10.1016/S0007-8506(07)61035-X.

Sharma, R. (2013). The 3D printing revolution you have not heard about. *Forbes.*

Shirazi, S. F., et al. (2015). A review on powder-based additive manufacturing for tissue engineering: Selective laser sintering and inkjet 3D printing. *Science and Technology of Advanced Materials*, 16. https://doi.org/10.1088/1468-6996/16/3/033502.

Skowyra, J., Pietrzak, K., & Alhnan, M. A. (2015). Fabrication of extended-release patient-tailored prednisolone tablets via fused deposition modelling (FDM) 3D printing. *European Journal of Pharmaceutical Sciences, 68*, 11–17. https://doi.org/10.1016/J.EJPS.2014.11.009.

Sood, A. K., Ohdar, R. K., & Mahapatra, S. S. (2010). Parametric appraisal of mechanical property of fused deposition modelling processed parts. *Materials and Design, 31*(1), 287–295. https://doi.org/10.1016/j.matdes.2009.06.016.

Tian, X., et al. (2016). Interface and performance of 3D printed continuous carbon fiber reinforced PLA composites. *Composites Part A: Applied Science and Manufacturing, 88*, 198–205. https://doi.org/10.1016/j.compositesa.2016.05.032.

Tian, X., et al. (2017). Recycling and remanufacturing of 3D printed continuous carbon fiber reinforced PLA composites. *Journal of Cleaner Production, 142*, 1609–1618. https://doi.org/10.1016/j.jclepro.2016.11.139.

Tiwari, R. V., Patil, H., & Repka, M. A. (2016). Contribution of hot-melt extrusion technology to advance drug delivery in the 21st century. *Expert Opinion on Drug Delivery*. https://doi.org/10.1517/17425247.2016.1126246.

Tumbleston, J. R., Shirvanyants, D., Ermoshkin, N., Janusziewicz, R., Johnson, A. R., Kelly, D., Chen, K., Pinschmidt, R., Rolland, J. P., Ermoshkin, A., Samulski, E. T., DeSimone, J. M. (2015). Additive manufacturing. Continuous liquid interface production of 3D objects. *Science, 347*(6228), 1349–1352. https://doi.org/10.1126/science.aaa2397.

van Lith, R., et al. (2016). 3D-printing strong high-resolution antioxidant bioresorbable vascular stents. *Advanced Materials Technologies, 1*(9), 1–7. https://doi.org/10.1002/admt.201600138.

Vasquez, G. M., Majewski, C. E., Haworth, B., & Hopkinson, N. (2014). A targeted material selection process for polymers in laser sintering. *Additive Manufacturing*, 127–138. https://doi.org/10.1016/j.addma.2014.09.003.

Verbelen, L., et al. (2016). Characterization of polyamide powders for determination of laser sintering processability. *European Polymer Journal*. https://doi.org/10.1016/j.eurpolymj.2015.12.014.

Voit, W., *et al.* (2016). *Cartridge-based 3D printing system.* Retrieved from https://patents.google.com/patent/US20160200044.

Wang, X., et al. (2017). 3D printing of polymer matrix composites: A review and prospective. *Composites Part B: Engineering, 110*, 442–458. https://doi.org/10.1016/j.compositesb.2016.11.034.

Wohlers, T. T. (2005). *Early Research & Development, Wohlers Report 2005: Rapid prototyping, tooling and manufacturing state of the industry worldwide progress report.* Wohlers Associates, Inc.

Wudy, K., Drummer, D., & Drexler, M. (2014). Characterization of polymer materials and powders for selective laser melting. *AIP Conference Proceedings, 1593*(May), 702–707. https://doi.org/10.1063/1.4873875.

Yao, X., et al. (2017). Evaluation of carbon fiber-embedded 3D printed structures for strengthening and structural-health monitoring. *Materials and Design, 114*, 424–432. https://doi.org/10.1016/j.matdes.2016.10.078.

Yun, Y., et al. (2018). Effect of printing parameters on tensile, dynamic mechanical, and thermoelectric properties of FDM 3D printed CABS/ZnO composites. *Materials, 11*(4), 466. https://doi.org/10.3390/ma11040466.

Zaharin, H. A., et al. (2018). Additive manufacturing technology for biomedical components: A review. *IOP Conference Series: Materials Science and Engineering, 328*, 012003. https://doi.org/10.1088/1757-899X/328/1/012003.

Zamarayeva, A., (EECS/UC B), et al. (2018). Custom, 3D sprayed MRI receive coils. *International Society for Magnetic Resonance in Medicine, 16-21 June, Paris, France.*

Zhang, P., et al. (2015). Placement suitability criteria of composite tape for mould surface in automated tape placement. *Chinese Journal of Aeronautics, 28*(5), 1574–1581. https://doi.org/10.1016/j.cja.2015.06.002.

Chapter 2
Design for Additive Manufacturing

Eoin P. Hinchy

Abstract Additive manufacturing has evolved over the past number of decades from a rapid prototyping technology into a viable net-shape manufacturing technology. Additively manufactured components are now being used for medical device applications all over the world, from custom splints and casts to implanted orthopaedic joints which improve patient experience quality of life.

Traditional design for manufacturing and assembly (DFMA) guidelines have moulded the design transition from product concept to the final product, allowing manufacturers to produce products more efficiently and cost-effectively. In traditional manufacturing, as product complexity increases, both cost and manufacturing time also increases; thus, products are often simplified to improve manufacturability. Simplifying products, however, reduces both designer intent and often product performance.

Additive manufacturing has many advantages over traditional manufacturing which will be discussed in this chapter, such as topology optimisation, geometrical design freedom and product consolidation as well as the ability to disrupt traditional supply chains. Additive manufacturing is not restrained by traditional DFMA guidelines, so as a result, design for additive manufacturing (DFAM) guidelines are instead used to guide product fabrication using this manufacturing technology.

Additive manufacturing is suited to low-volume, complex components, and thus careful consideration should be given to the selection of this technology over traditional manufacturing technologies. Product design considerations such as build orientation and support structures will be presented which affect manufacturing cost and time, post-processing requirements as well as mechanical properties of the final component. Some case studies will be presented at the end of this chapter which use additive manufacturing to produce lighter, more geometrically complex or consolidated products.

E. P. Hinchy (✉)
Confirm Centre Smart Manufacturing, School of Engineering, University of Limerick, Limerick, Ireland
e-mail: Eoin.Hinchy@ul.ie

D. M. Devine (ed.), *Polymer-Based Additive Manufacturing*,
https://doi.org/10.1007/978-3-030-24532-0_2

Keywords Design consideration · Design guidelines · Build orientation · Product optimisation · Topology optimisation

Abbreviations

ABS	Acrylonitrile butadiene styrene
ATP	Advanced turboprop
CAD	Computer-aided design
CFD	Computational fluid dynamics
CT	Computed tomography
DFAM	Design for additive manufacturing
DFMA	Design for manufacture and assembly
DMLS	Direct metal laser sintering
FEA	Finite element analysis
FFF	Fused filament fabrication (aka fused deposition modelling)
MRI	Magnetic resonance imaging
OEM	Original equipment manufacturer
PLA	Polylactic acid
PVC	Polyvinyl alcohol
SLA	Stereolithography
SLS	Selective laser sintering

2.1 Introduction

Since primitive man first walked the earth, humans have been making tools. The first tools for hunting were made from stone and wood, while metal appeared circa 6000 years ago. As man advanced, so did manufacturing technologies. New materials and new manufacturing techniques enabled man to produce tools and weapons which were stronger and more functional than previous versions. For centuries, product design has been driven by designer intent; however, the final design has always been limited by the constraints of the manufacturing technology. Product designers and design engineers need to strike a balance between product functionality and quality while also minimising cost. This has traditionally been done by following guidelines called design for manufacturing and assembly (DFMA), which focuses on reducing product complexity to reduce manufacturing costs. With additive manufacturing, however, product complexity is no longer linked to manufacturing costs, which has liberated designers, allowing for designer intent to be achieved affordably.

Additive manufacturing has somewhat broken away from the guidelines laid out by DFMA and instead requires its own guidelines, referred to as design for additive manufacturing (DFAM). Additive manufacturing has grown from a novel process traditionally used for rapid prototyping to an established process which is used for manufacturing net-shape devices such as medical devices and structural aircraft components. As a result of this, it has proven to be a disruptive technology in the manufacturing space, calling for designers to rethink the conventional product development life cycle, to maximise the advantages and capabilities of additive manufacturing. Traditionally, designers are taught how to design parts for ease of manufacturing, but with additive manufacturing, all parts can be made relatively easily, and thus, geometry is not a limiting factor (Hague 2006). The layerwise construction of additive manufacturing has given designers new design freedom which could not be really explored using conventional manufacturing technologies due to high costs and manufacturing complications.

In this chapter, an overview of conventional design for manufacturing and assembly (DFMA) will be presented highlighting traditional standard manufacturing guidelines. Following this, the advantages of additive manufacturing as a production process will be highlighted and explored, while considerations regarding the suitability of this manufacturing technology will also be addressed. Product design considerations for additive manufacturing will next be presented, including considerations regarding fabrication-related factors. Some post-processing considerations will be presented next, while some case studies using additive manufacturing for part consolidation and weight savings in the medical device and aerospace sectors will close this chapter.

2.2 Design for Manufacturing and Assembly

Everyday, we use products which have undergone extensive design and testing and have been manufactured cost-effectively. Design for manufacturing and assembly (DFMA) is a fundamental methodology which is traditionally used from design concept, all the way to manufacturing. This methodology has been used to develop high-volume products from the millions of cars on our roads to the billions of smartphones in our pockets, which humans have manufactured globally. The design and manufacturing engineers responsible for these products have struck a balance between product functionality, quality and cost—which form the three corners of the manufacturing process triangle (Tempelman et al. 2014). Ideally, the manufacturer would like to maximise quality and function, while minimising cost; however, to minimise cost, often product design needs to be simplified to boost manufacturability. Simplified designs, which substitute complex geometries or features for simpler ones, can reduce the cost per unit without having a significant impact on quality or functionality. However, simplification comes at the expense of designer intent. Successful cost reduction can have a significant impact on profit for high-volume production. For example, if Apple saved an additional $5 on every iPhone sold in 2017, then Apple would have made more than $1 billion additional profit.

Before discussing design for additive manufacturing (DFAM) in detail, it is important to touch on the traditional DFMA, as DFAM finds its roots in this methodology. DFMA is the process where an object being designed is evaluated critically for manufacturability. By using a set of established guidelines, design engineers can ensure a product can be manufactured quickly and efficiently at low-cost, but also provide the end-user with the quality and function expected of that product.

Traditionally, DFMA is conducted in a series of sequential steps. Firstly, customer requirements are established. This guides the functionality requirements of the product, and thus guides the next step—conceptual design, which includes product sketches. The design concept is critically examined with a view to easing manufacturability and assembly. Design engineers examine ease of assembly, identify potential materials for construction, as well as different manufacturing technologies which could be used. The costings for each of the material/manufacturing technology combinations are then critically analysed, where the most cost-effective combination is identified. The next step of this sequential design process is detailed design for minimum manufacturing costs, where the finalised product is designed. This includes drawings, bill of materials, identification of sub-components, etc. Once the product design is finalised, it is prototyped to ensure it meets design requirements (Boothroyd et al. 2001).

The DFMA process allows for decisions to be made which make manufacturing faster, cheaper and more efficient. For example, if a product has an intricate design, then perhaps it would be cheaper to assemble it from more simple sub-components rather than manufacturing it as one single component using traditional manufacturing methods. If the product is to be assembled from several sub-components, then the design engineer will need to choose a fastener technology which is cost-effective, but also robust. For example, mechanical fasteners such as bolts, screws or rivets may be used, or perhaps brazing, welding or adhesive joints. These are all decisions which need to me made during the design stage of the product (Boothroyd et al. 2001).

The selection of the manufacturing processes has a direct effect on cost. In fact, a product design engineer can influence 70% of the cost of the product, so by making smart choices early in the design process, costs can be minimised (Boothroyd et al. 2001). In traditional manufacturing, once a design suitable for manufacturing and assembly has been reached, the next step is to manufacture the associated tooling. If the manufacturing technology is plastic injection moulding, for example, then appropriate moulds would need to be designed and manufactured with a very high geometrical tolerance and surface finish. This leads to massive upfront costs for the manufacturer, who in turn needs to produce products in high volumes in order to offset the high tooling cost (Hague 2006). Thus, it can be very costly to manufacture once-off or low-volume customised products using conventional manufacturing technologies.

When considering DFMA, there are some fundamental guidelines which a design engineer should follow to ensure manufacturing costs are kept to a minimum:

Reduce handling time: Handling of parts may be considered as loading a component into a machine or processing a part. For example, a part might be loaded into machine A for one process, then machine B for a second process and then machine C for a final process. Time is wasted at each of the loading and unloading steps, and there is an associated cost incurred. If machine A could do all three process steps, then handling times are reduced, as are overall costs.

Ease of insertion: During manufacturing, products can be manufactured quicker if the assembly process is simple and straightforward. When designing an assembly, it is good practice to design parts which are easy to assemble; difficult or tricky assembly processes should be avoided. Designing features into parts such as guide hole or guide rails can speed up assembly. The addition of self-selecting features, where parts can only fit together one way, is also a good practice for reducing assembly time. The Japanese term poka-yoke, which broadly means "mistake proofing", describes this practice. An example of a self-selecting feature is a UK electrical plug, which has only one specific orientation in which the plug can mate with the electrical outlet, preventing misuse. Simple assembly processes also suit robotic assembly lines, which means assembly processes can be highly automated—saving time and long-term manufacturing costs.

Reducing part count: Reducing part count is a crucial guideline in DFMA which in turn reduces costs. By consolidating several discrete parts into fewer parts, manufacturing and assembly time can be reduced, as well as costs. In fact, by reducing part count, fewer fasteners are required, fewer sub-components need to be ordered and tracked, while there is less chance for an assembly error. This also reduces the number of operation steps and thus, reduces the number of operators, tools and manufacturing equipment required.

Modular design and multifunctional parts: By employing a modular design, costs can be reduced, the subcomponents which are used in assembly are less diverse. There can also be savings in terms of design time, where modules can be swapped out or changed based on the design requirements. Using multifunctional parts is also crucial in reducing the number of sub-components. For example, a metallic structural part in a device may also double as an electrical connector. A structural component may also have integrated guide rails, which serves a function in the assembly process, meaning additional guide rails do not need to be fitted to a part.

Avoiding common manufacturing defects: In terms of designing the manufacturing process, common manufacturing defects are well understood. For example, for injection moulding, if mould walls are too thin, the material may not fill completely—causing the product to fail. Similarly, with casting metals, if mould walls are too cold, molten metal may freeze, causing typical defects such as cold shut. There are standard rules for different manufacturing technologies which when followed, reduce the likelihood of failure. In the design stage, once a manufacturing technology and materials are selected, it is good practice for design engineers to re-evaluate their design to ensure it will not cause manufacturing defects, or else modify the design slightly to avoid features known to cause common defects.

Some of these guidelines apply mainly to traditional manufacturing processes, while there are others which apply to both DFMA and DFAM. In the next section of this chapter, the unique advantages of additive manufacturing will be discussed, which has lifted many design constraints faced by design engineers using traditional manufacturing technologies.

2.3 Advantages of Additive Manufacturing as a Production Process

With DFMA well understood, one must ask the question as to what benefits does additive manufacturing have to offer to design engineers as a production process? Why uptake this relatively new manufacturing technology when other manufacturing technologies such as casting, injection moulding and machining are so well established in the medical device sector? There are several key areas where additive manufacturing holds advantages over conventional manufacturing processes, which will be discussed in detail in this section. Firstly, additive manufacturing works well with digital product design and topology optimisation. It also gives the product designer the freedom to produce virtually any geometric shape, including personalised, patient-specific products and is very cost-effective for this kind of low-volume production. The ability to consolidate products and produce lightweight structures has seen tremendous benefits in both the medical device and aerospace industries, while additive manufacturing can also disrupt traditional manufacturing supply chains, which can save costs at a business enterprise level. Thus, it will be shown that these unique capabilities of additive manufacturing require designers to shift their way of thinking from the traditional DFMA approach.

2.3.1 Product Digitisation and Rapid Prototyping

With the ever-growing dominance of computer-aided design (CAD), including computational design simulations such as finite element analysis (FEA) and computational fluid dynamics (CFD), the digitisation of product manufacture is enabling design engineers to produce very effective and efficient products, rapidly. Design engineers can develop digital models of complex systems and subsystems, incorporating product and manufacturing information data such as kinematics, materials, tolerances, etc. The performance of these complex models can be tested using powerful simulation tools, such as thermal simulation, fluid dynamics and mechanical and dynamic simulations. These simulations can be run using cloud-based processors, thus, rapidly reducing simulation time compared to traditional local computing. Product digitisation has also allowed for design engineers to email digital model files or store on a cloud-based platform, allowing multinational companies to co-develop products at several different sites around the world simultaneously.

When digital testing and simulation of components is completed, and a satisfactory design is established, additive manufacturing can be used to fabricate a functional prototype which can be used for physical testing. Producing prototypes using conventional manufacturing can be slow and costly. However, additive manufacturing can be used to produce a net-shape or near-net-shape prototype rapidly—hence the term rapid prototyping. This accelerates product development and reduces time to market. It also enables manufacturers to produce components in final form, which can be comprehensively tested. Should design changes be needed, these can be updated digitally, and a new prototype can be rapidly manufactured.

2.3.2 Topology Optimisation

One key advantage of digital product manufacturing is topology optimisation. Topology optimisation is a process where components are structurally optimised using mathematical and computer modelling, resulting in higher strength-to-weight ratios, but also a more complex geometry. Topology optimisation optimises the material distribution across a component, given set boundary conditions and loading conditions, while also ensuring that the component performs as expected. This allows a design engineer to maximise the strength-to-weight ratio of a component, while also meeting specified loading conditions. CAD and FEA are two key technologies used in topology optimisation. In order to optimise component topology, loading conditions and interfacing (anchoring) points must be established. FEA can then be used to determine zones of maximum predicted stress/strain and minimum stress/strain. Material identified as being minimum stress/strain can be removed, topology optimising the component. Consider the simple bracket illustrated in Fig. 2.1a. This bracket has been designed using CAD to be manufactured using conventional manufacturing techniques. The bracket geometry consists of a triangular base plate with three mounting holes, while a loading hole is located on the perpendicular vertical fin. FEA was conducted on this bracket, where a 15 kN load was applied to one side of the loading hole in the direction away from the bracket, as arrowed in Fig. 2.1a. FEA results showed that the areas of high stress within the part were around the holes, as illustrated by the red and green colour contours in Fig. 2.1b. The colour contours also show that the dark blue-coloured material is surplus to requirements and serves no significant mechanical advantage to the bracket. Thus, topology optimisation uses this FEA data to eliminate material which was of no significant mechanical importance to the bracket—i.e. the blue material in Fig. 2.1b. The resulting topology optimised component shown in Fig. 2.1c, d has an organic internal structure, similar to that of bone. The smooth flat sides along the outer surfaces align with the original conformal shape of the bracket, ensuring the optimised structure does not increase the overall dimensions. This optimised bracket design showed a weight reduction of over 90% compared to the original bracket design, while still being able to meet loading conditions comfortably. From a DFMA perspective, the complex geometry of the optimised bracket would prove very difficult, slow and expensive to manufacture. However, additive manufacturing is

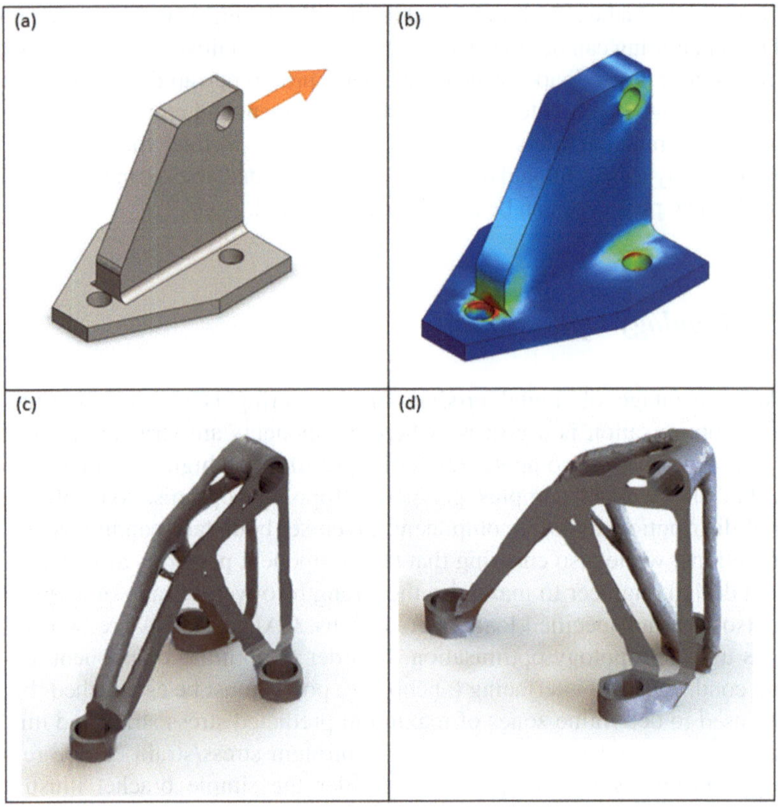

Fig. 2.1 (**a**) Schematic of a simple bracket, showing the direction of a 15 kN applied load; (**b**) FEA results showing zones of high stress (red) and low stress (blue); (**c**) and (**d**) rendering of topology-optimised structure front and rear, showing organic internal structure

renowned for its design freedom, and thus, these structures can be readily produced with this technology at no additional expense.

2.3.3 Geometrical Design Freedom at Low Cost

For conventional manufacturing technologies, there is a direct link between component complexity and cost: as component complexity increases, manufacturing cost also increases (Hague 2006). This cost increase is primarily attributed to both tooling costs and extended machining time. As components become more intricate, tooling needs to become more complex, and more fixtures may be required. In fact, some complex geometries are just impossible to manufacture using conventional subtractive manufacturing, irrespective of cost. In addition to these points, further assembly may be required to complete very complex and intricate components, again driving costs upwards. This means producing a medical device such as a patient-specific anatomically correct model of a human joint is difficult and costly to do using conventional manufacturing.

Additive manufacturing, on the other hand, offers almost complete geometrical design freedom, which can be attributed to the layer-by-layer fabrication process. In addition to this, component complexity is independent of cost for additive manufacturing. This means that design engineers can produce virtually any shape imaginable, without the associated increase in manufacturing costs that would be incurred using conventional manufacturing technologies.

One of the primary areas where additive manufacturing holds an average over traditional manufacturing is the lack of tooling requirements and associated costs. By this, it enables design engineers to produce virtually any geometric shape from a spectrum of materials, including polymers, metals and ceramics. This has enabled design engineers to produce complex, organic shapes, such as customised human bone structures, as well as topology-optimised aerospace components. Multi-material components can be designed and fabricated, with embedded functions and internal features. The ability to rapid prototype these during development and testing also reduces time to market, as complex tooling does not need to be designed nor fabricated. It also enables manufacturers to produce components in final form, which can be comprehensively tested and should design alterations be required; these can be done digitally without the need to update tooling.

2.3.4 Product Customisation

Due to the fundamental design freedom of additive manufacturing, customised, personalised products are now readily achievable. Not only that, but mass product customisation is also readily achievable. This has been a key driver for the uptake in additive manufacturing in the medical device field. Customers who receive a product which more closely suits their personal requirements experience increased customer satisfaction, which increases business value (Hague 2006).

Humans come in all different sizes and shapes, and although most medical devices may cover 90% of the population, these products are not always a best fit for the remaining 10%. However, additive manufacturing has opened new doors, where customised hearing implants can be 3D printed and custom dental crowns are additively manufactured, while humans across the globe are now using titanium orthopaedic knee replacements fabricated using additive manufacturing (Lipson and Kurman 2013).

In fact, traditional medical imaging techniques lend themselves well to additive manufacturing. Computed tomography (CT) and magnetic resonance imaging (MRI) stack slices of information, layer by layer, in order to develop a 3D image. Additive manufacturing produces solid objects layer by layer, in a similar fashion. Thus, CT and MRI data can be segmented and reproduced as a digital model, which can readily be 3D printed. The reproduction of 2D medical imaging data into 3D models has a huge advantage in terms of education. For example, one report looking at patient-specific additive manufactured models of congenital heart defects models of children suggested that cardiologists received much better engagement from parents when discussing the condition using the 3D models (Biglino et al. 2015). Additively manufactured cardiac models have also successfully been used in cardiac surgery planning, with reported potential for reduced operation times (Valverde et al. 2015). MRI and CT data can also be used to produce patient-specific implants, such as craniomaxillofacial (CMF) implants. These implant devices are an ideal example of the patient-specific nature of medical device manufacturing as the skull is one of the most varied structure through all homo sapiens despite ethnicity. The ability to design and additively manufacture custom CMF plates and 3D models of patients skulls can allow surgeons to test the implants for fit prior to surgery, reducing surgery time, and as a result, a more rapid patient recovery (Parthasarathy 2014).

2.3.5 Product Consolidation

One of the guidelines for DFMA is to reduce the number of sub-components used to make a product. This is referred to as part or product consolidation, where the number of discrete sub-components in a product is reduced by designing and fabricating these sub-components into one part, reducing cost and assembly time. Using traditional manufacturing processes, the consolidation of sub-components may be financially or technically challenging to achieve, which may negatively impact the final cost per unit. As a result, manufacturers have favoured the cheaper alternative of assembling many simple sub-components compared to the more expensive process of producing a very complex, consolidated component.

Additive manufacturing lends itself very well to product consolidation due to the layerwise construction during the manufacturing process. The enablement of part consolidation through additive manufacturing means that manufacturing requires less assembly, reducing part count and thus reducing the number of fasteners required. This also means that slow or difficult assembly processes can be avoided. Parts are also lighter, as components do not need to be reinforced with thicker material around fasteners. From an enterprise level, fewer sub-components also mean less storage requirements, less tracking, fewer orders, less sub-suppliers to manage, as well as less tooling.

2.3.6 Lightweight Structures

The design freedom of additive manufacturing has allowed designers to produce micro lattice structures, or scaffolds, which have a range of exciting applications. These lattices, like the scaffolding seen on the exterior of a building being constructed, consist of a series of interconnecting struts and nodes. Typically, a

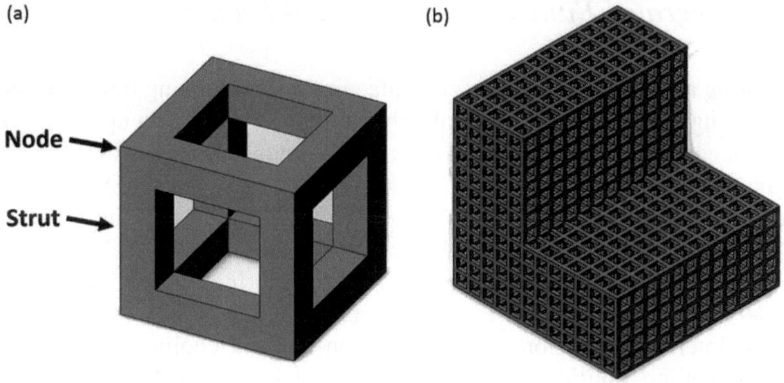

Fig. 2.2 Schematic illustrations showing (**a**) cubic lattice unit cell, highlighting node and strut regions and (**b**) an L-shaped component comprised of a homogeneous lattice structure

homogeneous lattice structure would consist of a unit cell (see Fig. 2.2), which is reproduced throughout the conformal volume where the lattice is required (Yang et al. 2015). The strength, stiffness and weight of these lattice structures are dependent on the size of the unit cell and the thickness of the lattice struts. Additive manufacturing allows for designers to alter the unit cell at different areas of a component producing heterogeneous lattice structures, which means that variable stiffness and weight can be obtained throughout a component. This kind of practice allows designers to lighten components, but also means strength can be increased in localised areas where there is concentrated loading. This maximises the strength-to-weight ratio of a component—ideal for aerospace and automotive applications where component weight needs to be minimised.

Although stiffness is one metric by how these lattice structures are evaluated, another is impact energy absorption. The unit cell and strut thickness have a marked effect on energy absorption, where lightweight structures can be produced for additive manufacturing which can readily absorb impact through deformation of the lattice structure. One such application envisioned for this would be sports personal protective equipment, which is lightweight for the user to perform, but tough enough to absorb impact during sport (Brennan-Craddock et al. 2012). Another application for these lattice structures is the medical device sector, where randomised lattice structures can be produced with a pore size of less than 300 μm mimicking cancellous bone (Wang et al. 2016). This has a huge benefit for orthopaedic devices such as knee and hip implants, where lattices at the bone-implant interfaces can be additively manufactured from biocompatible alloys such as Ti-6Al-4V and CoCr, mimicking the bone structure. These structures enable osteointegration—where the bone grows into the implant, improving implant fixation (Shah et al. 2016). Not only this but because the lattice structure is less stiff than the alloy implant, there is less of stiffness mismatch between the bone, which helps to reduce stress shielding and bone resorption, an issue with traditional orthopaedic implants (Shah et al. 2016).

2.3.7 Integrated Functions and Internal Features

One exciting application of additive manufacturing is integrating functions within a product being fabricated. For example, the ability to produce complex geometries means that complex internal features can also be produced. This allows designers to integrate functions into components, such as including an internal wiring conduit, or complex, free-form fluid channels which have been optimised using CFD simulation. Consider a mould for an injection moulding process. Traditionally, these would be fabricated using traditional subtractive manufacturing, where internal cooling channels would be drilled into the mould plate where possible, often resulting in undesired internal T-junctions and overall non-uniform cooling. Using CFD and thermal modelling, optimised cooling channels can be designed into a mould plate, which can then be produced using additive manufacturing. This has shown to reduce moulding cycle time, improve process efficiency and reduce internal corrosion (EOS GmbH 2013).

Articulating mechanisms are another kind of integrated function which can be produced using additive manufacturing. Articulating parts can be produced in situ, such as interlocking gears, sliders, hinges, ball joints, etc. The ability to include the mechanised functions into additively manufactured components not only reduces (or eliminates) assembly time but also eliminates the need to purchase, stock and trace small, intricate sub-components around a factory floor or warehouse. These mechanisms can also be fully sealed within a component such as a medical device during manufacturing, allowing designers to produce intricate internal systems without the risk of internal contamination during use.

2.3.8 Multiple Material Builds

Advances in additive manufacturing have allowed for multiple material builds to be produced, which has opened further doors for design engineers. Design engineers can now actively select what materials are required, and where. Multiple materials may be used to produce dissolvable support structures, reducing post-processing time. Multiple material may also be used to improve the mechanical properties of the component, such as the addition of high-strength materials at load-critical locations. One multi-material combination achievable by additive manufacturing is thermoplastic and carbon fibre. Components can be printed using thermoplastic, while carbon fibre strands can be selectively added to the material during the print process to increase component strength in desired locations (van der Klift et al. 2016). This means that external load-bearing devices such as splints or braces can produced stronger and lighter, resulting in a more comfortable user experience for the patient.

However, difficulties may arise when trying to combine different material using additive manufacturing. Firstly, the additive manufacturing hardware needs to be capable of providing more than one material feedstock to the build interface. This kind of hardware, such as a multi-head extrusion nozzle in FFF, increases the cost of the manufacturing equipment and requires more maintenance. In addition to hardware issues, there can be differences in thermal expansion/contraction in multiple material builds. As heat is used to additively manufacture parts in most cases, different rates of thermal contraction upon cooling can lead to residual stress, component warping and delamination and void formation.

2.3.9 Optimisation of Supply Chain and Inventory

From a supply chain perspective, the fact that parts can be readily produced on demand means that supply chains will be transformed. Large inventories of parts can be eliminated and replaced with 3D scanners, a CAD file repository and some additive manufacturing equipment. This will allow for decentralisation of manufacturing sites. Taking manufacturing decentralisation to one extreme, the International Space Station has a 3D printer on board for manufacturing general purpose devices like tools and brackets, but in 2017 it was used to manufacture a finger splint for an astronaut—the first medical device ever to be manufactured in space (Scott 2018). Back on earth, additive manufacturing supply chain optimisation is set to have a major impact on the future of healthcare. Instead of patient-specific implants being manufactured at one central manufacturing facility, it is more likely that large hospitals will carry their own 3D scanners, 3D printers and sterilisation equipment. This will allow the patient to come to the hospital for a 3D scan and then wait for their personalised device which is additively manufactured in the room next door.

Production of parts on demand is particularly suited for spare part supply chains, such as those for older transport systems like planes, trains and cars where subcomponents are no longer being produced by original equipment manufacturers (OEM). Due to the set-up time and tooling costs of conventional manufacturing, it would not be economically viable to reproduce these parts once the OEM ceases manufacturing, so instead, OEM parts would be purchased and stockpiled in a warehouse in case of future failure. This practice drives cost upwards, as there is an upfront purchase cost, and then a cost associated with long-term storage. Some products may even become obsolete over time and are eventually scrapped.

The alternative to stockpiling spare parts for years or decades is to reproduce these parts on demand. Low-volume production of spare parts using conventional manufacturing technologies can be very expensive due to tooling and setup costs. However, additive manufacturing is very suited this, and does not require additional tooling. A study examining the use of metal additive manufacturing to produce replacement air ducts for F-18 Hornet jet aircraft was conducted by Khajavi et al. (2014). Although it was highlighted that the slow production of the spare parts was a challenge compared to delivery from a warehouse, the author's model predicted that in the future, where additive manufacturing costs decrease and manufacturing speed increases, there would be localised production facilities producing spare parts. The economic benefits identified include a reduction in distribution costs, inventory storage costs, component obsolesce costs and less enterprise resource planning requirements (Khajavi et al. 2014).

2.4 Suitability of Additive Manufacturing

An essential precursor to DFAM is to take a holistic perspective on the product, to ensure that additive manufacturing is a suitable and cost-effective manufacturing technique. As is the case with DFMA, the selection of manufacturing technology should be made in the early product design phases of the design process to ensure the product design can be achieved by the manufacturing technology. Some considerations which must be made include product function, product performance, geometrical complexity, product personalisation, sales volumes, production times and ramp rates for launch to market. Additive manufacturing is a strong candidate for the manufacture of complex, intricate parts, patient-specific products, in low volumes, but other manufacturing techniques can be cheaper and more suitable when these design considerations change. For example, if a plastic product like a syringe is to be produced in high volumes, with little complexity and no customisation, then an injection moulding process would be more suited. Similarly, if a metallic component is to be produced in high volume, such as an aluminium walking frame, then perhaps an extrusion process is more suited than additive manufacturing. However, if a customised patient-specific implant is required for example, then additive manufacturing is ideally suited for such a low-volume, complex product.

Regarding material performance, components fabricated using fused filament fabrication (FFF) or direct metal laser sintering (DMLS) have been reported to have anisotropic performance in the Z-direction, that is, lower ultimate tensile strength (Frazier 2014; Redwood et al. 2017). This means that components may be mechanically weaker in one orientation compared to another. It is important to bear this in mind when designing and producing products using additive manufacturing (Frazier 2014). In order to minimise the effect of anisotropy, components should be orientated in such a way that the direction of stress is parallel to the layers of the build, i.e. perpendicular to the Z-direction (build direction).

2.5 Product Design Considerations

Like any design process, design engineers need to design a component which meets customers functional requirements, of high quality and can be manufactured cost-effectively. As part of this design process, the design engineer must select what manufacturing technologies will be used, as different manufacturing technologies have their own individual constraints which will affect the final design. Once a manufacturing technology is selected, the design engineer must develop a product or component which can meet the function, quality and cost requirements, but it also must avoid certain design features which result in common defects for the selected manufacturing technology. It has been already highlighted that not all parts are suited to additive manufacturing; however, the decision to select additive manufacturing as the fabrication process needs to be made early in the design stage.

As with traditional manufacturing, there are specific design considerations which should be made, as well as features which should be avoided when designing for additive manufacturing (DFAM). The first crucial decision which needs to be made is a decision of what type of additive manufacturing technology is to be selected—as different additive manufacturing technologies have different constraints regarding residual stress, minimum wall thicknesses, the necessity for support structures and many more. The layer height, or Z-height resolution, will need to be defined, where there is a trade-off between build time and surface finish. Dependent on the additive technology selected, the necessity and placement of support structures as well as component orientation need to be considered, as this influences manufacturing costs, post-processing, aesthetics and component performance. Careful attention also needs to be paid to design features such as overhangs and unsupported features, small holes, horizontal holes, thin walls and internal hollow sections, as incorrect component design can result in failures during the build process, or failure of the final component. Additive manufacturing also has its limitations with regard to geometrical tolerances and surface quality, and thus, post-processing may need to be completed to ensure product meets design requirements (Redwood et al. 2017).

2.5.1 Additive Technology Selection

A critical decision which needs to be made when addressing DFAM is the selection of additive manufacturing technology. There are several different technologies which are available to design engineers, and each of these technologies has their own advantages and drawbacks. The selection of the additive manufacturing technology can be guided by cost (material cost, energy cost and capital cost of hardware), material type, surface finish or geometric accuracy for example. Fused filament fabrication (FFF), for example, is a cost-effective printing technique, but the surface finish and geometrical accuracy of the part would not be as good as other additive manufacturing processes such as stereolithography (SLA) and powder bed selective laser sintering (SLS). Components fabricated using SLA can be transparent and have exceptional surface finishes compared to other additive manufacturing technologies. The materials used in SLA however are brittle and are not suited for mechanical or load-bearing parts. SLS powder bed additive manufacturing can produce robust nylon components which do not need support structures; unlike FFF, however, the manufacturing powder can become trapped in hollow surfaces, which is not an issue with FFF. Hardware costs for SLS equipment are considerably higher than FFF hardware. When selecting additive manufacturing technology, the design engineer should weigh up what features are critical to quality and select the appropriate technology which can meet financial constraints as well as component performance.

2.5.2 *Material Selection*

Material selection is essential for all product design and all manufacturing technologies. The correct material selection will strike a balance between user requirements, function, product quality and manufacturing cost. Often, product performance will drive material selection. However, visual appearance may also be a consideration when using additive manufacturing. For example, if a product or component needs to be functional with low-strength requirements, then polylactic acid (PLA) or acrylonitrile butadiene styrene (ABS) produced using FFF would be suitable and cheap to produce. For medium-strength applications (30–85 MPa), nylon components produced using SLS would be a welcome choice, but at a higher cost than FFF. Higher strength applications would most likely require metallic parts produced using DMLS. If a product needs to be biodegradable, then PLA is a good choice. If the component is required for medical purposes, such as a load-bearing orthopaedic implant, then the material would need to be biocompatible, while also resistant to degradation in the body. It should also have adequate mechanical performance. Thus, a titanium alloy (Ti-6Al-4V) would be suited for this application. If a product needs to be transparent, or have a smooth external surface, then SLA using a photopolymer resin should be selected. Overall, the selection of material should be decided on a case-by-case basis, where the most critical user requirements guide both material selection and additive technology selection (Redwood et al. 2017).

2.5.3 *Layer Height*

The layer height, sometimes referred to as Z-height resolution, has a direct effect on the surface finish of the part. Consider Fig. 2.3, which shows a schematic cross-section of two additively manufactured hourglass components. The component on the left has been manufactured with a layer thickness which is double that of the component on the right. This schematic shows that as layer thickness doubles, the surface becomes rougher, resulting in a pronounced "staircase" effect on curved surfaces. As the layer thickness decreases, the surface finish improves, and the staircase effect becomes less pronounced. The trade-off of surface finish is fabrication time. If the layer thickness halves, then the number of layers double, which approximately doubles the build time. Thus, to build parts fast, larger layers should be used; however, if a smoother surface finish is required, then it is recommended that the layer height is minimised. It is worth noting, however, that the maximum quality of surface finish is constrained by minimum layer height, meaning that very smooth surface finishes are not possible with most additive manufacturing technologies. In

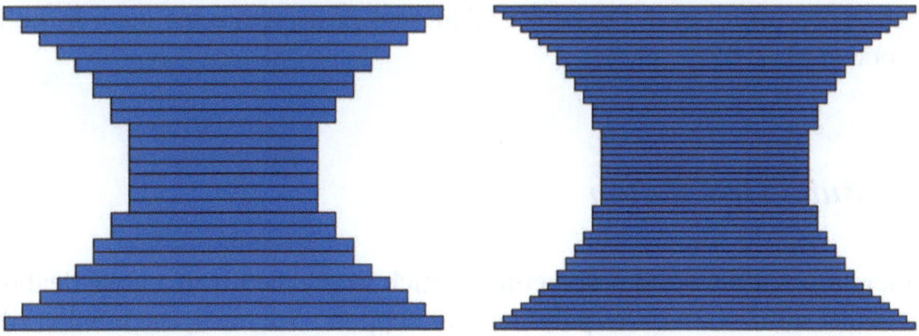

Fig. 2.3 Schematic illustrating the effect of layer thickness on surface finish, where the component on the left was fabricated with twice the layer thickness of the component on the right, and thus, the staircase effect is more pronounced

these cases, a post-manufacturing surface treatment such as machining or polishing may be required.

2.5.4 Support Structures

Support material, or support structures, are essential in some additive manufacturing technologies and serve multiple functions. These are required in polymer FFF and SLA manufacturing, and DMLS, but not required in polymer SLS additive manufacturing. There are several reasons why these structures are required. For example, in polymer additive manufacturing, these structures provide a build platform for overhangs to be built onto. They also ensure that the component is securely fixed to the building platform—which is essential to avoid residual stress-induced warpage. They have an additional function in DMLS, which is to act as a heat sink and pull heat from the melt surface into the build plate to aid solidification and cooling (Redwood et al. 2017).

The material used to fabricate support structures can be considered as material wastage, as once the component is manufactured, the support material is cut away and scrapped. As a result of this careful consideration needs to be made when designing support structures. Ideally, a balance needs to be struck where minimum material is used to produce these structures, but the structure also needs to be strong enough to resist support delamination where the support structures fracture due to excessive residual stress. This is undesirable, as it can result in component warping during manufacture, and potential damage to the additive manufacturing equipment.

Another design consideration which should be made is where the support structure should be applied to the component. Support structures should be avoided on thin walls, or on "cosmetic surfaces", which are aesthetic areas where surface finish is critical. The removal of support structures in a post-processing step can cause thin walls to fracture or leave a poorly textured surface finish on a cosmetic surface of a component, resulting in the need for additional post-processing such as polishing (Redwood et al. 2017).

2.5.5 Build Orientation

The orientation of the build is something which should be given a great deal of consideration. It influences the requirement for support structures, the number of parts which can be manufactured on the same build plate at once, overall height of the build, the building time and the component mechanical properties.

Consider a simple "T-" shaped component shown in Fig. 2.4a, which contains three holes. There are several different ways a design engineer could orientate this component, with advantages and disadvantages of each. Figure 2.4b shows the "T-" shaped component upright. In this case, however, for FFF and DMLS, it is not possible to

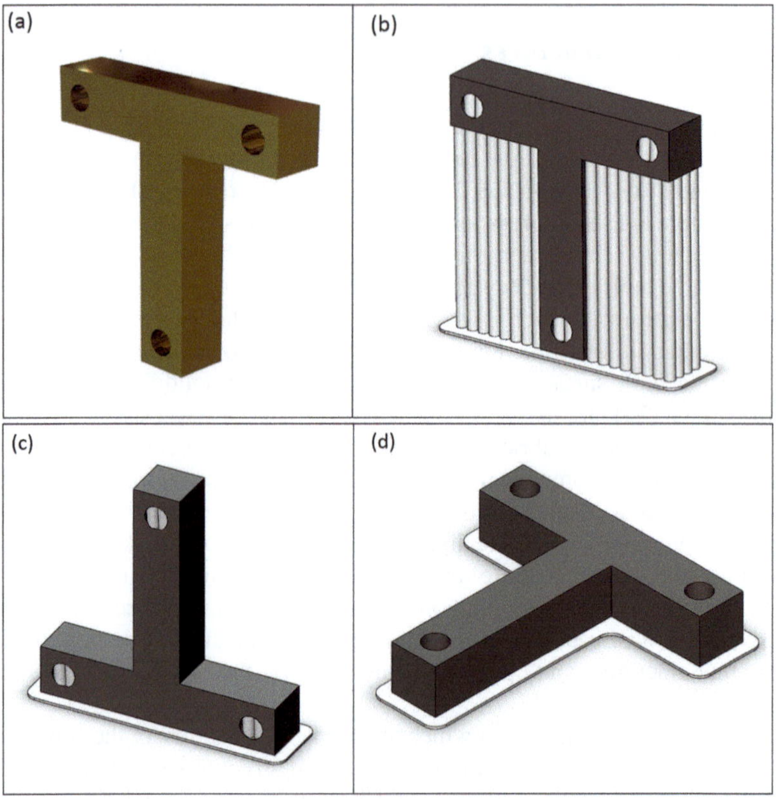

Fig. 2.4 Schematic of simple "T-" shaped bracket (**a**) final form; (**b**) upright build orientation showing light grey support structures; (**c**) upside-down orientation showing fewer support structures and (**d**) lying flat on the build plate

manufacture the overhangs without building support structures on either side. Thus, light grey support structures are seen in this figure. In addition to this, support structures are also added to the holes, which prevents the top of each hole collapsing during manufacture. By flipping the build orientation 180°, so that the horizontal portion is on the base of the build, support structures can be minimised, as shown in Fig. 2.4c. Using this orientation, a significant amount of material (45%) and fabrication time (40%) can be saved compared to the orientation used in Fig. 2.1b. By lying the component flat on the build plate, as shown in Fig. 2.4d, there is little difference in terms of build time and material volume for FFF when compared to Fig. 2.4c; however, it is good practice to keep the build height low where possible. In addition to this, as the axes of the holes align with the Z-direction in Fig. 2.1d, internal support structures are not required, saving both materials, but also post-processing. This orientation, however, requires a large footprint, so if the quantity of brackets to be produced is significant, more components will fit onto the build plate using an upright orientation.

Regarding build orientation for maximum mechanical properties, the component should be orientated so that the direction of predicted maximum stress aligns closest to the plane of maximum strength. In the case of additive manufacturing, the plane of maximum strength is in the *X–Y* plane. Thus, if the predicted maximum stress in the component was going to be along the leg of the "T", then the flat build orientation shown in Fig. 2.4d would produce the strongest component out of the three orientations.

2.5.6 Overhangs and Unsupported Features

Additive manufacturing techniques such as FFF, SLA and DMLS require a build platform beneath each layer—leading to the requirement for support structures. This means that overhangs, such as the arms in the T-shaped component presented in Fig. 2.4b, require support structures. In the case of this component, the angle between the underside of the overhang and the build plate is 0°. As this angle increases to 45°, support structures are required for FFF, SLA and DMLS; however, as the angle increases beyond 45°, support structures are not necessarily required for FFF (Redwood et al. 2017). For horizontal bridges, support structures are not always required for FFF and DMLS. In the case of FFF, bridge lengths of up to 10 mm can be achieved without supports. However, the surface finish of the down-skin (material facing down towards the build plate) can become rough. In the case of DMLS, the maximum bridge length recommended is only 2 mm (Redwood et al. 2017). For overhangs or longer bridges which are not supported, the build material is likely to fall into the void, or droop, which means there will be inadequate geometry resulting in product failure.

2.5.7 Hole Design

Producing accurate, concentric holes can be tricky to do with additive manufacturing, especially if holes are less than 1 mm in diameter. Technologies such as FFF has a dimensional accuracy in the region of ±0.5 mm (Redwood et al. 2017). For holes less than 2.0 mm in diameter, it is recommended that these holes are undersized, and a post-process drilling operation is used to increase the hole to the desired diameter. For long holes and cylindrical channels, it is recommended that the axes of these holes align with the Z-direction for technologies which require support structures such as FFF and DMLS, as illustrated in Fig. 2.4d. In this case, the "T-" shaped component is lying flat on the build plate, and thus the hole axes in the component do not need support structures. If a hole axis cannot be aligned with the Z-direction, the hole will likely require internal support structures to ensure it does not collapse in on itself during manufacture (see hole support structures in Fig. 2.4b, c). These support structures will need to be drilled out post-manufacturing, adding time and expense to the manufacturing process. If using a powder bed additive manufacturing process, blind holes will fill with powder. Depending on the aspect ratio of the hole, there may be difficulties removing loose manufacturing powder directly after printing, meaning further post-processing may be required.

2.5.8 Hollow Sections and Escape Holes

The design freedom of additive manufacturing means that designers can incorporate hollows into components for various purposes. These hollows may be used as fluidic channels or reservoirs, or perhaps used to reduce component weight. When a component is being additively manufactured using powder bed technology, manufacturing powder may become entrapped within these enclosed hollow sections of the component. This is undesirable, as it adds unnecessary weight to the part, but also wastes manufacturing powder which can be recovered and reused. To allow loose manufacturing powder to exit these hollow sections, escape holes should be incorporated into the design of the component. These holes are intentionally added to walls of hollow sections to allow loose internal powder to flow out of the component. The size and location of these holes are selected on a case-by-case basis. However, it is good practice to increase the size of the hole proportional to the volume of the powder-filled hollow section. Tiny escape holes combined with a large reservoir of entrapped powder will have an hourglass timer effect, resulting in longer cycle times as powder is restricted exiting the component. The location of escape holes should be away from load-critical areas, as these will act as load concentrators. In addition to this, a plug can be fabricated during the build, which can be glued in place to fill the escape hole once entrapped powder has been removed.

2.5.9 Thin Features

Thin walls and other thin features can be challenging to produce accurately using additive manufacturing, and different additive manufacturing technologies have their lower limits concerning the minimum thickness of these features. For example, it is recommended that features printed using FFF are 0.8 mm wide or greater. This increases to a minimum wall thickness for SLA of 1.0 mm (Redwood et al. 2017). Walls thinner than the recommended values are prone to damage and failure, and geometrical inaccuracies. These excessively thin walls may also suffer from thermal warpage or delamination. For powder bed additive manufacturing processes which use re-coater blades, the orientation of thin walls with regard to the direction of travel of the re-coater blade (coating direction) is something which needs to be considered carefully. Friction between the re-coater blade can cause thin walls to bend, and thus, the length of the thin wall should be aligned near to parallel (5° to 10° offset) to the direction of re-coater blade travel to ensure the blade has minimum effect on the wall.

2.5.10 Geometric Tolerances and Surface Quality

As discussed previously, the surface of a component manufactured using additive manufacturing may need to be modified for geometric tolerance or aesthetic reasons. It has already been discussed how the layer-by-layer building of components using additive manufacturing results in a staircase effect, where the step height corresponds to the layer thickness. Typically, the larger the build volume and the faster the build process, the less accurate parts are. Loose powder may also be bonded onto the surface, which needs to be removed to improve surface finish. Witness marks resulting from the removal of support structures may also cause aesthetic defects on surfaces (Gibson et al. 2010). To improve geometric tolerances and surface quality, components may be post-processing in the form of machining, abrasive blasting, laser machining, or polishing (Kumbhar and Mulay 2016).

2.6 Post-processing

DFAM not only considers the additive manufacturing process but also post-processing steps which ensure the final product meets customer requirements. Additive manufacturing can produce net-shape, or near-net-shape products, depending on the final requirements of the end product. In most cases, however, the fabricated product will need to undergo some form of post-processing step. Although additive manufacturing technology is very effective at making custom products in low volumes and of high complexity, it has intrinsic limitations regarding final product quality which can be improved with post-processing techniques. For example, depending on the product requirements, post-processing may need to be considered to remove support structures or other residual manufacturing material, improve surface finish and improve geometrical tolerances.

2.6.1 Material Removal

Material removal is a broad term, which relates to the removal of unwanted manufacturing material from additively manufactured products which is present after manufacturing but is not to be included in the final product. The two primary material removal processes are typically support structure removal and residual manufacturing powder removal. It has already been discussed how support structures are an essential part of the additive manufacturing process which ensures that complex geometries can be achieved, but these structures are unwanted in the final product and thus need to be removed during post-processing. In cases where these support structures are made from the same material as the component, these are removed using either a cutting process or an abrasive process, which can be laborious for components with complex geometries. Support structure removal may also leave blemishes or witness marks on the surface of the part, which may be undesirable for aesthetic requirements. Thus, in order to avoid tricky or time-consuming support structure removal, careful consideration should be made regarding the placement of these structures. Perhaps a component could be orientated in such a way during manufacturing so that support structures can be minimised or avoided altogether. Multi-material FFF hardware can allow for the fabricating of support structures using a soluble material which is different to the material fabricating the component. For example, polyvinyl alcohol (PVA) can be used to construct the support structures for a part being fabricated from ABS. PVA is water soluble, meaning that once the component is finished, it can be submerged in warm water, which entirely dissolves the support structure from the component—ideal for automating support structure removal from complex parts (Redwood et al. 2017).

Regarding SLS and DMLS, a critical post-processing step is residual powder removal. During the manufacturing process, the components become submerged in powder beds, meaning that all cavities—closed or open—are filled with un-sintered manufacturing powder. For closed hollow sections, it is recommended that escape holes are added in order to ensure powder can escape from the component. Vibration can be used to help agitate trapped manufacturing powder trapped aiding it to exit. Water-based bulk powder removal should be avoided for components with powder trapped in hollow sections. Water moisture is drawn into the powder-filled hollow by capillary action and causes the loose manufacturing powder to bind together like wet sand—making it very difficult to remove once wet. Fabricated lattice structures are also prone to being filled with loose powder. This loose powder can be mostly removed using a combination of vibration and pressurised streams of air or other inert gases.

In addition to loose powder, components manufactured in powder beds also have partially bonded manufacturing powder adhered to its surfaces. Some of these powder particles may be barely sintered to the surface, while some may be tightly fused. This poses a risk for components which may be sensitive to detachment of these beads during operation—such as medical implants or precision engine components. One such method for removing externally bonded powder is grit, sand or bead blasting, where an abrasive material is blasted at the surface of the part in a high-pressure air stream. This abrasive material knocks off any loosely adhered manufacturing powder and can also texture the surface of the component as desired (Redwood et al. 2017; Townsend et al. 2016). For internal features such as intricate cooling channels, where grit blasting is not readily achievable, abrasive slurry or paste can be injected through these channels, which helps cut any loosely bonded manufacturing powder off internal surfaces.

2.6.2 Surface Finishing and Improving Geometrical Tolerances

Surface finishing is an essential step in manufacturing—not just additive manufacturing. Indeed, the surface finish of an additive manufactured component is dependent on aspects such as additive manufacturing technology, layer height and support structure placement. Depending on the desired level of surface roughness of a final component, it may be necessary to improve the surface finish using surface treatment post-processing steps. These surface treatments are used to reduce surface roughness and the staircase effect of additively manufactured components. The geometrical accuracy of the component may also be increased with post-processing steps like CNC surface machining which can create a better surface finish and more accurate geometrical tolerances. In this case, components are manufactured using the additive manufacturing technology of choice, and then traditional machine cutting is used to ensure accurate geometrical tolerances. For this subtractive manufacturing process to work, however, there needs to be additional material stock added during the additive manufacturing process, where components are fabricated slightly oversized, and holes are fabricated undersized. Undersized small holes allow for these to be drilled out to required dimensions. Similarly, laser machining may also be used to finish external surfaces to ensure they meet both surface roughness and geometrical tolerances (Kumbhar and Mulay 2016).

Machining can sometimes result in machine lines, which are small marks left on the surface finish which show the path taken by the tool. If these are not desired, then grit blasting can be used to create a homogeneous surface finish through abrasive erosion. The roughness of the finish can be controlled by the size of the grit media. Grit blasting has a consequential impact on surface dimensions, however, where sharp features may become rounded, while complex surface features may also become worn away. Caution should also be taken during grit blasting of components which have surface openings for internal features, or exposed lattice structures. Grit can enter these regions and prove very difficult or impossible to remove. It is best practice to avoid blasting these components or to mask off these sensitive areas using appropriate masking material which can withstand blasting.

If very smooth or polished surface finish is required, then an additional polishing step will need to be completed after blasting. Vibro polishing can be used to reduce surface roughness, where parts are placed into a vibrating tumbler with small ceramic chips. The vibration in the tumble causes the hard chips to rub along the surface of the components, removing micro roughness peaks. Like blasting, this abrasive process may also cause further rounding of sharp features. Caution should be taken when polishing components which have complex surface features or open lattice structures, which may become damaged or entrap polishing media.

2.7 Product Consolidation and Weight Saving Using Additive Manufacturing

One of the key philosophies in conventional DFMA is to reduce the number of components in a product, in order to reduce weight, costs and manufacturing time. Additive manufacturing has evolved from being a rapid prototyping tool into a viable production process, capable of producing end-user product, becoming a key enabling technology for part consolidation. The ability to produce almost any geometrical shape in a final product at no additional cost has started a manufacturing design revolution. Designers can now produce components which have been topology optimised with weight savings above 90% or consolidate multi-component parts into single units which are lighter and more rigid than original versions. This design ability has many applications in heavily regulated manufacturing sectors such as medical devices and aerospace.

The fabrication of supports and braces is one key area in the medical device sector which lends itself to both part consolidation and weight savings. For example, the traditional fabrication of a plaster cast to protect a broken wrist requires several materials such cotton padding, plaster or fibreglass sheets, bandages, adhesive tape and instruments. These traditional plaster casts tend to be heavy, cumbersome, non-ventilated, un-washable and non-removable, meaning the patient experience can be poor (Kim and Jeong 2015). Additive manufacturing has been used with 3D scanning to produce lightweight patient-specific casts. These casts are consolidated into a single, additively manufactured structure, meaning the slow and often messy process of traditional cast fabrication can instead be replaced with a cleaner manufacturing process. These additively manufactured casts offer a variety of benefits to the patient experiences, such as lightweight yet rigid design, the ability to remove for bathing, and also offer good ventilation that can prevent formation of skin irritations and diseases which can form when using traditional casts (Fitzpatrick et al. 2017). The patient-specific design of these casts also means better fitting and thus better comfort for the patient during recovery.

Similar to wrist supports, additive manufacturing is also being used to produce lightweight, flexible bespoke back braces for children who suffer from scoliosis. Using DFAM principles, medical device manufactures have been able to consolidate these corrective braces with integrated lightweight lattice structures to reduce weight and improve flexibility. These topology-optimised, patient-specific devices are so discrete that children can wear them under their clothes without drawing attention from their peers—massively improving patient experience (Krassenstein 2014). Foot and ankle-foot orthoses devices used to realign lower legs is another area which is seeing the advantages of additive manufacturing as a production method. These devices are typically bespoke, and traditionally fabrication is laborious, requiring plaster cast moulds of patient's legs, reproducing a physical plaster model of the leg and then manually fabricating the orthoses to fit the model leg, before adapting it to fit the patient. With additive manufacturing and 3D scanning, scans of patient's legs can be captured in minutes, which are then used to design bespoke orthoses devices digitally. These devices are then topology optimised, where unnecessary material is removed, producing a lighter component compared to traditional manufacture. In addition to this, lightweight scaffold structures can be integrated into the load-bearing sites of the device, which elastically deform under loading to absorb impact and cushion the user—again improving patient experience (Pallari et al. 2010).

Product consolidation of cementless implants such as those used in total knee or total hip arthroplasty is something which additive manufacturing is also very capable of. Cementless implants do not require bone cement and, instead, have a porous osteointegration coating along the bone-cut surface. Traditionally, these porous coatings on titanium implants are manufactured using sintered titanium beads, with a typical diameter in the region of 400–700 μm (Carlsson et al. 2005). It is not possible to manufacturing these coatings during the forging or casting process; thus, these sintered coatings are applied to the implant in a post-processing step, adding time and cost to the manufacturing process. Additive manufacturing however can reproduce more effective porous coatings, which are fabricated during the manufacturing of the complete part, meaning a post-manufacturing sintering process is not required. However, metal parts fabricated by DMLS do require a heat treatment step post manufacture (Frazier 2014). The presence of these lattice structures after manufacturing also means that these may absorb any post-processing materials or liquids such as polishing media, so caution should be taken during the manufacturing process to limit exposure of lattice structures.

In addition to the medical device sector, the aerospace sector is another area which is using DFAM principles to revolutionise aircraft manufacturing. For example, general electric (GE), an American multinational company, has been using additive manufacturing to revolutionise the manufacture of gas turbine aircraft engines. GE has been pushing the boundaries with metal additive manufacturing in a bid to consolidate turbine components achieving reduced weight, improved durability and improved efficiency. One of GE's big success stories is the fuel nozzle of the CFM LEAP-1A gas turbine engine, which is used by the Airbus A320neo and Boeing 737 MAX. The traditional manufacture of this component required the assembly of around 20 sub-components, supplied by 20 sub-suppliers, and included high-temperature joining processes such as brazing and welding. However, using additive manufacturing and part consolidation, GE fabricated the fuel nozzle as a single unit, saving both time and cost. Not only was the final component 25% lighter but also it was also more than five times as durable than its assembled predecessor, resulting in an estimated saving of $3 million per aircraft per year (Kellner 2017; Rao 2015).

Taking product consolidation one step further, GE explored the possibility of using additive manufacturing and part consolidation to produce a complete gas turbine engine. This presented an entirely new design challenge for GE's design engineers, such as task had not been completed before. Using DFAM principles, GE was able to produce the advanced turboprop (ATP) engine using primarily additive manufacturing. The ATP engine was consolidated from 855 components down to just 12. The ability to produce complex internal geometries and consolidated subcomponents meant that GE could simplify the overall engine design to just a dozen parts, eliminating the need to design and tolerance hundreds of sub-components. The results were impressive, where the ATP engine showed a reduction in weight, an improvement in fuel efficiency of up to 20%, with an increase in power of 10%. In addition to these benefits, GE also highlights that the consolidated engine also shows reduced wear and leakage compared to predecessor engines (Kellner 2018).

2.8 Chapter Summary

Unlike traditional manufacturing technologies, additive manufacturing has the ability to produce products with very complex geometries, without the requirement of expensive tooling. This has started somewhat of a design revolution, as design simplification is no longer required by design engineers. Design engineers can produce complex, topology-optimised components, with integrated features, fabricated from several materials, and incorporating lightweight structures. Indeed, it has been discussed that additive manufacturing is not suited in all cases, and during the design process, design engineers need to identify what product traits and customer requirements are driving the manufacturing technology decision. Custom, patient-specific devices are very well suited to additive manufacturing due to low volume and high complexity, while simple, high-volume components may be more suited to an alternative traditional manufacturing process.

There are several design considerations which need to be made during the DFAM process. These included the selection of manufacturing technology, material selection, layer thickness, placement of support structures (if required) and component manufacturing orientation. The latter two considerations will be dictated by the geometrical features of the component, such as overhangs, unsupported features, hole size and hole orientation. Additive manufacturing has limitations in terms of minimum feature size, geometrical accuracy and surface finish. These can be improved in post-processing steps such as grit blasting or sanding, laser polishing, machining or vibro polishing.

Based on the case studies presented throughout this chapter, it is evident that additive manufacturing has presented a huge opportunity to the medical device and aerospace sectors. Surgeons can produce patient-specific 3D models to help visualise surgery and test implants before the patient ever enters the operating theatre. Personalised implants can be manufactured with intricate osteointegration coatings which encourage bone ingrowth. Custom splints and orthoses can be manufactured with integrated lightweight structures, simplifying the manufacturing process, while also producing lightweight, rigid devices. The opportunities for additive manufacturing in the medical device sector have shown encouraging growth for the technology. DFAM principles are key to manufacturing safe and effective medical devices using additive manufacturing.

Acknowledgments This work was supported in part by financial support Science Foundation Ireland (SFI) under Grant Number 16/RC/3918, co-funded by the European Regional Development Fund.

References

Biglino, G., Capelli, C., Wray, J., Schievano, S., Leaver, L.-K., Khambadkone, S., Giardini, A., Derrick, G., Jones, A., & Taylor, A. M. (2015). 3D-manufactured patient-specific models of congenital heart defects for communication in clinical practice: Feasibility and acceptability. *BMJ Open, 5*, e007165.

Boothroyd, G., Dewhurst, P., & Knight, W. A. (2001). *Product design for manufacture and assembly, revised and expanded.* Boca Raton: CRC Press.

Brennan-Craddock, J., Brackett, D., Wildman, R., & Hague, R. (2012). The design of impact absorbing structures for additive manufacture. *Journal of Physics Conference Series, 382*, 012042. https://doi.org/10.1088/1742-6596/382/1/012042.

Carlsson, Å., Björkman, A., Besjakov, J., & Önsten, I. (2005). Cemented tibial component fixation performs better than cementless fixation: A randomized radiostereometric study comparing porous-coated, hydroxyapatite-coated and cemented tibial components over 5 years. *Acta Orthopaedica, 76*, 362–369.

EOS GmbH (2013). Plastic manufacturer FWB achieves major savings in production time and cost using additive manufacturing, Technical report, Retrieved January 11, 2019, from https://cdn2.scrvt.com/eos/public/d6021a2d744738ce/d359c096085bce7316bf56d5c9341a4d/fwb.pdf

Fitzpatrick, A. P., Mohanned, M. I., Collins, P. K., & Gibson, I. (2017). Design of a patient specific, 3D printed arm cast. *KnowledgeE Engineering, 2*, 135–142.

Frazier, W. E. (2014). Metal additive manufacturing: A review. *Journal of Materials Engineering and Performance, 23*, 1917–1928. https://doi.org/10.1007/s11665-014-0958-z.

Gibson, I., Rosen, D. W., & Stucker, B. (2010). Design for additive manufacturing. In *Additive manufacturing technologies* (pp. 299–332). Boston, MA: Springer.

Hague, R. (2006). Unlocking the design potential of rapid manufacturing. *Rapid Manufacturing: An Industrial Revolution for the Digital Age*, pp. 5–18.

Kellner, T. (2018). Fired up: GE successfully tested its advanced turboprop engine with 3D-printed parts—GE reports (WWW document). GE Rep. Retrieved January 11, 2019, from https://www.ge.com/reports/ge-fired-its-3d-printed-advanced-turboprop-engine/.

Kellner, T. (2017). How 3D printing will change manufacturing—GE reports [WWW document]. GE Rep. Retrieved January 11, 2019, from https://www.ge.com/reports/epiphany-disruption-ge-additive-chief-explains-3d-printing-will-upend-manufacturing/.

Khajavi, S. H., Partanen, J., & Holmström, J. (2014). Additive manufacturing in the spare parts supply chain. *Computers in Industry, 65*, 50–63.

Kim, H., & Jeong, S. (2015). Case study: Hybrid model for the customized wrist orthosis using 3D printing. *Journal of Mechanical Science and Technology, 29*, 5151–5156. https://doi.org/10.1007/s12206-015-1115-9.

Krassenstein, B. (2014). 3D systems unveils 3D printed bespoke back brace to treat chronic scoliosis. 3DPrintcom voice 3D print. *Additive Manufacturing*, Retrieved January 11, 2019, from https://3dprint.com/5596/3dprinted-bespoke-back-brace/

Kumbhar, N. N., & Mulay, A. V. (2016). Post processing methods used to improve surface finish of products which are manufactured by additive manufacturing technologies: A review. *Journal of the Institution of Engineers (India): Series C*, 1–7.

Lipson, H., & Kurman, M. (2013). *Fabricated: The new world of 3D printing*. New York: John Wiley & Sons.

Pallari, J. H. P., Dalgarno, K. W., Munguia, J., Muraru, L., Peeraer, L., Telfer, S., & Woodburn, J. (2010). Design and additive fabrication of foot and ankle-foot orthoses. In *Proceedings of the 21st Annual International Solid Freeform Fabrication Symposium—An Additive Manufacturing Conference, 9–11 August 2010, Austin, TX, USA*.

Parthasarathy, J. (2014). 3D modeling, custom implants and its future perspectives in craniofacial surgery. *Annals of Maxillofacial Surgery, 4*, 9–18. https://doi.org/10.4103/2231-0746.133065.

Rao, R. (2015). How GE is using 3D printing to unleash the biggest revolution in large-scale manufacturing in over a century [WWW Document]. TechRepublic. Retrieved January 6, 2019, from https://www.techrepublic.com/article/how-ge-is-using-3d-printing-to-unleash-the-biggest-revolution-in-large-scale-manufacturing/.

Redwood, B., Schffer, F., & Garret, B. (2017). The 3D printing handbook: Technologies, design and applications. 3D Hubs. Vancouver

Scott, C. (2018). The impact of 3D printing on the medical device industry | 3DPrint.com | The Voice of 3D Printing/Additive Manufacturing [WWW Document]. 3DPrint.com. Retrieved January 16, 2019, from https://3dprint.com/220137/the-impact-of-3d-printing-on-the-medical-device-industry/.

Shah, F. A., Omar, O., Suska, F., Snis, A., Matic, A., Emanuelsson, L., Norlindh, B., Lausmaa, J., Thomsen, P., & Palmquist, A. (2016). Long-term osseointegration of 3D printed CoCr constructs with an interconnected open-pore architecture prepared by electron beam melting. *Acta Biomaterialia, 36*, 296–309. https://doi.org/10.1016/j.actbio.2016.03.033.

Tempelman, E., Shercliff, H., & van Eyben, B. (2014). *Manufacturing and design*. Oxford: Elsevier.

Townsend, A., Senin, N., Blunt, L., Leach, R. K., & Taylor, J. S. (2016). Surface texture metrology for metal additive manufacturing: A review. *Precision Engineering, 46*, 34–47.

Valverde, I., Gomez, G., Gonzalez, A., Suarez-Mejias, C., Adsuar, A., Coserria, J. F., Uribe, S., Gomez-Cia, T., & Hosseinpour, A. R. (2015). Three-dimensional patient-specific cardiac model for surgical planning in Nikaidoh procedure. *Cardiology in the Young, 25*, 698–704.

van der Klift, F., Koga, Y., Todoroki, A., Ueda, M., Hirano, Y., & Matsuzaki, R. (2016). 3D printing of continuous carbon fibre reinforced thermo-plastic (CFRTP) tensile test specimens. *Open Journal of Composite Materials, 6*, 18–27. https://doi.org/10.4236/ojcm.2016.61003.

Wang, X., Xu, S., Zhou, S., Xu, W., Leary, M., Choong, P., Qian, M., Brandt, M., & Xie, Y. M. (2016). Topological design and additive manufacturing of porous metals for bone scaffolds and orthopaedic implants: A review. *Biomaterials, 83*, 127–141. https://doi.org/10.1016/j.biomaterials.2016.01.012.

Yang, S., Tang, Y., & Zhao, Y. F. (2015). A new part consolidation method to embrace the design freedom of additive manufacturing. *Journal of Manufacturing Processes, 20*, 444–449.

Chapter 3
Mechanics Modeling of Additive Manufactured Polymers

Pu Zhang, Yiqi Mao, and Xin Shu

Abstract Polymers will continue to dominate the global market of additive manufacturing for at least a few decades. Investigating the mechanics of additive manufactured polymers not only creates fundamental knowledge of their mechanical behaviors but also guides the design and additive manufacturing of engineering or biomedical components. Unfortunately, the mechanical responses of these polymers are strongly dependent on the manufacturing process, material composition, printing orientation, temperature, etc., which bring about great challenges for the modeling. To overcome such challenges, this chapter introduces two high-fidelity nonlinear constitutive models devised specifically for additive manufactured photopolymers to facilitate the computational design process. The first model is developed to take into account the intrinsic anisotropy of photopolymers caused by the layer-wise manufacturing feature, while the second model is targeting the thermomechanical property of photopolymers across the glass transition temperature, which can be used to simulate the shape memory effect. Note that both models are applicable to the finite deformation scenario and inelastic deformation. At the end of this chapter, we also provide our humble opinions on the research demands and trends regarding the mechanics and modeling of additive manufactured polymers.

Keywords Photopolymer · Additive manufacturing · Constitutive model · Anisotropy · Shape memory

P. Zhang (✉)
Department of Mechanical Engineering, State University of New York at Binghamton, Binghamton, NY, USA
e-mail: pzhang@binghamton.edu

Y. Mao
School of Mechanical and Vehicle Engineering, Hunan University, Changsha, China
e-mail: myq_1984@hnu.edu.cn

X. Shu
Department of Mechanical Engineering, State University of New York at Binghamton, Binghamton, NY, USA

School of Mechanical and Vehicle Engineering, Hunan University, Changsha, China
e-mail: xinshu@hnu.edu.cn

D. M. Devine (ed.), *Polymer-Based Additive Manufacturing*,
https://doi.org/10.1007/978-3-030-24532-0_3

3.1 Introduction

Additive manufactured (AM) polymers are usually fabricated from extrusion, photopolymerization, and laser sintering processes. They have gained wide applications in both engineering and biomedical areas as introduced in many other chapters. Early development of the AM polymers was mainly targeting the rapid prototyping application which has been proven to be very successful and still serves as one of the major applications of the AM technology nowadays. In recent years, AM polymers are also used for structural usage when the load-bearing property is not a very critical concern, e.g., tooling, fixture, lattice structures, multifunctional structures, etc. In particular, AM polymeric composites have been commercialized for both the extrusion and laser sintering processes; e.g., the MarkForged Onyx 3D printer utilizes composite filaments with a variety of fibers, and the EOS incorporation offers polyamide/carbon fiber composite powders for laser sintering fabrication. Owing to the high stiffness and high strength, these polymeric composite materials are promising candidates to replace many of the metallic components for future additive manufactured structures. Besides the engineering applications, AM polymers are also used for biomedical applications like bioscaffolds, bioreactors, stents, and implants. Compared to the metallic or ceramic counterparts, the polymer structures usually have advantages like lighter weight, lower price, and better biocompatibility, which make them more competitive than the metallic and ceramic materials for biomedical applications. Although non-polymeric materials, especially metals and alloys, have gained dramatic advancement in the AM market, polymers and polymer printers will continue dominating the market in the near future due to the relatively low price, high technology readiness level, and wide applications.

Exploring the mechanical behaviors of AM polymers is a crucial task in order to facilitate the design and fabrication of engineering and biomedical components made of polymers. Firstly, these components need to satisfy certain design criteria for stiffness, strength, actuation, service life, and among others so that mechanical testing and mechanics modeling/simulation are necessary for the structural design. Secondly, material testing and modeling also need to be carried out to understand the fundamental material behaviors during the processing of polymers since the processing parameters will affect the microstructure evolution and mechanical properties, which in turn influence the deformation and distortion of the parts during the fabrication process. Hence, modeling the material response during processing will enable engineers to choose optimal processing parameters to achieve enhanced properties or reduced building failure. Certainly, the mechanical responses of polymers are distinct for different processing techniques and polymer types. Some of the major processing techniques are remarked below in brief.

For the fused deposition modeling (FDM), thermoplastics are usually melted and extruded layer by layer to form a part, which shows anisotropic property that depends on the orientation and bonding of the beads. Besides the stiffness and strength properties that are usually considered for FDM parts, the processing-induced distortion is also a critical concern since it directly affects the building accuracy of the engineering parts or prototypes, which can normally be predicted through thermomechanical modeling and simulation by conducting finite element analysis (FEA). The modeling of FDM plastics is a relatively matured technique given that lots of research works have been published in this area (Zhang and Chou 2008; Domingo-Espin et al. 2015). However, the modeling for other extrusion processes is rarely considered. For example, researchers have initiated the efforts to fabricate thermoset/thermoplastic composites (Ning et al. 2015; Compton and Lewis 2014) and hydrogels (Gladman et al. 2016) through the extrusion process, but the mechanical properties are not thoroughly studied and well modeled yet. Thus, it is envisioned that more research works will be conducted for these newly developed materials.

The photopolymerization process fabricates the polymeric parts by cross-linking the polymer chains through UV lights or laser. Normally both elastomers and plastics can be fabricated from these processes. These processes include, but are not limited to, the stereolithography (SLA), PolyJet, continuous liquid interphase printing (CLIP), two-photon lithography, etc. Nowadays, these processes can usually achieve quite high resolution and precision down to the micron scale or even submicron scale. In addition, researchers also overcome a lot of obstacles on the brittleness of photopolymers by developing new types of elastomer or plastic with remarkable resilience or ductility. Although most of the photopolymer parts from AM exhibit anisotropic properties due to the layer-wise processing feature, the CLIP technique can manufacture parts that are almost isotropic. Note that the photopolymerization techniques are the most dynamic and disrupting area for AM polymers in recent years. Significant progress has also been achieved on the modeling of these photopolymers. For example, advanced mechanics models have been developed by a number of researchers on their finite deformation behaviors (Zhang and To 2016; Liu and Li 2018), shape memory effect (Mao et al. 2015a), anisotropic properties (Zhang and To 2016; Barclift et al. 2012), etc. Very recently, researchers also conduct experimental and modeling studies to the polymer behavior during the photopolymerization process (Wu et al. 2018), which provides insights into the effect of processing parameters on the mechanical behaviors of the as-built parts. Note that the processing study is still at a nascent stage and a lot more works should be carried out in the future to fully understand the photopolymers' behavior.

Although the laser-sintered AM polymers have been investigated for decades, they are not as common as the extrusion or photopolymerization techniques due to the relatively high cost. At present, both thermoplastics and their relevant composites can be fabricated. The most popular material is polyamide (i.e., nylon) (Goodridge et al. 2012) due to its high strength, high impact resistance, and low density. Usually these polyamide particles have spherical shapes with diameters around 50 μm. The laser can either fully or partially melt the particles layer by layer to form the final products. Laser-sintered polyamide materials are semicrystalline, showing spherulite-like microstructure, dependent strongly on the processing parameters like laser power and scanning speed. In addition, laser-sintered polyamide materials also show anisotropic mechanical behaviors (Zarringhalam et al. 2009; Ajoku et al. 2006) due to the printing-orientation effect and the existence of micropores. These mechanical properties have been reported in the literature, albeit advanced thermomechanical modeling is still scarce. There does have a strong demand on the thermomechanical modeling of the laser sintering process to predict the residual stress and distortion so that the process can be optimized to improve the building accuracy.

The modeling of AM polymers is in fact similar to that for conventional elastomers and plastics (Bergstrom 2015), although some new features arise for the AM-oriented design and fabrication. A common feature for AM polymers is the material anisotropy due to the layer-wise processing characteristics. Hence, the anisotropy will inevitably aggravate the challenge to model the polymer's response. Another common feature is that the processing of AM polymers usually involves phase transitions so the polymer responses at different thermal stages need to be modeled. Note that some AM processes also involve thermo-chemo-mechanical coupling which will not be discussed further in this chapter. The major aim of this chapter is to present the advanced mechanics modeling of some AM polymers so that researchers can extend these models to other materials. In Sect. 3.2, the finite strain constitutive models for photopolymers including both plastics and elastomers are introduced by considering the processing-induced anisotropy. In Sect. 3.3, the shape memory effect of the photopolymer plastics is introduced as an example for the thermomechanical modeling. In Sect. 3.4, we will discuss about the mechanics modeling needs and future research trends for both engineering and biomedical applications. This chapter intends to present a more inspiring rather than comprehensive introduction to the mechanics modeling of AM polymers given that the research in this area is still very limited to date.

3.2 Nonlinear Modeling of Additive Manufactured Photopolymers

Photopolymerization is a very popular additive manufacturing technology for polymer components such as plastics and elastomers. Two typical examples of these polymers are the VeroWhite plastic and TangoBlack elastomer fabricated from the Stratasys Objet 3D printers. These UV or laser cross-linked plastics and elastomers usually have amorphous microstructure and exhibit anisotropic properties to some extent. Due to the layer-wise fabrication feature, most of these polymers show transversely isotropic behavior with the printing direction as the anisotropic axis. Therefore, this section aims at introducing two nonlinear constitutive models for UV cross-linked plastics and elastomers, respectively, that are modified from the hyperelastic and viscoplastic models developed by Arruda and Boyce (1993a, b). The polymers are considered to be transversely isotropic for the two models in this section. The readers are referred to the reference (Zhang and To 2016) for more details on the numerical implementation of the models.

3.2.1 *Finite Strain Anisotropic Model for Plastics*

Here we consider the finite deformation scenario for the inelastic deformation behavior of the glassy polymers. Figure 3.1 illustrates different configurations of a continuum solid body. Based on continuum mechanics theory, the finite deformation of the solid body is described by the deformation gradient **F**, which maps an arbitrary material point **X** in the reference configuration $\mathscr{B}_0$ in the 3D Euclidean space to a corresponding spatial point **x**(**X**) in the current configuration $\mathscr{B}$, as

$$\mathbf{F} = \frac{\partial \mathbf{x}}{\partial \mathbf{X}} \tag{3.1}$$

The deformation gradient is usually decomposed into two parts for finite deformation involving plasticity (Lee 1969), as

$$\mathbf{F} = \mathbf{F}^{e}\mathbf{F}^{p} \tag{3.2}$$

Note that the relaxed configuration $\mathscr{B}_p$ is nonunique. In order to simplify the analysis, we adopt a relaxed configuration such that $\mathbf{F}^e$ is symmetric, i.e., $\mathbf{F}^e = \mathbf{V}^e$, where $\mathbf{V}^e$ is the left stretch tensor of $\mathbf{F}^e$ (Boyce et al. 1988). Further, the plastic deformation gradient $\mathbf{F}^p$ is decomposed into two parts, as

$$\mathbf{F}^{p} = \mathbf{R}^{p}\mathbf{U}^{p} = \mathbf{V}^{p}\mathbf{R}^{p} \tag{3.3}$$

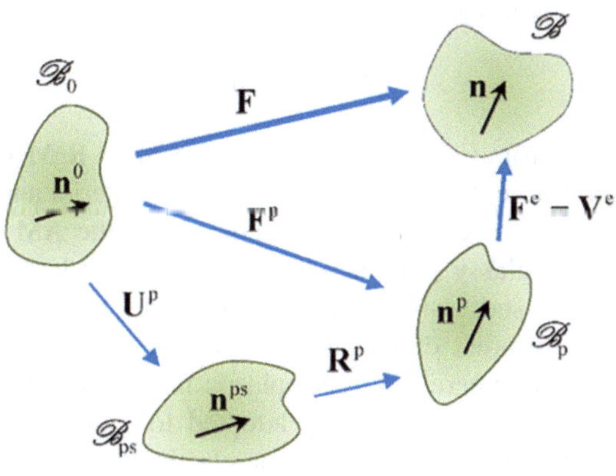

Fig. 3.1 Multiplicative decomposition $\mathbf{F} = \mathbf{F}^e\mathbf{F}^p$ of the elastoplastic deformation gradient of a continuum solid body. The plastic deformation can be further expressed as the product of a stretching part and a rotation part, as $\mathbf{F}^p = \mathbf{R}^p\mathbf{U}^p$. The symbols $\mathscr{B}_p$, $\mathscr{B}_{ps}$, $\mathscr{B}_0$, and $\mathscr{B}$ indicate the reference configuration, current configuration, relaxed configuration, and plastic stretching configuration, respectively. The privileged axes in these four configurations are denoted by $\mathbf{n}^0$, **n**, $\mathbf{n}^p$, and $\mathbf{n}^{ps}$, respectively

where $\mathbf{R}^p$ is the rotation tensor, $\mathbf{U}^p$ is the right stretch tensor, and $\mathbf{V}^p = \mathbf{R}^p\mathbf{U}^p\mathbf{R}^{pT}$ is the left stretch tensor. The decomposition in Eq. (3.3) is also illustrated in Fig. 3.1 by introducing an intermediate configuration $\mathscr{B}_{ps}$ with plastic stretch deformation only.

The AM photopolymers are assumed to be transversely isotropic in the reference configuration given the fact that they usually form a lamellar or layer-wise structure. Therefore, it is natural to choose the printing direction as the material's privileged axis, i.e., $\mathbf{n}^0$ in the reference configuration $\mathscr{B}_0$ of Fig. 3.1. The evolution of the privileged axis $\mathbf{n}^0$ is essential to the constitutive modeling in the finite deformation scenario. As shown in Fig. 3.1, the evolution of the privileged axis $\mathbf{n}^0$ is usually assumed to be unchanged by the plastic stretch $\mathbf{U}^p$ during the plastic stretch process, i.e., $\mathbf{n}^{ps} = \mathbf{n}^0$. Hence, the privileged axis $\mathbf{n}^p$ is only affected by the plastic rotation $\mathbf{R}^p$ in the relaxed configuration $\mathscr{B}_p$, as

$$\mathbf{n}^p = \mathbf{R}^p\mathbf{n}^0 \tag{3.4}$$

In addition, the corresponding rate form of Eq. (3.4) is derived as

$$\dot{\mathbf{n}}^p = \dot{\mathbf{R}}^p\mathbf{n}^0 = \tilde{\mathbf{W}}^p\mathbf{n}^p \tag{3.5}$$

where $\tilde{\mathbf{W}}^p = \dot{\mathbf{R}}^p\mathbf{R}^{pT}$ is called the material spin tensor. The material anisotropy of transversely isotropic solids is characterized by the structural tensor, which evolves with the deformation. In the reference configuration $\mathscr{B}_0$, the structural tensor is defined as

$$\mathbf{M}^0 = \mathbf{n}^0 \otimes \mathbf{n}^0 \tag{3.6}$$

where $\otimes$ is the dyadic product of tensors. Correspondingly, in the relaxed configuration $\mathscr{B}_p$, the structural tensor is defined as

$$\mathbf{M}^p = \mathbf{n}^p \otimes \mathbf{n}^p \tag{3.7}$$

It is easy to verify from Eqs. (3.6) and (3.7) that $\mathbf{M}^p = \mathbf{R}^p\mathbf{M}^0\mathbf{R}^{pT}$. The structural tensors will exhibit different forms for other types of material anisotropy.

At room temperature, the deformation behavior of the AM photopolymers is similar to many other glassy polymers, which usually undergoes elastic regime, yielding, softening, and successive hardening due to the stretching and alignment of molecular chains at different deformation stages. In order to model all of these behaviors, we present a hyperelastic-viscoplastic model in Fig. 3.2. The 1D rheological model in Fig. 3.2a is frequently adopted to model the glassy polymers (Arruda and Boyce 1993b), which consists an elastic spring, a hyperelastic spring, and a viscoplastic element. The hyperelastic spring is used to model the backstress evolution due to the hardening of the glassy polymer under large inelastic strain, as shown in Fig. 3.2b. Finally, the equilibrium of the stress state is formulated in the relaxed configuration $\mathscr{B}_p$ in this work.

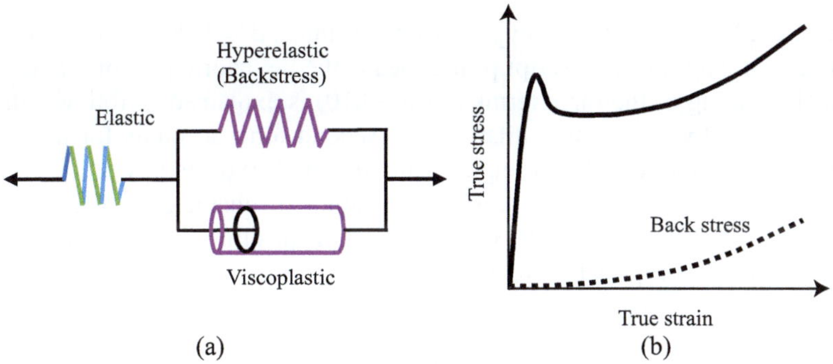

Fig. 3.2 (**a**) Schematic illustration of the rheological model of the transversely isotropic photopolymer. (**b**) A representative stress-strain curve of glassy polymers

The elastic deformation of the model is represented by the elastic spring shown in Fig. 3.2a. Note that the elastic deformation is usually quite small compared to the inelastic deformation. Thus, the elastic stress $\boldsymbol{\sigma}^{\mathrm{e}}$ in the relaxed configuration $\mathscr{B}_{\mathrm{p}}$ is determined from the following formula, as (Zhang and To 2016)

$$\sigma^{\mathrm{e}} = J_{\mathrm{p}}^{-1}\{\lambda\left(\mathrm{tr}\mathbf{E}^{\mathrm{e}}\right)\mathbf{I} + 2\mu_{\perp}\mathbf{E}^{\mathrm{e}} + \alpha\left[\left(\mathbf{M}^{\mathrm{p}} : \mathbf{E}^{\mathrm{e}}\right)\mathbf{I} + \left(\mathrm{tr}\mathbf{E}^{\mathrm{e}}\right)\mathbf{M}^{\mathrm{p}}\right] \\ + 2\left(\mu_{\parallel} - \mu_{\perp}\right)\left(\mathbf{M}^{\mathrm{p}}\mathbf{E}^{\mathrm{e}} + \mathbf{E}^{\mathrm{e}}\mathbf{M}^{\mathrm{p}}\right) + \beta\left(\mathbf{M}^{\mathrm{p}} : \mathbf{E}^{\mathrm{e}}\right)\mathbf{M}^{\mathrm{p}}\} \tag{3.8}$$

where λ, $\mu_{\parallel}$, $\mu_{\perp}$, α, and β are five independent elastic constants, $\mathbf{E}^{\mathrm{e}} = (\mathbf{F}^{\mathrm{eT}}\mathbf{F}^{\mathrm{e}} - \mathbf{I})/2$ is the elastic Green strain tensor, $\mathbf{I}$ is the second order identity tensor, $J_{\mathrm{p}} = \det \mathbf{U}^{\mathrm{p}}$ accounts for the plastic volume change, and ":" is the double dot product of the tensors.

In addition, the Cauchy stress $\boldsymbol{\sigma}$ in the current configuration $\mathscr{B}$ can be determined from Eq. (3.8) by conducting a push-forward transformation, as

$$\sigma = J_{\mathrm{e}}^{-1}\mathbf{F}^{\mathrm{e}}\sigma^{\mathrm{e}}\mathbf{F}^{\mathrm{eT}} \tag{3.9}$$

where $J_{\mathrm{e}} = \det \mathbf{V}^{\mathrm{e}}$ denotes the elastic volume change. The Cauchy stress in Eq. (3.9) is usually used to implement the model in commercial finite element software like ABAQUS.

The material hardening and anisotropy evolution in the post-yielding stage are modeled by introducing a backstress $\boldsymbol{\sigma}^{\mathrm{b}}$ derived from a hyperelastic model (Arruda and Boyce 1993b). Note that the polymer is transversely isotropic in the reference configuration $\mathscr{B}_0$. The backstress $\boldsymbol{\sigma}^{\mathrm{b}}$ in the relaxed configuration $\mathscr{B}_{\mathrm{p}}$ is defined as (Zhang and To 2016)

$$\sigma^{\mathrm{b}} = \kappa\left(J_{\mathrm{p}} - 1\right)\mathbf{I} + \frac{\bar{\mu}\sqrt{N}\beta_{ch}}{3J_{\mathrm{p}}^{5/3}\lambda_{ch}}\mathbf{B}^{\mathrm{p}'} + \tilde{\mu}J_{\mathrm{p}}^{-1}\left(I_4^{\mathrm{p}} - 1\right)\mathbf{M}^{\mathrm{p}} \tag{3.10}$$

where κ, $\overline{\alpha}$, N, and $\tilde{\alpha}$ are four hyperelastic constants, $\mathbf{B}^{p} = \mathbf{F}^{p}\mathbf{F}^{pT}$ is the left Cauchy-Green tensor, and the superscript prime means the deviatoric part of a tensor. The physical meaning of the three terms in Eq. (3.10) is explained as follows: the first term is caused by the volume dilation, the second term accounts for the volume distortion that is modeled by the eight-chain model (Arruda and Boyce 1993a), and the third term is the standard reinforcing model to characterize the anisotropy (Chagnon et al. 2014; Ogden 2003) induced by the printing process. Moreover, the terms λ_{ch} and β_{ch} in Eq. (3.10) are defined as

$$\lambda_{ch} = \sqrt{\frac{I_1^{\mathrm{p}}}{3J_{\mathrm{p}}^{2/3}}}$$

$$\beta_{ch} = \mathcal{L}^{-1}\left(\frac{\lambda_{ch}}{\sqrt{N}}\right) \tag{3.11}$$

where $\mathcal{L}(\beta_{ch}) = \coth\beta_{ch} - 1/\beta_{ch}$ is the Langevin function. Note that other forms of hyperelastic models can also be chosen depending on the available experimental data.

Experimental data indicates that the inelastic deformation of AM photopolymers is pressure-sensitive, rate-dependent, and orientation-dependent. Therefore, the viscoplastic element is modeled to incorporate all these factors. The driving stress $\tilde{\sigma}$ for the viscoplastic flow is derived from the equilibrium of the 1D rheological network in Fig. 3.2a, as

$$\tilde{\sigma} = \sigma^{\mathrm{e}} - \sigma^{\mathrm{b}} \tag{3.12}$$

where $\boldsymbol{\sigma}^{\mathrm{e}}$ and $\boldsymbol{\sigma}^{\mathrm{b}}$ are defined in Eqs. (3.12) and (3.10), respectively. In the case that the material is transversely isotropic and pressure sensitive, the equivalent stress is defined in a modified Hill form by incorporating the Bauschinger effect (Hill 1948; Dafalias and Rashid 1989; Hütter et al. 2013), as

$$\tau = \sqrt{a_1\mathrm{tr}\left(\tilde{\sigma}'^2\right) + a_2\mathrm{tr}^2\left(\mathbf{M}^{\mathrm{p}}\tilde{\sigma}'\right) + a_3\mathrm{tr}\left(\mathbf{M}^{\mathrm{p}}\tilde{\sigma}'^2\right)} + a_4\mathrm{tr}\left(\mathbf{M}^{\mathrm{p}}\tilde{\sigma}'\right) + a_5\mathrm{tr}\tilde{\sigma} \tag{3.13}$$

where a_i (i = 1, 2, ⋯, 5) are five dimensionless yielding constants and $\mathbf{M}^{\mathrm{p}}$ is the structural tensor in Eq. (3.7).

Since the material is pressure sensitive, a nonassociated flow rule is used, otherwise unphysical plastic volume dilatation will occur (Nemat-Nasser 2004; Gao et al. 2011). The plastic flow potential is given as

$$\psi = \tau + g\left(J_{\mathrm{p}}, p\right) \tag{3.14}$$

where g is an unknown function to be determined and p is the equivalent pressure defined as

$$p = \frac{1}{3a_5}\left[a_4 \mathrm{tr}\left(\mathbf{M}^{\mathrm{p}}\tilde{\sigma}'\right) + a_5 \mathrm{tr}\tilde{\sigma}\right] \tag{3.15}$$

where the term related to a_4 is introduced to account for the anisotropy of the polymer. Note that the pressure in Eq. (3.15) is positive in tension and negative in compression.

The evolution of the plastic deformation rate is

$$\mathbf{D}^{\mathrm{p}} = \dot{\gamma}^{\mathrm{p}}\tilde{\mathbf{N}} \tag{3.16}$$

where $\dot{\gamma}^{\mathrm{p}}$ and $\tilde{\mathbf{N}}$ are the amplitude and direction of the viscoplastic flow, respectively. More specifically, the flow direction tensor $\tilde{\mathbf{N}}$ is obtained from the derivative of the nonassociated flow potential in Eq. (3.14), as (Zhang and To 2016)

$$\begin{aligned}\tilde{\mathbf{N}} &= \frac{\partial \psi}{\partial \tilde{\sigma}} \\ &= \frac{a_1\tilde{\sigma}' + a_2 \mathrm{tr}\left(\mathbf{M}^{\mathrm{p}}\tilde{\sigma}'\right)\mathbf{M}^{\mathrm{p}'} + 0.5a_3\left(\mathbf{M}^{\mathrm{p}}\tilde{\sigma}' + \tilde{\sigma}'\mathbf{M}^{\mathrm{p}}\right)'}{\sqrt{a_1 \mathrm{tr}\left(\tilde{\sigma}'^2\right) + a_2 \mathrm{tr}^2\left(\mathbf{M}^{\mathrm{p}}\tilde{\sigma}'\right) + a_3 \mathrm{tr}\left(\mathbf{M}^{\mathrm{p}}\tilde{\sigma}'^2\right)}} \\ &+ \left(1 + \frac{1}{3a_5}\frac{\partial g}{\partial p}\right)\left(a_4\mathbf{M}^{\mathrm{p}'} + a_5\mathbf{I}\right)\end{aligned} \tag{3.17}$$

In addition, the volume change rate of the plastic deformation is assumed to be

$$\dot{J}_{\mathrm{p}} = \frac{J_{\mathrm{p}}\dot{\gamma}^{\mathrm{p}}}{\kappa_{\mathrm{p}}} p \tag{3.18}$$

Moreover, the equivalent plastic flow rate $\dot{\gamma}^{\mathrm{p}}$ is taken as the following form.

$$\dot{\gamma}^{\mathrm{p}} = \dot{\gamma}_0 \exp\left[-\frac{\Delta G}{k_{\mathrm{B}}\vartheta}\left(1 - \frac{\tau}{s}\right)\right] \tag{3.19}$$

where $\dot{\gamma}_0$ is a pre-exponential factor, ΔG is the zero stress level activation energy, and s is the athermal shear strength, a parameter characterizing the resistance to the viscoplastic shear deformation in the material. The strain softening is modelled by introducing a preferred state s_s so that the athermal threshold stress evolves in the following way, as (Boyce et al. 1988)

$$\dot{s} = h\left(1 - \frac{s}{s_s}\right)\dot{\gamma}^p \tag{3.20}$$

where h is the softening slope and the initial condition is $s = s_0$ when $\gamma^p = 0$.

This nonlinear viscoplastic model can be implemented into commercial finite element software ABAQUS to conduct numerical simulation. A backward Euler algorithm is suggested to compute the evolution of the stress state and elastic/plastic deformation as shown in (Zhang and To 2016).

3.2.2 Anisotropic Hyperelastic Model for Elastomers

The photopolymerization-based AM techniques can produce compliant elastomers with rubber-like mechanical properties. The mechanics modeling of these AM elastomers will help design the engineering parts. Normally the elastomers are modeled by hyperelastic models. For a systematic introduction to this subject, please refer to (Holzapfel 2001). In this section we consider a transversely isotropic hyperelastic model that takes into account the printing orientation effect.

The key problem in developing hyperelastic models is to define the hyperelastic strain energy function. The deformation gradient is defined similarly to the viscoplastic model in Fig. 3.1. However, the deformation gradient $\mathbf{F}$ is the same as the elastic deformation gradient $\mathbf{F}^e$ since there is no plastic deformation in this case. For transversely isotropic hyperelastic materials, the strain energy function has a general form of

$$\Phi\left(\mathbf{C},\mathbf{M}^0\right) = \Phi\left(I_1, I_2, I_3, I_4, I_5\right) \tag{3.21}$$

where $\mathbf{C} = \mathbf{U}^2$ is the right Cauchy-Green tensor, I_i $(i = 1, 2, \cdots, 5)$ are five invariants defined as (Spencer 1971)

$$\begin{aligned}
I_1 &= \mathrm{tr}\mathbf{C} \\
I_2 &= \left[I_1^2 - \mathrm{tr}\left(\mathbf{C}^2\right)\right]/2 \\
I_3 &= \det\mathbf{C} = J^2 \\
I_4 &= \mathbf{M}^0 : \mathbf{C} \\
I_5 &= \mathbf{M}^0 : \mathbf{C}^2
\end{aligned} \tag{3.22}$$

Numerous strain energy functions have been developed for Eq. (3.21) in the literature (Chagnon et al. 2014). For AM elastomers, we adopt the following hyperelastic strain energy, as

$$\Phi = \frac{\kappa}{2}(J-1)^2 + \bar{\mu}\sqrt{N}\left[\beta_{\mathrm{ch}}\lambda_{\mathrm{ch}} - \sqrt{N}\ln\left(\frac{\sinh\beta_{\mathrm{ch}}}{\beta_{\mathrm{ch}}}\right)\right] + \frac{\tilde{\mu}}{4}(I_4-1)^2 \tag{3.23}$$

where κ, $\bar{\propto}$, N, and $\tilde{\propto}$ are four material constants related to the hyperelastic behavior. The first term of Eq. (3.23) is the volume dilation energy, the second term is the distortion energy modeled by the eight-chain model (Arruda and Boyce 1993a), and the last term is the energy contribution from the material anisotropy (Chagnon et al. 2014; Ogden 2003) . Similar to Eq. (3.11), the terms λ_{ch} and β_{ch} in Eq. (3.23) are defined as

$$\begin{aligned} \lambda_{\mathrm{ch}} &= \sqrt{\frac{I_1}{3J^{2/3}}} \\ \beta_{\mathrm{ch}} &= \mathcal{L}^{-1}\left(\frac{\lambda_{\mathrm{ch}}}{\sqrt{N}}\right) \end{aligned} \tag{3.24}$$

Note that one can also choose other forms of the strain energy functions in Eq. (3.23) depending on the available data. For example, the I_2 related term often needs to be determined by biaxial testing (Holzapfel 2000). In addition, the I_5 term is also disregarded for simplicity since it is usually less important than I_4 and does not have a clear physical meaning (Chagnon et al. 2014). Therefore, only three invariants appear in Eq. (3.23) for the current hyperelastic model.

The derivation of the Cauchy stress $\boldsymbol{\sigma}$ is straightforward once a proper hyperelastic strain energy is given. Following a procedure outlined by Ogden (2003), the Cauchy stress is determined as

$$\begin{aligned} \sigma &= J^{-1}\left(2\frac{\partial\Phi}{\partial I_1}\mathbf{B} + 2I_3\frac{\partial\Phi}{\partial I_3}\mathbf{I} + 2\frac{\partial\Phi}{\partial I_4}\mathbf{M}^0\right) \\ &= \kappa(J-1)\mathbf{I} + \frac{\bar{\mu}\sqrt{N}\beta_{\mathrm{ch}}}{3J^{5/3}\lambda_{\mathrm{ch}}}\mathbf{B}' + \tilde{\mu}J^{-1}(I_4-1)\mathbf{M}^0 \end{aligned} \tag{3.25}$$

where $\mathbf{B} = \mathbf{F}\mathbf{F}^{\mathrm{T}}$ is the left Cauchy-Green tensor and the superscript prime means the deviatoric part of a tensor. Note that Eq. (3.10) satisfies the stress-free condition when $\mathbf{F} = \mathbf{I}$.

The model presented here is modified from the eight-chain model proposed by Arruda and Boyce (1993a). Other types of models can also be used, e.g., the neo-Hookean model and Gent model (Boyce and Arruda 2000). The neo-Hookean model is a first order approximation of the Arruda-Boyce model. On the other hand, comparison study has indicated that the Arruda-Boyce model is very close to the

Gent model. Hence, the adoption of the hyperelastic models mainly depends on the available data and users' preference. The hyperelastic models can be implemented into ABAQUS easily.

3.3 Modeling of Shape Memory Photopolymers

3.3.1 Background of Shape Memory Polymers

Shape memory polymers (SMPs) are capable to "memorize" a permanent shape. After being fixed to a temporary shape, an SMP can recover its original permanent shape upon activation by an external stimulus (Mao et al. 2015b). However, SMPs can offer large shape change (Guang Yang et al. 2014; Zhao et al. 2015) but generally only in one-way actuation. A notable exception is reported in the work by Mather et al. (Chung et al. 2008), where a two-way actuation is realized by utilizing a semicrystalline polymer, while an external biasing load had to be maintained. This condition was later released by integrating an external bias into a composite architecture, where an SMP programmed strip was filled inside another polymer matrix; this concept was recently applied to design interpenetrating network polymers, with one group of polymers capable of melt-crystallization transition under temperature change (Yakacki et al. 2015). Since these approaches depend on the applied mechanical loading, they are applicable only in simple shape changes, where the simple tension is limited. Environmentally responsive hydrogels (Westbrook and Jerry Qi 2008; White 2013; Naficy et al. 2014) can reversibly swell or shrink under activation of environmental conditions, such as temperature or PH value, and correspondingly give rise to the two-way actuation. However, hydrogels has a low modulus, typically of a few tens to hundreds of kPa, which limits their applicability, e.g., for some biomedical applications. Generally main efforts to the development of active materials concentrate on the sophisticated multi-physics constitutive behavior of the materials, although it has been integrated in the pursuit of composite concepts with controlled material architecture in fibrous or layered forms.

As an advanced manufacturing technique, 3D printing (or additive manufacturing) allows materials to be deposited in a layer-by-layer manner to form a 3D component. The recently developed multi-material 3D printing techniques (such as the polyjet technique) enable printing of digital materials with a continuously varied properties with a large range at almost any predefined spatial location. With such capabilities of printing complicated geometries and digital materials, 3D printing is readily integrated with novel designs to create materials or components with unprecedented properties, such as metamaterials with negative Poisson's ratio (Babaee et al. 2013), ultralight and ultra-stiff metamaterials (Zheng et al. 2014), and medical devices (Sanchez-Dehesa et al. 2013; Khalyfa et al. 2007).

To understand the operative phenomena and empolder this in design, we firstly develop a theoretical model integrating the sophisticated constitutive behaviors of active materials (both the SMP and the hydrogel) with the controlled spatial distributions of these materials in 3D space to discuss the shape change and to probe the design space. And then, we create and demonstrate the reversible shape change behavior of several structures by applying the design principles. More detailed introduction to this model is shown in (Mao et al. 2015a).

3.3.2 Model Descriptions

Here, a thermodynamic viscoelastic constitutive model is established for shape memory polymer. A thermal expansion element is considered in series with the mechanical part. A multibranch model (shown as Fig. 3.3), following the way in (Yu et al. 2014), is applied to capture the temperature-dependent viscoelastic behavior. In current multibranch model, totally 16 nonequilibrium branches as well as 1 equilibrium branch are adopted.

The deformation gradient of the model $\mathbf{F}$ is multiplicity decomposed into the mechanical part, $\mathbf{F}^M$, and thermal deformation part, $\mathbf{F}^T$, as

$$\mathbf{F} = \mathbf{F}^M \mathbf{F}^T \tag{3.26}$$

Then, the total Cauchy stress $\boldsymbol{\sigma}$ can be written as combination of contributions from equilibrium part and nonequilibrium part,

$$\sigma = \sigma_{\text{eq}} + \sum_{i=1}^{m} \sigma^i \tag{3.27}$$

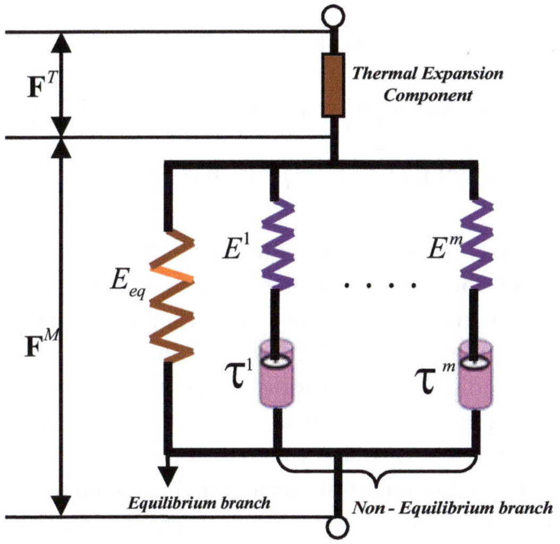

Fig. 3.3 1D rheological representation of the multibranch model. Top block indicates the thermal deformation. The equilibrium spring represents the entropic elasticity of elastomer matrix. The multibranches indicate the nonequilibrium properties. Here multi-physical mechanisms of viscoelastic properties are considered

here $\boldsymbol{\sigma}_{\mathrm{eq}}$ and $\boldsymbol{\sigma}^i$ are the Cauchy stresses for the equilibrium and the i-th ($1 \leq i \leq m$) nonequilibrium branch (as shown in Fig. 3.3), respectively. In current analysis, $m = 16$, i.e., 16 nonequilibrium branches are considered.

The isotropic thermal expansion or contraction of the constructed component is assumed,

$$\mathbf{F}^T = J_T \mathbf{I} \tag{3.28}$$

where $\mathbf{I}$ is the second order identity tensor, J_T is the volume change due to thermal expansion/contraction that is defined as

$$J_T = \frac{V(T,t)}{V_0} = \left[1 + 3\alpha_r \left(T - T_0\right)\right]\left(1 + \delta\right) \tag{3.29}$$

where $V(T, t)$ is the volume at time t and temperature T. V_0 is the reference volume at the reference temperature T_0. α_r is the linear coefficient of thermal expansion (CTE) of the rubbery state, and δ characterizes the deviation of volume from equilibrium volume and

$$\delta = \frac{V(T,t)}{V_{eq}(T)} - 1,\; V_{eq}(T) = \left[1 + 3\alpha_r \left(T - T_0\right)\right] V_0 \tag{3.30}$$

where δ is defined by the KAHR 33-parameter.

The Cauchy stress tensor in the equilibrium branch uses Arruda-Boyce eight chain model, i.e.,

$$\begin{aligned} \sigma_{\mathrm{eq}} &= \frac{nk_B T}{3J_{\mathrm{M}}} \frac{\sqrt{N}}{\lambda_{\mathrm{ch}}} \mathcal{L}^{-1}\left(\frac{\lambda_{\mathrm{ch}}}{\sqrt{N}}\right) \bar{\mathbf{B}}' + \kappa\left(J_{\mathrm{M}} - 1\right)\mathbf{I}, \\ \bar{\mathbf{B}}' &= \bar{\mathbf{B}} - 1/3\mathrm{tr}\left(\bar{\mathbf{B}}\right)\mathbf{I},\; \bar{\mathbf{B}} = \bar{\mathbf{F}}^M \bar{\mathbf{F}}^{MT},\; \bar{\mathbf{F}}^M = J_M^{-1/3}\mathbf{F}^M, \\ \lambda_{\mathrm{ch}} &= \sqrt{tr\left(\bar{\mathbf{B}}\right)/3},\; \mathcal{L}(\beta) = \coth\beta - 1/\beta. \end{aligned} \tag{3.31}$$

where n is the cross-linking mole concentration, i.e., mole number of cross-linked points per unit volume, k_B is Boltzmann's constant, and $_N$ is the number of Kuhn segments between two cross-link sites (and/or strong physical entanglements). The shear modulus $\mu_r(T)$ of the elastomer in the equilibrium state is defined as nk_BT, which is an indication of entropic elasticity; κ is the bulk modulus and is typically orders of magnitude larger than μ_r to ensure material incompressibility.

We assumed a unified viscous flow rule for all nonequilibrium branches, and the temperature-dependent relaxation time formation is defined for each branch. For the *i-th* nonequilibrium branch ($1 \leq i \leq m$), the deformation gradient is further decomposed into an elastic part, $\mathbf{F}_i^v$, and a viscous part, $\mathbf{F}_i^e$, as

$$\mathbf{F}_i^M = \mathbf{F}_i^e \mathbf{F}_i^v \tag{3.32}$$

Therefore, the Cauchy stress can be calculated as

$$\begin{gathered}\sigma^i = \frac{1}{J_e^i}\left[\mathbf{L}_i^e(T) : \mathbf{E}_i^e\right], \text{for } 1 \le i \le m \\ J_e^i = \det\left(\mathbf{F}_i^e\right),\ \mathbf{E}_i^e = \ln \mathbf{V}_i^e,\ \mathbf{V}_i^e = \mathbf{F}_i^e \mathbf{R}_i^{eT}\end{gathered} \tag{3.33}$$

where $\mathbf{L}_i^e(T)$ is the fourth-order isotropic elasticity tensor in the i-th nonequilibrium branch ($1 \le i \le m$), which is taken to be temperature independent in general, i.e.,

$$\mathbf{L}_i^e(T) = 2G_i\left(I - \frac{1}{3}\mathbf{I} \otimes \mathbf{I}\right) + K_i \mathbf{I} \otimes \mathbf{I}, \text{for } 1 \le i \le m, \tag{3.34}$$

where $\mathbf{I}$ is the fourth order identity tensor, G_i and K_i are shear and bulk moduli for each nonequilibrium branch ($1 \le i \le m$), respectively.

It is assumed that all the rubbery branches in the nonequilibrium branches ($1 \le i \le m$) have the same shear modulus, i.e.,

$$G_i(T) = n_R k_B T \text{ for } 1 \le i \le m \tag{3.35}$$

with n_R the cross-linking density. Then the elastic modulus in each nonequilibrium branch is calculated as

$$E_i(T) = \frac{G_i(T)}{2(1 + v_i)} \text{ for } 1 \le i \le m \tag{3.36}$$

The temperature-dependent relaxation times of nonequilibrium mechanism are calculated according to the thermorheological simplicity principle as

$$\tau_i(T) = \tau_{i0} \alpha_T(T) \text{ for } 1 \le i \le m \tag{3.37}$$

where $\alpha_T(T)$ is the time-temperature superposition (TTSP) shift factor and τ_{i0} is the relaxation time at the reference temperature. At temperatures around or above T_s, the WLF equation is adopted,

$$\log \alpha_T(T) = -\frac{C_1(T - T_M)}{C_2 + (T - T_M)} \tag{3.38}$$

here C_1 and C_2 are material constants and T_M the WLF reference temperature. When the temperature is below T_s, $\alpha_T(T)$ is defined following the Arrhenius-type behavior,

$$\ln \alpha_T(T) = -\frac{AF_c}{k_B}\left(\frac{1}{T} - \frac{1}{T_g}\right) \tag{3.39}$$

where A and F_c are material constants.

3.3.3 Additive Manufactured Shape Memory Structures

Based on the 3D nonlinear models for understanding the interplay of geometry, material, and environmental stimuli parameters, we design reversibly actuating components and sequential self-folding structures realized by thermal activation of spatially variable patterns which are 3D printed by digital shape memory polymers. There digital materials have different shape memory behaviors. Measurements of the spatial and temporal configurations of these structures agree well with the companion finite element simulations.

The finite element simulations of the folding and unfolding processes are presented in Fig. 3.4a. During the unfolding process, the structure is successively unfold by rigid surfaces one panel after the other at a high temperature. After lowering the temperature and unloading, the structure is fixed into the flat configuration. By heating the structure uniformly, it recovers after sequentially folding into the original configuration when hinges are sequentially activated due to their different thermally activated response time. The simulated recovery ratios for representative hinges (H-1, H-4, and H-7) are presented in Fig. 3.4b and compared with measurements and results from the simulation by Matlab. The FEA simulations agree well with the measurements, although they predict slightly faster folding than the experiment, which is maybe due to the neglect of the water resistance, an obvious effect observed in the experiment. Figure 3.4 also shows the Matlab simulations agree well with the FEA simulations and experiments.

Combing thermally activated SMP and water-activated hydrogel, we designed another example of 3D printed reversible shape memory component. A flower-shaped 3D structure composed of three types of petals is designed and then experimentally demonstrated that the folding of petals is a function of geometrical sizes. By elaborating structural size, we can realize the sequence folding of petals like a flower folding. The elastomer, the hydrogel, and the SMP are designed with different thickness ratios for the three layers as shown in Fig. 3.5a. The inner layer was designed with the thicknesses of the elastomer, the hydrogel, and the SMP layers as 0.2 mm, 0.4 mm, and 0.3 mm.

The second layer was designed with the thicknesses of 0.2 mm, 0.3 mm, and 0.4 mm for the elastomer, the hydrogel, and the SMP layers; and the outer layer was

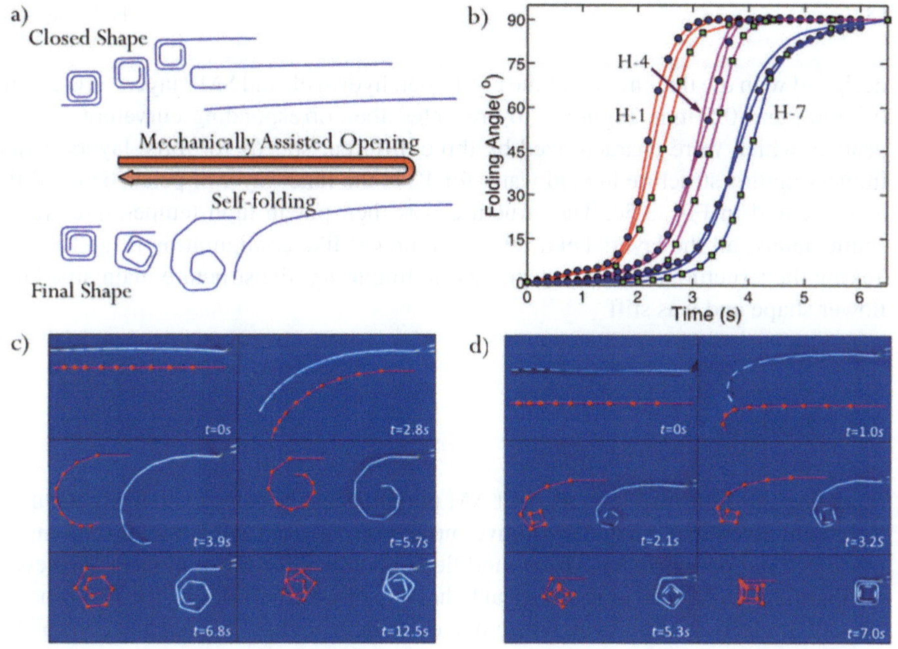

Fig. 3.4 (**a**) The finite element simulation of shape memory recovery activated sequential self-folding after the strip is firstly programmed at high temperature. (**b**) The simulations of folding angles of three representative hinges by Matlab simulation and compared with those by FEA and experiments. Solid line only, FEA; round dot, Matlab simulation; square, experiments. Red color, H-1; magenta color, H-4; blue color, H-7. Comparison between Matlab simulation and experiments for (**c**) Sample 1 and (**d**) Sample 2, where red lines indicate the simulation results

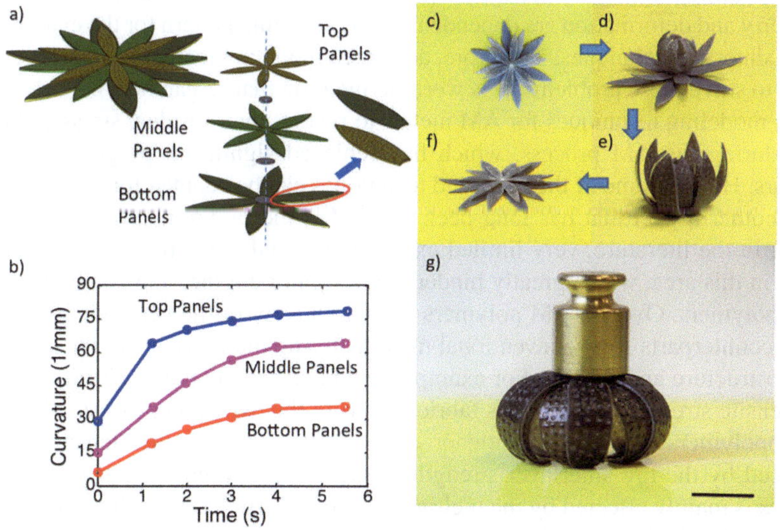

Fig. 3.5 A self-folding/unfolding flower. (**a**) Schematic of two activated shape memory petal-like structure. (**b**) The curvature changes under thermal activation as functions of time. (**c–f**) The sequence of reversible actuation. (**g**) The dried configuration is stiff and can carry a load of 25 g. The scale bar in (**g**) is 12.5 mm

designed with the thicknesses of the elastomer, hydrogel, and SMP layers as 0.2 mm, 0.2 mm, and 0.5 mm. Figure 3.5b presented the corresponding curvature change profiles which were characterized by three different speeds for three layers. After immersing this structure in cold water for 12 h, the inner layer of petals bent a little as presented in Fig. 3.5c. The structure was then put in high temperature water. Immediately, all the layers bent, forming a flower-like configuration (Fig. 3.5d, e). Taking the structure out of hot water and letting it dry, the structure maintained the flower shape and was stiff.

3.4 Summary

Modeling the deformation of as-built AM polymers has become a relatively matured technique given that many constitutive models (Bergstrom 2015) of conventionally manufactured polymers have been modified or adapted for AM polymers. However, the thermal-mechanical or rheological behaviors of polymers during the AM process are still not fully explored yet, which makes it challenging to predict and control the distortion of the polymer parts precisely during the fabrication process. This is a critical problem that will tackle many of the building failures for AM processes. Admittedly, the material deformation during the AM process is very complicated, which is affected by phase transition and multi-field coupling, e.g., thermal, chemical, and mechanical coupling. Therefore, researchers not only need to develop high-fidelity models but also have to investigate the fundamental material properties during the AM process. Moreover, the processing feature of each AM technique may impose difficulties to model the material deformation. For example, the material property and deformation are dependent on the scanning pattern for the extrusion and laser sintering techniques. Therefore, advanced modeling techniques must be developed to solve these problems. It is worth to mention that researchers have developed some modeling techniques for AM metals to predict their residual stress and distortion during the AM process, which has achieved significant progress in the last 5 years. Hence, some of the proposed techniques can be adapted for polymers.

Another aspect that has long been ignored is the failure analysis of AM polymers. In the literature, very limited experimental and theoretical works have been done in this area, which greatly hinder the design of engineering components from AM polymers. Overall, AM polymers show relatively brittle behavior compared to their counterparts from conventional manufacturing methods due to the anisotropic microstructure and defects. For example, the inter-bead bonding condition dictates the tensile strength of AM parts fabricated from the extrusion process. Most of the photopolymers also show brittle or quasi-brittle feature under tension partially induced by the low interlayer strength. While for the laser-sintered polymers, the failure is mainly affected by the melting degree of particles, porosity, and interlayer strength. Predicting the failure of these AM polymers will need more research works on the damage or fracture properties and their modeling development. The modern image-based failure characterization techniques like micro-CT, ultrasonic, and digital image correlation (DIC) will help track the failure process of AM polymers.

Compared to polymers and metals, the modeling research on AM composites has been very rare in the literature, which is potentially because they are not well commercialized yet. It is anticipated that AM composites, reinforced either by fibers or particles, will expand rapidly in the near future since they usually have improved mechanical properties than polymers and lighter weight compared to metals. In addition, the AM composites can be designed to exhibit multiple functionalities like actuation, heat control, etc. Thus, this is an open area with enormous research opportunity in the near future. Certainly, the modeling for AM composites is very challenging no matter for the as-built materials or material deformation during the AM process. We envision significant development toward this direction soon.

We have introduced a kind of active polymers in Sect. 3.3 for shape memory polymers, which are useful to fabricate shape-changing structures or robotics. Some researchers call this process 4D printing, which adds a time dimension to the conventional 3D printing technology. In fact, researchers also tried other techniques to fabricate active polymers and polymeric composites in the literature. For example, hydrogel with the anisotropic expansion behavior can be used to fabricate structures that change shape under swelling (Gladman et al. 2016). Research works have also been reported for magnetic and piezoelectric composites in the literature (Wang et al. 2017). Modeling the material responses of these active polymers and polymeric composites is a very exciting area that will facilitate the design and control of active structures fabricated from AM.

Finally, mechanics modeling of AM polymers used in the biomedical field has important implications on the functionality of the parts. This is obvious for implants, which need to be designed and tailored to satisfy the stiffness and strength requirements. As a result, the design-optimization of these AM polymeric implants depends on high-fidelity models of the AM polymers. In addition, some implants are biodegradable, which will require constitutive models to monitor the degradation of the mechanical properties for a lifetime prediction. Besides the implants, AM polymeric scaffolds are also used in tissue engineering to serve as frames for cell growth and differentiation. The topology and stiffness of the scaffolds will influence the growth and differentiation of cells. So the mechanical properties of the scaffolds and cell-scaffold interactions should be investigated and modeled to regulate the growth of tissues or organs. This area is still emerging, so it will take years to fully understand how scaffolds facilitate the growth of tissues or organs. This is an active field in 3D bioprinting.

References

Ajoku, U., Saleh, N., Hopkinson, N., Hague, R., & Erasenthiran, P. (2006). Investigating mechanical anisotropy and end-of-vector effect in laser-sintered nylon parts. *Proceedings of the Institution of Mechanical Engineers, Part B: Journal of Engineering Manufacture, 220*, 1077–1086.

Arruda, E. M., & Boyce, M. C. (1993a). A three-dimensional constitutive model for the large stretch behavior of rubber elastic materials. *Journal of the Mechanics and Physics of Solids, 41*, 389–412.

Arruda, E. M., & Boyce, M. C. (1993b). Evolution of plastic anisotropy in amorphous polymers during finite straining. *International Journal of Plasticity, 9*, 697–720.

Babaee, S., Shim, J., Weaver, J., Chen, E., Patel, N., & Bertoldi, K. (2013). 3D soft metamaterials with negative Poisson's ratio. *Advanced Materials, 25*, 5044–5049.

Barclift, M. W., & Williams, C. B. (2012). Examining variability in the mechanical properties of parts manufactured via polyjet direct 3D printing. In *International Solid Freeform Fabrication Symposium* (pp. 6–8). Austin, TX: University of Texas at Austin.

Bergstrom, J. S. (2015). *Mechanics of solid polymers: Theory and computational modeling*. Norwich, NY: William Andrew.

Boyce, M. C., & Arruda, E. M. (2000). Constitutive models of rubber elasticity: A review. *Rubber Chemistry and Technology, 73*, 504–523.

Boyce, M. C., Parks, D. M., & Argon, A. S. (1988). Large inelastic deformation of glassy polymers. Part I: Rate dependent constitutive model. *Mechanics of Materials, 7*, 15–33.

Chagnon, G., Rebouah, M., & Favier, D. (2014). Hyperelastic energy densities for soft biological tissues: A review. *Journal of Elasticity, 120*, 1–32.

Chung, T., Romo-Uribe, A., & Mather, P. (2008). Two-way reversible shape memory in a semicrystalline network. *Macromolecules, 41*, 184–192.

Compton, B. G., & Lewis, J. A. (2014). 3D-printing of lightweight cellular composites. *Advanced Materials, 26*, 5930–5935.

Dafalias, Y., & Rashid, M. (1989). The effect of plastic spin on anisotropic material behavior. *International Journal of Plasticity, 5*, 227–246.

Domingo-Espin, M., Puigoriol-Forcada, J. M., Garcia-Granada, A.-A., Lluma, J., Borros, S., & Reyes, G. (2015). Mechanical property characterization and simulation of fused deposition modeling polycarbonate parts. *Materials & Design, 83*, 670–677.

Gao, X., Zhang, T., Zhou, J., Graham, S. M., Hayden, M., & Roe, C. (2011). On stress-state dependent plasticity modeling: Significance of the hydrostatic stress, the third invariant of stress deviator and the non-associated flow rule. *International Journal of Plasticity, 27*, 217–231.

Gladman, A. S., Matsumoto, E. A., Nuzzo, R. G., Mahadevan, L., & Lewis, J. A. (2016). Biomimetic 4D printing. *Nature Materials, 15*, 413.

Goodridge, R., Tuck, C., & Hague, R. (2012). Laser sintering of polyamides and other polymers. *Progress in Materials Science, 57*, 229–267.

Guang Yang, W., Lu, H., Min Huang, W., Jerry Qi, H., Lian Wu, X., & Yuan Sun, K. (2014). Advanced shape memory technology to reshape product design, manufacturing and recycling. *Polymers, 6*, 2287–2308.

Hill, R. (1948). A theory of the yielding and plastic flow of anisotropic metals. *Proceedings of the Royal Society A, 193*, 281–297.

Holzapfel, G. A. (2000). *Nonlinear solid mechanics*. Chichester: Wiley.

Holzapfel, G. A. (2001). *Nonlinear solid mechanics: A continuum approach for engineering*. New York: Wiley.

Hütter, M., Senden, D., & Tervoort, T. (2013). Comment on the use of the associated flow rule for transversely isotropic elasto-viscoplastic materials. *International Journal of Plasticity, 51*, 132–144.

Khalyfa, A., Vogt, S., Weisser, J., Grimm, G., Rechtenbach, A., Meyer, W., et al. (2007). Development of a new calcium phosphate powder-binder system for the 3D printing of patient specific implants. *Journal of Materials Science: Materials in Medicine, 18*, 909–916.

Lee, E. H. (1969). Elastic-plastic deformation at finite strains. *Journal of Applied Mechanics, 36*, 1–6.

Liu, L., & Li, Y. (2018). Predicting the mixed-mode I/II spatial damage propagation along 3D-printed soft interfacial layer via a hyperelastic softening model. *Journal of the Mechanics and Physics of Solids, 116*, 17–32.

Mao, Y., Yu, K., Isakov, M. S., Wu, J., Dunn, M. L., & Qi, H. J. (2015a). Sequential self-folding structures by 3D printed digital shape memory polymers. *Scientific Reports, 5*, 13616.

Mao, Y., Robertson, J. M., Mu, X., Mather, P. T., & Jerry Qi, H. (2015b). Thermoviscoplastic behaviors of anisotropic shape memory elastomeric composites for cold programmed non-affine shape change. *Journal of the Mechanics and Physics of Solids, 85*, 219–244.

Naficy, S., Spinks, G., & Wallace, G. (2014). Stimuli-responsive hydrogel actuators (presentation video). *Proc. SPIE 9056, Electroactive Polymer Actuators and Devices (EAPAD)*, 90560V.

Nemat-Nasser, S. (2004). *Plasticity: A treatise on finite deformation of heterogeneous inelastic materials*. Cambridge: Cambridge University Press.

Ning, F., Cong, W., Qiu, J., Wei, J., & Wang, S. (2015). Additive manufacturing of carbon fiber reinforced thermoplastic composites using fused deposition modeling. *Composites Part B: Engineering, 80*, 369–378.

Ogden, R. W. (2003). Nonlinear elasticity, anisotropy, material stability and residual stresses in soft tissue. In G. Holzapfel & R. Ogden (Eds.), *Biomechanics of soft tissue in cardiovascular systems* (pp. 65–108). Vienna: Springer.

Sanchez-Dehesa, J., Garcia-Chocano, V., Climente, A., Cervera, F., Gomez-Lozano, V., Sanchis, L., et al. (2013). Three-dimensional axisymmetric cloak based on the cancellation of acoustic scattering from a sphere. *Physical Review Letters, 110*, 124301.

Spencer, A. (1971). Theory of invariants. In A. Eringen (Ed.), *Continuum physics* (pp. 239–352). New York: Academic Press.

Wang, X., Jiang, M., Zhou, Z., Gou, J., & Hui, D. (2017). 3D printing of polymer matrix composites: A review and prospective. *Composites Part B: Engineering, 110*, 442–458.

Westbrook, K., Jerry Qi, H. (2008). Actuator designs using environmentally responsive hydrogels. *Journal of Intelligent Materials Systems and Structures, 19*, 597–607.

White, M. E., Yatvin, J., Joe III, B. G., Bilbrey, J., & Locklin, J. (2013). Advances in smart materials: Stimuli-responsive hydrogel thin films. *Journal of Polymer Science Part B, 51*, 1084–1099.

Wu, J., Zhao, Z., Hamel, C. M., Mu, X., Kuang, X., Guo, Z., et al. (2018). Evolution of material properties during free radical photopolymerization. *Journal of the Mechanics and Physics of Solids, 112*, 25–49.

Yakacki, C., Saed, M., Nair, D., Gong, T., Reed, S. M., & Bowman, C. N. (2015). Tailorable and programmable liquid-crystalline elastomers using a two-stage thiol-acrylate reaction. *RSC Advances, 5*, 18997–19001.

Yu, K., McClung, A. J. W., Tandom, G. P., Baur, J. W., & Qi, H. J. (2014). A thermomechanical constitutive model for an epoxy based shape memory polymer and its parameter identifications. *Mechanics of Time Dependent Materials, 18*, 453–474.

Zarringhalam, H., Majewski, C., & Hopkinson, N. (2009). Degree of particle melt in Nylon-12 selective laser-sintered parts. *Rapid Prototyping Journal, 15*, 126–132.

Zhang, Y., & Chou, K. (2008). A parametric study of part distortions in fused deposition modelling using three-dimensional finite element analysis. *Proceedings of the Institution of Mechanical Engineers, Part B: Journal of Engineering Manufacture, 222*, 959–968.

Zhang, P., & To, A. C. (2016). Transversely isotropic hyperelastic-viscoplastic model for glassy polymers with application to additive manufactured photopolymers. *International Journal of Plasticity, 80*, 56–74.

Zhao, Q., Qi, H. J., & Xie, T. (2015). Recent progress in shape memory polymer: New behavior, enabling materials, and mechanistic understanding. *Progress in Polymer Science, 49-50*, 79–120.

Zheng, X., Lee, H., Weisgraber, T., Shusteff, M., DeOtte, J., Duoss, E. B., et al. (2014). Ultralight, ultrastiff mechanical metamaterials. *Science, 20*, 1373–1377.

Chapter 4
Additive Manufacturing of Tooling for Use in Mass Production Processes

Suzan Gunbay, Michael Hopkins Jr, Guang Hu, Noel M. Gately, Declan M. Devine, and Conor Hayes

Abstract 3D printing is increasingly being used for the fabrication of customised parts for use in biomedical applications. However, the number and type of regularity approved resins is limited, and it cannot reach the economies of scale observed with traditional mass production processing techniques such as injection moulding and blow moulding. Nevertheless, it can still play a vital role in the development of biomedical products and tools for their mass production. Both academic and industry research teams have successfully utilised 3D-printed printing of tools for a variety of applications and have shown that it is a viable alternative to metallic tooling for short product runs. Using this hybrid system where the tools are 3D printed for use in mass production processes, trials can be run using the resin and process desired for the production of the final biomedical part. Its quick production times can simplify design iterations while ensuring the functionality of final parts. Due to the layer-by-layer nature of manufacture, 3D printing has also enabled optimisation of tools through the incorporation of conformal cooling channels which can increase manufacturing efficiencies. The current chapter give a brief outline to injection moulding and blow moulding, which are two of the most common production techniques for biomedical parts. It further discusses the importance of cooling of these parts and how conformal cooling can benefit cycle times. Finally, a number of case studies are outlined which highlight some of the current research in this field.

Keywords Rapid tooling · Prototyping · Conformal cooling · Injection moulding · Blow moulding

S. Gunbay · M. Hopkins Jr · G. Hu · N. M. Gately · D. M. Devine · C. Hayes (✉)
Materials Research Institute, Athlone Institute of Technology, Athlone, Ireland
e-mail: chayes@ait.ie

D. M. Devine (ed.), *Polymer-Based Additive Manufacturing*,
https://doi.org/10.1007/978-3-030-24532-0_4

4.1 Introduction

It has been established both in the industry and more broadly in the literature that incorporating rapid prototyping into the product design process can lead to reductions in production costs and lead time from initial design concept to end-product production (Mischkot et al. 2017). Particularly, in the medical technology sector, prototyping is an essential part of this process. Prototypes function by allowing for end-user feedback at early stages of development; they can show investors how the product looks and functions as well as their usefulness for testing safety, performance and overall suitability for use by humans (Tábi et al. 2016). Typically, traditional moulding methods allow for the production of highly accurate parts using specified materials that have their limitations (Rajaguru et al. 2015; Qayyum et al. 2017). With medical devices, the materials used to make components and the processes from which these components are manufactured require approval from regulatory bodies such as the US Food and Drug Administration (FDA) and the European Commission (EC). As such, medical device companies are incentivised to build prototypes using the specified materials intended to be used in the final product utilising the final fabrication method because this enables accurate verification and validation testing by regulatory authorities. In many cases, specific materials used in medical devices can be incompatible with many current AM technologies and may require more traditional manufacturing methods such as injection, micro and extrusion moulding (Vasco and Pouzada 2013), reaction injection moulding (RIM) (Udroiu and Braga 2017), vacuum forming, casting, compression moulding and blow moulding (Au and Yu 2013). While these methods are capable of producing uniform high-quality parts, production of steel moulds can take at least 6 weeks and thousands to tens of thousands of dollars to produce depending on the complexity of the component. This process can leave little room for design iteration and extending lead time significantly (Huang et al. 2017). As a result the use of additively manufactured insert tools for the production of medical device prototypes to speed up product-to-market time is emerging as a popular option, offering medical device developers the best of both worlds regarding speed, cost and product quality (Piotter et al. 2002). Stratasys is a global leader in 3D printing and additive solutions and is one such company that has been able to capitalise on the benefits of integrating additive manufactured insert tools into the product design process (PDP) in a variety of industries which includes the medical technology sector (Sella 2019; Volpato et al. 2016; Mendible et al. 2017). In this chapter, we briefly describe injection moulding and blow moulding as these are two production techniques frequently used for the manufacture of medical devices. Conformal cooling is discussed as this is a major advantage of this method of tool production. Finally, practical applications of AM tooling are discussed through the description of case studies and their major findings.

4.2 Technologies

4.2.1 Injection Moulding

Injection moulding is a process whereby polymeric material is heated into its molten state and injected into a metallic mould containing a specific design where it is cooled to create a three-dimensional component. Modern injection moulding machines have evolved from being hydraulically driven to become superior electrically driven, energy saving machines that provide precise, stable moulding conditions (Uezono et al. 2011). The injection moulding industry has revolutionised the way we create products over the years and has been used in countless industries around the globe including automotive, aerospace, medical, packaging, toys and much more. This single-step operation makes injection moulding economically viable despite the high cost of machinery and moulds. Injection moulding machines and moulds are very costly because of the high pressures required when injecting polymer melts and the complexity of the process controls (Ebnesajjad 2015). Additionally, long lead times are associated with the manufacture of steel tools. In some instances, softer metallic tools are employed to reduce material costs; however, lead times may still range from weeks to months depending on the complexity of the part and location of the tool manufacturer. Nonetheless, over 30% of all the plastic parts are manufactured by the injection moulding process and is one of the most preferred processes in manufacturing industry because it can produce complex-shaped plastic products and having good dimensional accuracy with short cycle times (Singh and Verma 2017).

To reduce start-up costs and lead times associated with short runs or pilot production injection moulding runs, many industries are seeking out alternative and advanced methods for creating mould and tooling inserts. One alternative currently under development is the use of AM technologies to produce functional mould prototypes. Polymeric-based materials used for 3D-printed tooling have gained a lot of attention in global markets as well as research fields. As delivery times and costs of tools are on a downward trend, the modern tool manufacturer is under pressure to produce tools quickly, accurately and at a lower cost. Reducing the time to produce prototypes is key to speed up the development of new products. Stereolithography (SLA) is a technology which utilises a photocurable polymeric resin to create each layer of a print build which is used to create injection mould tools. This technology is ideally suited to 3D printing tooling as it is the oldest and therefore best established 3D printing technique, fast processing times, high spatial resolution, ease of post-production finishing and the resins once cured do not melt. This technology can lead to savings in cost and time and is ideally used to build rapid injection mould tools for short-run production directly. Rahmati and Dickens (2007) researched SLA-printed tooling regarding the longevity of the tool due to it been exposed to such high pressures and temperatures. It was found that after 500 shot various failure mechanisms were identified. They concluded that the key point to the success of this technique is a very high conductivity of the SLA-printed material

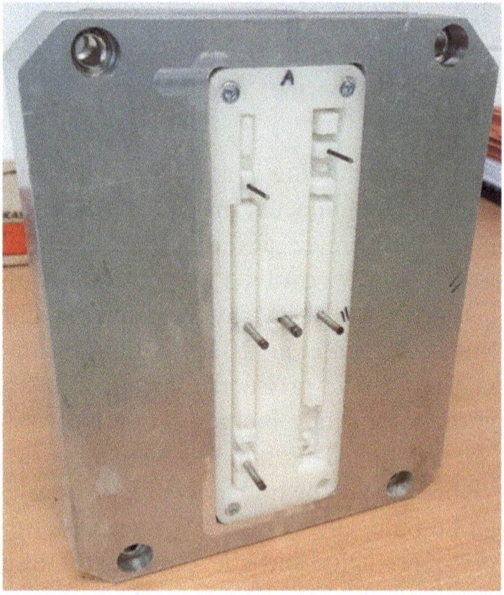

Fig. 4.1 SLA-printed insert tool with standard ejection system

and the short injection time. These two major factors are the key to the success of the SLA injection mould tooling, which is overlooked by many. The research team also identified such failure mechanisms in Athlone Institute of Technology. They have carried out extensive studies utilising SLA technology for printing hybrid production scale tooling for injection moulding. The first failure mode which was identified was gate degradation within the tool as seen in Fig. 4.1. This challenge was resolved through the creation of a hybrid tool steel/3D-printed mould base system which supported the 3D-printed tool while allowing easy interchanging of inserts.

However, due to the low thermal conductivity levels of the SLA printable resins, it was necessary to conduct trials to assess the properties of parts produced using this hybrid method, as cooling efficiencies would be affected due to the high insulating properties of the printed tooling. These trials were conducted using commercial polypropylene (PP) to evaluate the thermal and mechanical properties of mouldings via the 3D-printed tooling, with particular attention focused on the crystalline structure of the PP material. Results indicated that the insulating nature of the 3D-printed inserts did affect the crystallinity of the moulded parts. However, parts moulded using 3D-printed tools maintained a tensile strength and Young's modulus at least comparable to parts moulded using a traditional steel tool indicating that this system may be used for short and pilot runs in applications where these properties are critical parameters. Again failure mode was observed around areas where the printed tooling was subjected to high speed machined to locate ejector pins as seen in Fig. 4.2. These results indicate that this technology is more suitable for use om prototype and low production moulding trials rather than full production runs.

Lead times of mould production are another advantageous area which this technology can address. Traditionally it could take weeks if not months to get your concept part from paper to moulded component. Now with this technology,

Fig. 4.2 (**a**) Initial gate design prior to moulding (left) and gate degradation after 50 shots of PP at 200 °C (right), (**b**) Traditional tool steel design (left) compared to hybrid tool base system used in AITs study (right), (**c**) Failure mode of the 3D-printed tool after 50 shots with PP (left and right)

depending on the complexity of the part, lead times can be measured in hours or days. However, the major disadvantage with the technology is low numbers of parts produced before failure and post-processing requirements which need to be carried out. In this sense, polymer-based 3D printing has advantages over metal printing as both the cost and finishing costs are lower when compared to metal parts.

4.2.2 Blow Moulding

Blow moulding is a polymer processing technique commonly used for the production of hollow plastic parts. It is used for the production of packaging (bottles) for milk, carbonated drinks, other foodstuffs, household chemicals, personal toiletries and pharmaceuticals. These items account for approximately 70% of the blow moulding market (Brydson 1999). However, blow moulding is also used for the production of other important products such as drums, tool cases, toys, fuel tanks and air ducts for the automotive sector and catheter balloons and other biomedical applications. In this process, air is introduced through a 'blow pin' under pressure into a preformed 'parison'. The pressurised air forces the softened polymer to the edges of a mould thereby forming a hollow container with an outer shape replicating the shape of the mould. There are two main categories of blow moulding which are characterised by the way in which the parison is formed. These are injection blow moulding and extrusion blow moulding. In injection blow moulding, the parison is injection moulded prior to blow moulding in a two-step process. Extrusion blow moulding processes involve the parison been formed in a continuous extrusion process where two mould halves close on to the extrudate and air is blown through the blow pin to inflate the parison so that it takes up the shape of the mould. The former is used for applications when the blow moulded part must interface with another high-tolerance component such as high-pressure carbonated drink bottles interfacing with bottle caps, while the latter is used for low-pressure items such as milk bottles, toys, drill cases and automotive parts. Almost 75% of all blow moulded plastic materials are extrusion blow moulded (Wagner et al. 2014). In both processes, the parison must be heated or maintained at a temperature which allows the polymer to deform and fill the mould by air introduced into the part through a blow pin under pressure (Brydson 1999).

In blow moulding, the mould or tool used to form the part is relatively simple. However, it has two critical functions: (1) imparting the desired shape to the melted polymer and (2) cooling the moulded part. In its simplest form, this tool consists of two female cavities which close around the parison. The inner pressures in a blow mould are significantly lower than an injection mould, 8–10 bar compared to some hundred bars (Gilbert 2017). Hence, low clamping forces are required, and there are low shear loads at the mould surface compared to an injection mould due to the material not being pushed across the surface of the mould under high pressure. Therefore, traditionally in blow moulding most of the moulds were made of aluminium (McGrum et al. 1988). Hence, the main barriers to using polymer-based additive manufacturing are greatly reduced in blow moulding compared to injection moulding. However, there are areas within the mould where steel or other hard-wearing materials should be used. For instance, the pinch-off is the section of the mould where the parison is squeezed and welded together. These are often made out of a harder material like steel to extend the lifetime and to make the exchange of these inserts easier, and this would also be necessary in 3D-printed tooling as it must have the structural soundness to withstand the pressure of the plastic material and to withstand the repeated closing cycles of the mould. Additionally, heat dissipation from the part which enables fast cooling and short cycle times still remains a challenge, similar to what is encountered during the injection moulding part cooling process.

4.3 Cooling

Cooling is particularly important in mould design because it consumes much of the cycle time and, therefore, has a significant effect on product economics. Cooling can take two thirds or more of the entire mould-closed cycle time. Best part economics require cooling as quickly as possible. The cooling action occurs principally through mould cooling (cooling the outside of the blown part via conduction), with internal cooling of the blown part via convection contributing in the blow moulding process. Nevertheless in both injection moulding and blow moulding, conduction cooling via the mould wall is the most significant driver of cooling times.

There are three primary factors involved as a function of mould cooling:

- The coolant circulating rate.
- The coolant temperature.
- The efficiency of the overall heat transfer system.

The coolant temperature and circulating rate can be easily controlled by a chiller system, while only process and mould design can adequately affect the efficiency of the overall heat transfer system. Coolants are circulated through moulds in several ways. Often the simplest method is to machine out the back of the mould leaving an open area. This can then be enclosed with a gasketed back plate as seen in Fig. 4.3a. Using simple entrance and exit openings, the coolant is circulated in relatively large volumes. This is often known as flood cooling. A more efficient method, however, is to drill holes through the mould block from top to bottom and from side to side until they intersect as seen in Fig. 4.3b. Surplus openings are plugged, and hose connections are inserted for the inlet and outlet lines. This type of arrangement of cooling can provide extra cooling where needed in the mould such as at the neck and bottom.

Cooling water for blow moulding is often temperature controlled using a chiller, but sometimes is taken directly from the mains and so may be considered to be less effective and uncontrollable. The ideal water temperature needed for blow moulding of polyolefins is about 4–6 °C (Gilbert 2017). It is recommended that a chiller

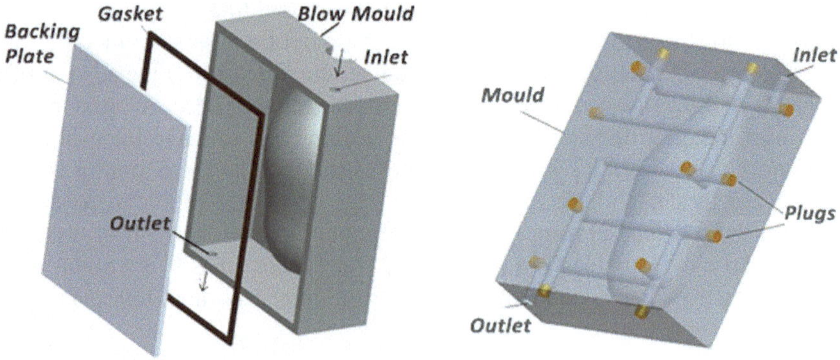

Fig. 4.3 Flood cooling system (left 1.4) and standard drill through cooling system (right 1.5)

must be laid out for a cooling capacity so that the temperature difference between water inlet and outlet does not exceed 2 °C at a pressure drop of not more than 5 bar. The water should be clean and should be running in a closed circuit. To reduce cycle times, extremely cold water can be used.

4.3.1 Benefits of Optimised Cooling System Design

The relevance of optimised cooling system design in blow moulding can be summarised into two areas: quality and productivity.

Quality is a major interest of emphasis in the plastic industry. The quality of products can depend on three major factors: the accuracy of fabrication of the mould, repeatability of the moulding machine being used and the correct design of the mould to produce the part. Of fundamental importance is the design of the cooling system that will extract heat from the hot plastic in the mould cavity at a maximum and uniform rate throughout the mould.

If uniform cooling is employed, numerous moulding quality issues can be circumvented. Warpage of parts is a large problem that is directly caused by the unsuitable design of cooling systems (Graham 2000). Another issue is the homogeneity of properties in parts moulded from crystalline resins. The morphology of the crystal structure of the final product is controlled by temperature history and its cooling time. Minor changes in temperature history and cooling time can lead to major changes in crystal formation which affects the mechanical properties of the finished part (Liu and Gehde 2016). Consequently, blow moulds with uneven cooling profiles can produce parts with unacceptable mechanical properties, possibly causing premature failure of the moulded product.

Cooling rate enhancement will reduce the cycle time but lead to defects like shrinkage and warpage. Warpage is due to residual stresses arising when shrinkage is differential and anisotropic across the part and part thickness. Difference in shrinkage may be due to (a) molecular and fibre orientation in fibre-filled materials, (b) non-uniform cooling of the final product and asymmetric cooling across part thickness from core side to cavity side, (c) part thickness variation and (d) asymmetry in part geometry (part has a specific feature on only one side of it). Earlier in two research contributions, Hassan et al. (2010) showed that an efficient cooling system is the one which achieves uniform temperature distribution on either side of the part to produce a defect-free product and also reduces the cooling time.

Further potential advantages of optimised cooling systems can include:

- Large energy savings on blow mould equipment can be gained by decreasing cycle time.
- Optimised cooling systems can result in the use of higher coolant temperatures which implies less chiller capacity requirements or less plant water usage.
- Capital outlays for new equipment can be minimised, since shorter cycle times reflect themselves in higher plant capacity through utilisation without an associated increase in the number of moulding machines.
- Less material wastage and downtime due to consistent production of quality parts.

4.3.2 Conformal Cooling

Conformal cooling is based on the principle that cooling channels follow the part's contours to facilitate faster and more uniform cooling. Until recently this simple concept has been difficult to execute. Some of the geometries required in conformal cooling are impossible to fabricate with traditional machining. However, the emergence of additive manufacturing technologies have allowed mould designers to increasingly utilise this technique. Additive manufacturing enables the design of complex cooling channels within the mould design phase rather than post-processed at suboptimal locations. Effective cooling achieved with a conformal cooling system has been developed in numerous studies for injection moulding (Au and Yu 2007, 2011). Tools with conformal cooling channels, which are fabricated by 3D printing, have demonstrated improvements in both the production rate and the part quality as compared with conventional production tools. Sachs et al. (2000) examined their conformal cooling tools in industrial applications; they reduce the cycle time by 15% and the part distortion by 9%. The high efficiency of a conformal channel in thermal control is useful not only for cooling purposes but also for mould heating (Xu and Sachs 2009).

Numerous studies have been carried out on the optimisation of cooling channels by numerical methods utilising various optimisation algorithms. Wu et al. (2015) developed a framework for optimising the design of injection moulds with conformal cooling for additive manufacturing. The proposed system consists of three modules, namely, process and material modelling, multiscale topology optimisation and experimental testing, calibration and validation. Wang et al. (2011) developed a method for designing conformal cooling circuits by investigating a geometric modelling algorithm. Kwon (1988) demonstrated the fundamentals of the cycle-averaged cooling modelling and some features of the boundary element method mould cooling design system in both two and three dimensions. Qiao (2006) performed a case study by replacing the mould part interface by the mid-surface of the part; the numerical quadrature over the surface of the cooling channels was reduced to integrate over the axis of the cylindrical segments; and modelling and meshing of the exterior surface was avoided. From this, a new systematic computer-aided approach was developed for cooling system optimal design with the advantage of being simple in concept and easy in implementation while showing high computational efficiency and quality.

Although these studies provide a viable method in optimising cooling channel position using computer-aided engineering (CAE), the design variables remained in the realm of the number of cooling channels and spacing between them with the constraints being defined as limits between the cooling channels themselves and between the surface and the cooling channels. Concerning the coolant, only its temperature was taken into account as a steady-state fixed condition.

According to Park and Pham (2009), there are three different techniques to be employed when designing conformal cooling channels: zigzag, parallel and spiral. Depending on the geometry of a part, these methods may be used in combination or on their own. The zigzag pattern, also known as a series cooling path, has part regions cooled one after the other rather than at the same time. Cooling in series is generally not preferred unless parts are small enough that the delay is negligible. The parallel channel design allows for different areas of the mould to be cooled at the same time. Park and Pham (2009) acknowledged that the main drawback for the parallel cooling method is that it requires a lot of coolant. The spiral conformal cooling channel design is often used with parts that have a curvature or spherical elements.

Although the design of the cooling system is essential, it remains, most of the time, based on the experience of the designer and not on a thorough heat transfer analysis in the mould and the plastic part. Today, the widespread availability of new rapid prototyping processes, like laser sintering or stereolithography, which allow the design of very complex channel shapes increases the number of possible cooling system configurations. Thanks to these techniques, the concept of conformal cooling, i.e. channels that conform to the surface of the moulding cavity, is of primary interest. However, achieving a uniform temperature field (i.e. quality) and having the shorter cycle time (i.e. productivity) are two antagonistic objectives. The optimal cooling system design is then compromise between uniformity of the temperature field in the part at the end of the cycle and the duration of the cycle itself. The control and/or design parameters to reach these objectives are then the location and the shape of the cooling channels, the temperature and the flow rate of the coolant fluid, and to a lesser extent the thermal properties of the mould material.

4.4 Comparison of UV Photocurable AM Resin Tools to Metal AM Tools

Metal AM tools are continually gaining traction as an alternative to standard injection moulding tools due to their ability to build complex conformal cooling channels which would be difficult or impossible using standard CNC tooling methods. Powder materials like maraging steel are available for metal tooling through technologies such as direct metal laser sintering (DMLS) and selective laser melting (SLM) which can perform in the high pressure, temperatures and surface wearing of injection moulding tools.

Metal AM injection moulding tools are being produced particularly as inserts where complex cores or critical conformal cooling requirements are needed for injection moulded parts to increase part quality and/or reduce the cooling time which in turn increases part throughput. LBC Engineering produces conformal cooled inserts for clients with metal AM technologies. By using DMLS they were able to optimise the thermal performance of a mould insert with the reduction of cycle time from 90 s to 40 s which corresponded to a productivity increase of 125 per cent per annum.

Salcomp,[1] a company that produces plugs and power supplies for mobile phones in large volumes, required optimisation of their injection moulding processes to compete on the market. To reduce the cooling time, they integrated a DMLS steel additive manufactured core insert with conformal cooling channels where standard tooling could not do so. The solution to the first part of the task was as follows: The cooling elements themselves were to be brought closer to the cavity. In this way, the heat from the layer of metal can pass through and exit the machine faster. At the same time, the fine points of the design should make sure that this evacuation takes place as effectively as possible. To bring the plan to fruition, Salcomp required a technology that made the precision positioning of the cooling channels possible, without causing any new problems. Since drilling, turning or other options could neither provide the necessary flexibility of design nor the required precision, additive manufacturing, in this case metal-based, entered the discussion: Salcomp commissioned a third party to utilise EOS technology in producing EOSINT M 270, applying Direct Metal Laser Sintering (DMLS) for the manufacture of its core inserts. Simulation models were used to establish an ideal form for use in the production process. The successful change to tool production had an effect: As planned the cooling channels were to be moved to the smallest possible distance from the edges of the core inserts, so they became conformal. At the same time, the quality of the end-product as a whole was improved, thanks to the use of additive manufacturing. The improvements are now showing positive effects in the daily production run at Salcomp: The time required for cooling was reduced from 14 s to just 8 s for each production cycle. The company was able to increase its monthly output through this efficiency gain by more than 56,000 units, without having to make adjustments

[1] https://www.eos.info/press/customer_case_studies/salcomp.

to their manufacturing machinery. The annual cost savings amount to 20,000 euros. In addition, the reject rate fell from 2% to 1.4%, bringing cost savings and a reduction in the environmental impact of the production process. The final result is increased production quality, in spite of shortening the production cycle.

Despite the successes shown in the use of metal additive manufacturing of tools, this approach is mainly suited to full production rather than prototyping low volumes as it requires a relatively high investment with significant development and processing stages to perfect the design of the parts. There are processing challenges with DMLS and SLM where thermal stresses induced after building can cause part warpage and/or cracking. Often optimisation of the additive manufacture build process is required with iterative build requirements and potentially with simulation software to identify the build failure issues. This is often a relatively slow process so not ideal for rapid development. Combined with high material costs and high investment for equipment and running costs, it can lead to a prohibitively expensive tooling part for prototyping. This tooling insert can also have limitations with surface finishes and tolerances due to the nature of additive manufacture with layer lines and the aforementioned warpage issue. This can lead to secondary processing requirements such as CNC lathe machining and EDM to attain the surface finishes and tolerance requirements. Indeed there are specific hybrid CNC/additive manufacture machines such as Hybridmanutech's Ambit system that combine CNC with powder fed Laser Metal Deposition to provide the means of adding and removing material within the same process. With the current cost and technical limitations, metal additive manufacture tooling would not be deemed suitable for early prototyping tools or bridge tooling as a standard CNC or EDM tooling process is still being recognised as an affordable approach to producing prototype tools in comparison.

The use of additive manufacturing to produce photocurable resin-based rapid tooling utilising technologies such as SLA has reduced in cost dramatically, and products such as the Formlabs Form 2 with a cost of less than €4000 and their availability of high-temperature resins have provided a low-risk/low-cost platform to develop prototype tooling. Industrial SLA equipment costs are still relatively high ranging from €50k to €250k depending on the machine specifications such as platform size and resolution. Metal additive manufacturing industrial machines such as Renishaw SLM or EOS DMLS are still at a higher cost ranging from €250k to €500k and require specialist ancillary equipment and space to run due to safety issues with the management of the powders. With SLA there are no high ancillary costs with the only requirement being UV ovens and potentially air extraction for solvents when cleaning.

4.5 Benefits of Using Resin-Based Rapid Tools for Injection Moulding

The window of opportunity that additive manufactured resin-based tooling can provide are within low volume needs, prototyping, bridge tooling or for performance validation of parts where standard polymer production material is a critical requirement that current standard additive manufacture part prototyping processes struggle to replicate accurately for an injection moulded component.

Due to the limitations of resin tools concerning the longevity of the tools, surface quality, etc., it would be deemed prudent to focus on specific features of the relative part to obtain critical information during prototyping and testing rather than focus on achieving a fully formed comparable part to a production tool which is currently difficult. This process can have the potential to minimise late tooling design changes which are expensive and time-consuming. By verifying critical functionality in the design phase with additive manufacturing resin, tool prototyping costs and often more importantly time to market can be reduced. Nevertheless, the approach should be focused on specific features within the component, so sometimes partial or sectional moulding of the part could be suitable rather than full part moulding, as when using standard resins, 20–50 parts are achievable.

For example, if snap-fits are crucial in the function of two enclosures, these features could be the focus with an additive manufacture resin tool for testing and prototyping before full production tooling is manufactured. Other features like surface quality, quantity, flash, etc. would be irrelevant for the part. In theory, the snap-fit can be developed to a stage with additive manufacture resin tools with the desired full production material which in turn minimises the chances of having to adjust the production tool. This can accelerate development time in the long run with optimising and performance verification based on actual production materials and moulding processes. A series of small resin insert tools with varying geometry can be built and moulded with the production material to test for comparison and/or DOE techniques.

Specific common features that resin-based rapid tooling can provide accurate test feedback on prior to entering full production include:

- Chemical resistance
- Thermal resistance
- Surface defects
- Moulding optimisation
- Snap-fit requirements
- Living hinge requirements
- Texturing
- Over moulding
- Insert moulding
- Fixings

As such resin-based rapid tooling is suitable for use during the development phase of tools to identify and eliminate processing issues such as weld lines and air traps, while potentially highlighting shrinkage and sink marks in the final part. The design of the gate shape, size and position can be trialled. This can provide valuable feedback in the part and tooling design development phase of complex parts or where these issues may be of critical value.

Our research group have previously developed a small PP component which was having issues with gate design leading to pressure loss and short shots. Using a high-temperature resin from Formlabs Form 2 to build an insert tool with a new gate design, we were able to quickly test the design and verify that the pressure issue was eliminated.

To verify the force requirement or functionality in terms of number of cycles of a living hinge, the only method of doing so is through actual injection moulding of parts with a suitable material, normally PP. FDM or SLS prototype parts can work as limited functioning living hinge features but do not represent the moulded state. Similarly, PU vacuum cast parts can resemble living hinges but are not functionally accurate. Potentially with a full understanding of the effects of resin tools in the cooling process, living hinge parts can be produced via resin additive manufactured tools to verify its function. Again our group have been able to produce living hinge features using Somos GP+ resin tools. Using this process a design flaw was quickly identified where the living hinge features was unable to function due to it being too thick. Subsequently, the living hinge design feature was adjusted to a minimum of 0.3 mm for production tooling. This simple trial saved the client company in terms of both tooling costs and lead time as well as time to market.

Surface textures and finishes are a common feature with injection and blow moulded parts. Polished, textured or complex patterns are often used to signify function such as a handle grip, improve aesthetics and add to the overall quality or "feel" of a part. Often texturing can provide a means to mask surface defects in injection moulded parts such as resin flow or swirl lines, sink, gates and weld lines. Textures are traditionally difficult to obtain with standard rapid prototyping of parts, e.g. vacuum casting, SLA, FDM, etc. without significant manual finishing of the parts. The rapid prototyping of parts intended for full production with injection or blow moulding still will not capture potential surface defects inherent to these processes. With resin-based rapid tooling, the production process, batch of polymer and colour allow the detection of surface defects early before production tooling has been produced. Suitable textures could then be applied to the resin tools in relevant areas to determine how best to mask defects or for functional and/or aesthetic reasons. This process could be done relatively quickly and cheaply which could provide valuable feedback for the decision-making process when going to production tooling. Texturing can be applied to the resin tool via bead blasting with various mediums, polishing via wet and dry paper, digital surface manipulation on the 3D CAD file and potentially plasma, ion, chemical and laser etching. Our team are investigating this area by determining specific procedures to match surface tooling finishes on additive manufactured resins with current standard injection moulded texturing and polished levels.

Another functional aspect of advanced prototype development that additive manufactured resin tools can bring is with overmoulding of two polymer materials. Over moulding is a process where a single part is created using two or more different materials in combination. Typically the first material, sometimes referred to as the substrate, is partially or fully covered by subsequent materials during the manufacturing process. Sometimes the two materials are chemically bonded only during the moulding process, sometimes mechanically interlocked or ideally both. By prototyping, this process using additive manufactured tools early in the design process, any issues with chemical adhesion or mechanical interlocking can be identified and designs to improve mechanical interlocking and/or adhesion trialled. Again this process could lead to getting the production tool "right first time" and reduce expensive tooling modifications late in the development process and ensure the production material will function as intended.

A similar potential high-value prototyping method where additive manufactured tooling may be beneficial is the area of insert tooling. In this process, usually metal parts are inserted into an injection moulding tool, and either a hard or soft resin-like thermoplastic elastomers (TPEs) are moulded around the metal part. This is a common practice with products like screwdrivers, and other hand tools are produced to provide soft grip surfaces. Another aspect of insert tooling is with enclosing fully or partially electronic components. This process can be quickly prototyped using the additive manufactured tooling processes and allow for early troubleshooting and identification of tooling design flaws.

Our team simulated this process to determine the lead time from concept to a final moulded part. In our investigation, it was found that using rapid tooling processes, a USB key demonstrator went from concept to final moulded parts in 3 days but with only 24 person-hours. The resin-based rapid tooling insert which was produced using Somos GP+ resin on a 3D Systems Viper was able to produce 15 parts before resin failure occurred with resin cracking resulting from the high pressure and temperatures utilised in moulding. Nevertheless, this process verified that the insert moulding was achievable and provided a fully functioning USB key (Fig. 4.4).

There are numerous methods for fixing plastic components together including screws, inserts, adhesives and welding. Again standard rapid prototyping of parts can partially verify such methods before committing to production tools but not there limitations. Resin-based rapid tooling can provide the basis to test such fixings with production material. For example, thread forming screw manufacturers provide approximate hole diameters for broad materials but advise on conducting tests on proposed applications before they are implemented. There is a balance between the thread stripping torque and minimum torsional strength of the screw which can be difficult to achieve without actual testing on the production material part. By producing the relevant boss features with resin-based rapid tooling using the relevant production material, a range of boss and hole diameters can be produced to test and find the optimal dimensions to inform final production tooling specifications. A similar approach can be applied to inserts.

For welding applications resin-based rapid tooling can allow for optimisation of features such as energy concentrators which can be crucial in testing to produce an

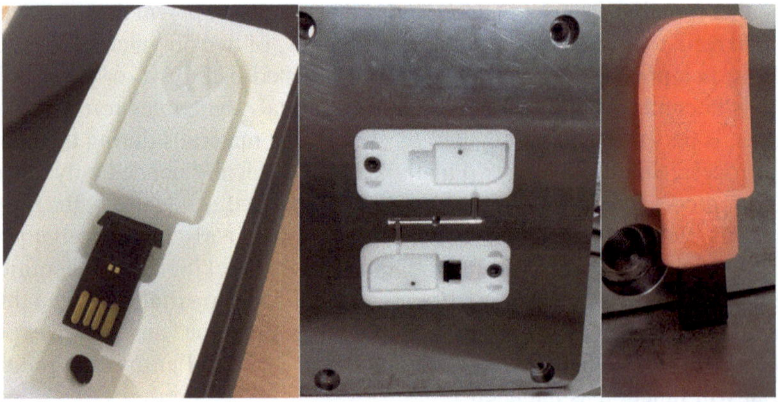

Fig. 4.4 Resin-based rapid tooling was used to produce USB keys from design to production in 3 days which represented 24 person-hours. (Left) the USB component loaded into the printed insert; (middle) inserts placed in tool; (right) moulded USB key

optimised bond. Again a number of energy concentrator designs could be produced and the feature tested verified prior to production. Surface area optimisation can be a similar approach with the use of adhesives.

4.6 Rapid Tooling: Case Studies

4.6.1 Design Verification Through the Use of Resin-Based Tooling

Outside of the medical technology sector, Digital ABS moulds have also been utilised to create realistic, precise production suitable parts. In 2015, HASCO tested Digital ABS tools and evaluated their ability to produce high-quality sealing plug parts for their A8001 clamping fixture. Following production of an initial part prototype using PolyJet technology, the company realised that the thickness of the part needed to be 12 mm thick in order to ensure a tight fit in the fixture. Immediately the design team were able to reduce the thickness of the part and incorporate this change into the design of the resin-based rapid tooling insert which enabled them to injection mould functional parts in as little as 4 days. Since their incorporation of resin-based rapid tooling insert, HASCO's development of a quick change injection mould system HASCO K3500 enables production of small series of products as well as prototypes quickly and cost-effectively (Wheeler 2019).

Coloplast, a Danish medical device company specialising in products related to urology, ostomy, continence and skin care were seeking to produce prototype TPE components for novel neurology devices that required functional testing prior to production in order to analyse the durometry and wall thickness. Coloplast knew building a soft steel mould to produce the TPE part was an expensive venture as several design iterations were likely. As a solution, Coloplast utilised rapid tooling technologies from Diversified Plastics to produce thermoplastic moulds that would be used to print prototype TPE parts. The mould was reported to take approximately 11 h to build including both halves and centre core. By using an existing mould base, further time savings were achieved with a pre-existing ejector system (3D Printed Mold Components 2019). Having evaluated the TPE prototype, Coloplast deemed the part too soft to meet compression resistance requirements and decided to revise and stiffen the design of the part by thickening the wall sections. As such, early incorporation of rapid tooling technologies enabled final mould production in 5 days as opposed to weeks and allegedly saved the company thousands of dollars (3D-printed medical mold components 2019).

4.6.2 Resin-Based Rapid Tooling to Reduce Costs and Lead Times

Global design firm Worrell Design Inc. who specialise in healthcare product solutions joined forces with Advanced Moulding Technologies (AMT) to develop rapid tooling for the production of injection moulding insert tools from Digital ABS resin (Stratasys). This resin is reported to have high strength and heat resistant. Combining the use of an Objet500 Connex (Stratasys), an 85-ton injection moulding press (Toshiba) and a 250-ton press from AMT for larger parts, Worrell used this set-up to produce high-functioning prototypes for medical device companies in a variety of thermoplastics including acrylonitrile butadiene styrene (ABS), a variety of nylons, Delrin (acetyl resin), TPEs, polyethylenes (PE) and PP. For example, Worrell designed and engineered a dual-flow needleless blood collection system for healthcare company MedTG that reduced the need for multiple injections, thereby increasing patient comfort and hospital efficiency. By utilising resin-based rapid tooling to prototype their device, the company was were able to reduce the costs associated with traditional tooling from $11,000 to $2600 as well as cutting overall lead times from 56 days to 2 days. Similarly, the design of a prototype for OBMedical's maternal foetal monitor allowed for prototyping in their specified end-product material without having to invest in steel tooling reducing overall cost and lead times from $12,000 to $4000 and from 84 days to 3 days, respectively. The use of resin-based rapid tooling allowed for clinical trials enabling OBMedical to submit clinical data to the FDA in a fraction of the time it would've taken without the use of the AM tools (Pomager 2019).

It is well known that process cycle time plays a crucial role in the injection moulding as it directly affects the efficiency of the processes. More than two thirds of the moulding cycle duration is taken up by part cooling time, while the cooling uniformity of moulded part has important influence on part quality (Mazur et al. 2017). However, most current cooling channels which are fabricated by traditional manufacturing techniques such as CNC machining might have restrictions in terms of geometry, cost, lead time and cooling fluid flow (Bue et al. 2018). Therefore, the application of conformal cooling channels made by solid freeform fabrication is becoming popular recently. In a study by Park and Dang (2017), spiral cooling channels made by 3D printing were evaluated to determine the quality and productivity of the moulding process by applying the conformal cooling channels. The proposed cooling system modified the existent cooling system by adding new spiral cooling channels. Moreover, the flow rate of the coolant was increased from 0.75 to 3 L/min. Results demonstrated that the higher flow rate decreased cycle time and that the applied conformal cooling channels reduced the cycle time by 23% (from 39 s to 30 s) compared to conventional cooling channels (Park and Dang 2017).

4.6.3 Ceramic-Polymer Tooling Inserts for Use in the Production of Electrical Switch Components

Somos® PerFORM is a ceramic-polymer composite resin that is also capable of producing strong, stiff, high-temperature resistant composite parts that are ideal for tooling and wind tunnel testing applications. It was reported that Wehl and Partner[2] tested the use of insert tools for production of electrical switch components fabricated from Somos PerFORM resin using PolyJet (Stratasys). Interestingly, the company had initially tested ABS-based resins but found that the resin lacked sufficient temperature tolerance which in turn had an effect mould reliability and part quality. It is reported that components produced using Somos PerFORM moulds could approximately 40 shots for 20% glass filled PC injected at 150 °C and 47 shots for Nylon 6 at injected at 270 °C. Overall, the company reported successful production of functional, testable parts in days as well as fabrication of 30 different moulds in less than a year, with a reported 70% reduction overall lead time compared to previous traditional methods of tool production.[2]

4.6.4 Comparison of Resin-Based Printed Tooling to Metal Tooling

A recent collaborative study between Laser Prototypes Europe, Belfast and our research group produced resin-based rapid insert tools for injection moulding from a variety of SLA resins including Somos PerFORM. The overall objective of the study was to investigate viability, time and cost effectiveness of using 3D-printed tools for short injection moulding production runs. The study compared the thermal and mechanical profiles as well as surface finish of the SLA moulds. A metal-based

[2] https://3dprint.com/122604/somos-perform-wehl/.

additive manufactured mould fabricated using metal powder bed fusion (MPBF) and a traditional CNC mould were also fabricated for comparative purposes. Somos PerFORM outperforms all of the other SLA resins tested which succumbed more quickly to temperature and pressure-related blistered, warpage and deterioration. The additive design of the SLA mould also allowed for easy incorporation of cooling channels that allowed for observation and a design iterative approach towards heat build-up and cooling profiles of the moulds during the injection process. It was also observed that the SLA and CNC mould cavities had smoother surface finish compared to the MPBF which had the issue of moulded parts sticking to the cavity upon ejection. All three moulds were found to produce more than 500 shots with appropriate holding pressure and cooling, while the gate life on both the SLA and MPBF moulds insert broke under the high shear stress of processing (McKay 2017).

4.6.5 Comparison of Service Life of Tools Using Different Resins

An extensive study by Boxtrom (2017) aimed to investigate different challenges and opportunities of using resin-based rapid processes for injection moulding tools in the product development process. Having designed and acquired a mould base for the testing of the insert tools from a variety of resins, three different products in different levels of development were chosen. These included a part already in production, a part in the detail design phase and a part that was a new variation of an existing production part. Initially, the mechanical performance and pressure tolerance of a mould with simple part geometry was fabricated on a ProX800 using high-temperature resin Bluestone Accura (3D Systems) was first studied. Moulded PP parts were selected due to their easy flow characteristics which enabled initial determination of a suitable shot size without damaging the mould. Despite visible layering on the surface of the SLA mould and some brittleness, the results of the study showed complete filling of the cavity with each shot but also significant wear on the SLA mould after 50 shots. After 30 shots mechanical failure of the mould was observed, flash effects were noted due to cracks in the core insert both during processing and before as a result of injector pinhole drilling which required aluminium reinforcement.

Following this, insert tools fabricated from FormLabs High Temp and an externally sourced DMS Somos tool exhibited failure after the first shot due to splitting of the cavity in the same location, and this was more extensive in the Formlabs insert. It was concluded that this is likely due to dimensional incompatibility of the cavity insert design which prevented a tight fit with the adjacent half of the core insert and this is a common problem with polymer-based rapid tooling. Following reinforcement with aluminium, 15 PP parts were produced, and one part was used to test dimensional accuracy of the part with respect to the CAD model. Warpage of moulded PP parts was attributed to the lack of sufficient holding pressure and inadequate cooling of the mould. Ejection force-related deformation surrounding the injector pin was also observed.

The author of this study concluded that early knowledge of a suitable resin as well as standard manufacturing conditions and mould fitting process would produce more favourable outcomes and longevity for resin-based rapid insert tools. Moreover, the biggest impact on the lead time of the production part was the design of the tool inserts. Lack of adequate cooling and conservative pressure profiles were shown to cause a variety of issues from part warpage, mould cracking and inadequate gas ventilation issues due to part geometry and lack of draft angles on the internal mould surface. It was observed that damage to polymer moulds caused by force of injection might be reduced by exerting a higher ejection force and additional mould compensatory tampering with metals such as aluminium (Boxtrom 2017).

4.6.6 Carbon Fibre-Reinforced Rapid Tooling Inserts

Hofstätter (Hofstatter et al. 2016) investigated the effect of carbon fibres on the lifetime and surface deterioration of rapid tooling. Tools were produced using DLP with short carbon fibres added as fillers. The inserts were used during injection moulding of low-density polyethylene until their failure. It was found that the lifetime of fibre-reinforced inserts was extended compared to resin-only inserts. The surface wear in terms of mean surface roughness was negligible when inspecting the surface without crack-like features. Minor cracks on the surface appeared from 300 shots but did not propagate to major cracks, and therefore did not result in fatal failure of the entire insert. Propagation of major cracks through the entire insert were found to spread slower compared to plain inserts. Thus, it can be concluded that the lifetime of the insert was increased by generating a composite using short carbon fibres (Hofstatter et al. 2016).

4.7 Limitations of Polymer-Based Rapid Tooling

Despite their cost and time-saving benefits from a product design perspective, the lifetime of thermoplastic AM insert tools in manufacturing capacity is limited by a variety of factors. Largely, these factors concern temperature, pressure and the mould materials' ability to withstand continuous cycles of harsh manufacturing environments over time (Kovacs and Bercsey 2005). Other factors include applied force, surface roughness and hardness, resin type, post-processing conditions as well as curing time (Vasco and Pouzada 2013).

Traditional tools manufactured from steel and aluminium can easily withstand many cycles of high-temperature and high-pressure environments (Griffiths et al. 2012). However, commercial thermoplastic resins and other polymer composite materials used to prototype rapid tooling generally have much lower melting temperatures and lower impact strengths depending on the curing treatments and polishing that are applied. Thermoplastics also tend to act as insulators, which make sufficient cooling of the mould difficult to predict and varies from resin to resin (Rajaguru et al. 2015). This has an overall effect on both the lifetime of the mould, which is prone to warpage (Asproiu and Străjescu 2007), cracking and even breakage as well as the structure and performance of the moulded parts. Moreover, the pressure profile for thermoplastic moulds looks very different from that of their steel counterparts. Often the time between cycles known as holding pressure is extended for as long as 60 s in order to give the mould sufficient time to cool down between each cycle before new molten material is passed through it (Boxtrom 2017). On average, a thermoplastic rapid tooling might produce anything from 25 to 100 parts, while its metal counterpart would likely easily produce at least multiple 1000s.

Another factor that has been shown to affect the quality of parts moulded from polymer-based rapid tools is the surface finish of the mould which is related to the quality and resolution (layer thickness) of the 3D printer. Brydson (1999) studied the hardness characteristics and surface finish of epoxy-based resins (Dupont Sonos 6110 and 7110) and reported that the layered nature of AM fabrication gives rise to a rough, uneven surface finish which if not subject to post-processing prior to use can prevent the ejection of the moulded parts or damage their quality during the process of removal. It was reported that polishing/sanding of the insert tools with abrasive tools of smaller particle sizes (approximately 30 μm) leads to better surface finish for the moulds and an overall increase in part quality. Bryan also found that further curing using a heat source as well as the traditional UV curing of the printed mould improves the overall heat distribution temperature of the mould as well as increased hardness and flexibility with the extra polymer bonding curing provides which in turn affects the strength and durability of the mould (Bryan 1998).

4.8 Summary

Polymer-based 3D-printed tooling currently cannot replace metallic tools for high volume production due to machine and material constraints. However, it has being shown repeatedly that the fabrication of polymer-based rapid tooling offers tool designers a fast, low-cost method to verify tool designs and test initial parts. It can also be used for short runs of parts which can be advantageous for biomedical sectors as enough parts can be produced for regularity trials using this process that allows the production of the final part design in the desired polymer resin using the desired production method which greatly reduces development costs.

Polymer-based tooling does have difficulties in terms of heat build-up due to their insulating properties. However, this issue can be addressed somewhat through the incorporation of conformal cooling channels, optimisation of machine settings and through the development of thermally conductive reinforced composite resins for use in the production of tooling.

Finishing of polymer tooling can also play a role in tool quality and longevity. This is relatively simple with polymer-based tooling and is the subject of Chap. 10 of the current book.

Acknowledgements This work was supported in part by financial support from Athlone Institute of Technology under the Presidents Seed Fund, Enterprise Ireland funding under the Technology Gateway program, grant number TG-2017-0114 and Science Foundation Ireland (SFI) under Grant Number 16/RC/3918, co-funded by the European Regional Development Fund.

References

3D-printed medical mold components (2019). In: Today's Medical Developments. https://www.todaysmedicaldevelopments.com/article/tmd1014-sample-thermoplastic-eleastomer/. Accessed 18 Jul 2019

3D Printed Mold Components (2019). In: Divplast.com. https://www.divplast.com/wp-content/uploads/2015/05/3D-Printed-Mold-Components-Case-Study2.pdf. Accessed 18 Jul 2019

Asproiu, S., & Străjescu, E. (2007). Influence of mold properties on the quality of molded parts. *UPB Scientific Bulletin, Series D: Mechanical Engineering, 69*(3), 39–46.

Au, K. M., & Yu, K. M. (2007). A scaffolding architecture for conformal cooling design in rapid plastic injection moulding. *International Journal of Advanced Manufacturing Technology*. https://doi.org/10.1007/s00170-006-0628-x.

Au, K. M., & Yu, K. M. (2011). Modeling of multi-connected porous passageway for mould cooling. *CAD Computer Aided Design*. https://doi.org/10.1016/j.cad.2011.02.007.

Au, K. M., & Yu, K. M. (2013). Conformal cooling channel design and CAE simulation for rapid blow mould. *International Journal of Advanced Manufacturing Technology*. https://doi.org/10.1007/s00170-012-4326-6.

Boxtrom J. (2017). Improving the product development process of injection molded parts by utilization of additively manufactured molding tools. Masters Thesis, Aalto University, Helsinki, Finland.

Bryan M. B. (1998). Post-build processing of stereolithography molds. Masters Thesis. *Georgia Institute of Technology*, Atlanta GA, USA.

Brydson, J. (1999). *Plastics materials* (7th ed.). Oxford: Butterworth-Heinemann, Elsevier.

Bue, D., et al. (2018). Evolution of additively manufactured injection molding inserts investigated by thermal simulations.

Ebnesajjad, S. (2015). Injection molding. *Fluoroplastics,* 236–281. https://doi.org/10.1016/B978-1-4557-3197-8.00010-9.

Gilbert, M. (2017). Chapter 1—Plastics materials: Introduction and historical development. In M. B. T.-B. P. M. Gilbert & E. Eighth (Eds.), *Brydson's plastics materials* (8th ed., pp. 1–18). Oxford: Butterworth-Heinemann. https://doi.org/10.1016/B978-0-323-35824-8.00001-3.

Graham, L. (2000). What is a mold? An introduction to plastic injection molding and injection mold construction. The Tech Group Inc.

Griffiths, C. A., et al. (2012). Micro-stereolithography tools for small-batch manufacture of polymer micro-parts. *Proceedings of the Institution of Mechanical Engineers, Part B: Journal of Engineering Manufacture, 226*(4), 708–721. https://doi.org/10.1177/0954405411429242.

Hassan, H., et al. (2010). 3D study of cooling system effect on the heat transfer during polymer injection molding. *International Journal of Thermal Sciences, 49*(1), 161–169. https://doi.org/10.1016/j.ijthermalsci.2009.07.006.

Hofstatter, T., et al. (2016). Evolution of surface texture and cracks during injection molding of fiber-reinforced, additively-manufactured, injection molding inserts. In *Proceedings—ASPE/Euspen 2016 Summer Topical Meeting: Dimensional Accuracy and Surface Finish in Additive Manufacturing* (pp. 38–43). Retrieved from https://www.engineeringvillage.com/share/document.url?mid=cpx_M18f3d89215704e76b47M59011017816317 1&database=cpx.

Huang, R., et al. (2017). Environmental and economic implications of distributed additive manufacturing: The case of injection mold tooling. *Journal of Industrial Ecology, 21*, S130–S143. https://doi.org/10.1111/jiec.12641.

Kovacs, J., & Bercsey, T. (2005). Influence of mold properties on the quality of molded parts. *Periodica Polytechnica Mechanical Engineering, 49*(2), 115–122.

Kwon, T. H. (1988). Mold cooling system design using boundary element method. *Journal of Engineering for Industry, 110*(4), 384–394. Retrieved from http://manufacturingscience.asmedigitalcollection.asme.org/article.aspx?articleid=1447114.

Liu, Y., & Gehde, M. (2016). Effects of surface roughness and processing parameters on heat transfer coefficient between polymer and cavity wall during injection molding. *The International Journal of Advanced Manufacturing Technology, 84*(5), 1325–1333. https://doi.org/10.1007/s00170-015-7816-5.

Mazur, M., et al. (2017). Numerical and experimental evaluation of a conformally cooled H13 steel injection mould manufactured with selective laser melting. *The International Journal of Advanced Manufacturing Technology, 93*(1–4), 881–900. https://doi.org/10.1007/s00170-017-0426-7.

McKay, W. (2017). LPE research investigates rapid tooling for injection moulding. *Engineers Journal, Engineers Ireland.*

McGrum, N. G., Buckley, C. P, & Bucknall, C. B. (1988). Principles of Polymer Engineering. (pp. 311–312). Oxford University Press, New York.

Mendible, G. A., Rulander, J. A., & Johnston, S. P. (2017). Comparative study of rapid and conventional tooling for plastics injection molding. *Rapid Prototyping Journal, 23*(2), 344–352. https://doi.org/10.1108/RPJ-01-2016-0013.

Mischkot, M., et al. (2017). Injection moulding pilot production: Performance assessment of tooling process chains based on tool inserts made from brass and a 3D printed photopolymer. *Proceedings of ANTEC, 2017*, 1–5.

Park, H. S., & Dang, X. P. (2017). Development of a smart plastic injection mold with conformal cooling channels. *Procedia Manufacturing, 10*, 48–59. https://doi.org/10.1016/j.promfg.2017.07.020.

Park, H. S., & Pham, N. H. (2009). Design of conformal cooling channels for an automotive part. *International Journal of Automotive Technology*. https://doi.org/10.1007/s12239-009-0011-7.

Piotter, V., et al. (2002). Microinjection moulding of medical device components. *Business Briefing—Medical Device Manufacturing and Technology, 2002*, 63–66.

Pomager, J. (2019). 3D-printed injection molding: The future of rapid prototyping. In: Meddeviceonline.com. *International Journal of Automotive Technology*. https://www.meddeviceonline.com/doc/3d-printed-injection-molding-the-future-of-rapid-prototyping-0001. Accessed 18 Jul 2019

Qayyum, J. A., et al. (2017). Performance of 3D printed polymer mold for metal injection molding process. *ARPN Journal of Engineering and Applied Sciences, 12*(22), 6430–6434. https://doi.org/10.3390/met8060433.

Qiao, H. (2006). A systematic computer-aided approach to cooling system optimal design in plastic injection molding. *International Journal of Mechanical Sciences*. https://doi.org/10.1016/j.ijmecsci.2005.11.001.

Rahmati, S., & Dickens, P. (2007). Rapid tooling analysis of Stereolithography injection mould tooling. *International Journal of Machine Tools and Manufacture, 47*(5 Special issue), 740–747. https://doi.org/10.1016/j.ijmachtools.2006.09.022.

Rajaguru, J., Duke, M., & Au, C. K. (2015). Development of rapid tooling by rapid prototyping technology and electroless nickel plating for low-volume production of plastic parts.

International Journal of Advanced Manufacturing Technology, 78(1–4), 31–40. https://doi.org/10.1007/s00170-014-6619-4.
Sachs, E., et al. (2000). Production of injection molding tooling with conformal cooling channels using the Three Dimensional Printing process. *Polymer Engineering and Science*. https://doi.org/10.1109/ROMAN.2005.1513780.
Sella, N. (2019). 3D Printed Injection Moulds. In: Usglobalimages.stratasys.com. http://usglobalimages.stratasys.com/Main/Secure/Webinars/InjectionMolding_PPT_Webinar.pdf?v=635804191840947776. Accessed 18 Jul 2019
Singh, G., & Verma, A. (2017). A brief review on injection moulding manufacturing process. *Materials Today: Proceedings, 4*(2), 1423–1433. https://doi.org/10.1016/j.matpr.2017.01.164.
Tábi, T., et al. (2016). Comparison of thermal, mechanical and thermomechanical properties of poly(lactic acid) injection-molded into epoxy-based Rapid Prototyped (PolyJet) and conventional steel mold. *Journal of Thermal Analysis and Calorimetry, 123*(1), 349–361. https://doi.org/10.1007/s10973-015-4997-y.
Udroiu, R., & Braga, I. C. (2017). Polyjet technology applications for rapid tooling. *MATEC Web of Conferences*, 1–6. https://doi.org/10.1051/matecconf/201711203011.
Uezono, H., Katsuhiro, F., & Ochi, K. (2011). History of development of injection moulding machine technologies and future perspectives. JSW Technical Review.
Vasco, J. C., & Pouzada, A. S. (2013). A study on microinjection moulding using moulding blocks by additive micromanufacturing. *International Journal of Advanced Manufacturing Technology, 69*(9–12), 2293–2299. https://doi.org/10.1007/s00170-013-5165-9.
Volpato, N., Solis, D. M., & Costa, C. A. (2016). An analysis of Digital ABS as a rapid tooling material for polymer injection moulding. *International Journal of Materials and Product Technology, 52*(1/2), 3. https://doi.org/10.1504/IJMPT.2016.073616.
Wagner, J. R., Mount, E. M., & Giles, H. F. (2014). 52—Extrusion blow molding. In J. R. Wagner, E. M. Mount, & H. F. B. T.-E. (Eds.) Extrusion (2nd ed., pp. 593–601). Oxford: William Andrew Publishing. https://doi.org/10.1016/B978-1-4377-3481-2.00052-1.
Wang, Y., et al. (2011). Automatic design of conformal cooling circuits for rapid tooling. *CAD Computer Aided Design*. https://doi.org/10.1016/j.cad.2011.04.011.
Wheeler, A. (2019). New HASCO/Stratasys Hybrid System 3D Prints Inserts For Injection Molded Prototypes. In: Engineering.com. https://www.engineering.com/3DPrinting/3DPrintingArticles/ArticleID/10699/New-HASCOStratasys-Hybrid-System-3D-Prints-Inserts-For-Injection-Molded-Prototypes.aspx. Accessed 18 Jul 2019
Wu, T., et al. (2015). A framework for optimizing the design of injection molds with conformal cooling for additive manufacturing. *Procedia Manufacturing*. https://doi.org/10.1016/j.promfg.2015.09.049.
Xu, R. X., & Sachs, E. (2009). Rapid thermal cycling with low thermal inertia tools. *Polymer Engineering and Science*. https://doi.org/10.1002/pen.21261.

Chapter 5
Current Market for Biomedical Implants

Aleksandra Foerster, Laura Ruiz Cantu, Ricky Wildman, and Christopher Tuck

Abstract The use of additive manufacturing (AM) for the fabrication of biomedical implants has grown in the recent years because it gives the flexibility to create patient-specific customised implants and on-demand production. AM has opened the possibility of the fabrication of implants in the clinical setting, so-called bedside manufacturing that gives the opportunity to print directly into the patient or in a nearby location.

In this chapter, an insight into the global 3D-printed medical devices market and more detailed discussion of the different applications including artificial skin; cardiovascular stents; ear, nose and throat (ENT); and dental are presented. The chapter reviews recent advancements in 3D printing technologies for biomedical implants. It describes recent developments in the 3D printing of artificial skin with a focus on the treatment of burn wounds. Additionally, the recent progress in the field of electronic, multifunctional skin is reviewed.

The chapter also discusses the methods and materials being used for 3D printing stents including computational-based 3D-printed metal, biodegradable polymer, self-expandable and drug-eluting stents.

ENT is a speciality area where AM has been also successfully utilised for the fabrication of implants. The examples of implants such as hearing aids which are already on the market and 3D printing techniques which can be used to create implants to replace the bones in the middle ear (ossicles) when damaged or to reconstruct the entire ear are discussed.

Finally, dental implants in the research and development stage is reviewed.

Keywords Biomedical implants · Artificial skin · Additive manufacturing of stent · 3D-printed ENT · Dental 3D printing

A. Foerster · L. R. Cantu · R. Wildman · C. Tuck (✉)
Faculty of Engineering, University of Nottingham, Nottingham, UK
e-mail: Christopher.Tuck@nottingham.ac.uk

D. M. Devine (ed.), *Polymer-Based Additive Manufacturing*,
https://doi.org/10.1007/978-3-030-24532-0_5

5.1 Introduction

Medical achievements using additive manufacturing (AM) technology are growing rapidly and are expected to change healthcare markedly (Schubert et al. 2014). The possibility of translating medical imaging of body parts done by X-rays, computerised tomography (CT) or magnetic resonance imaging (MRI) to digital models that can be used by AM machinery has opened a number of applications in the medical field. AM technology has been used to produce bones, skeletons, ears, trachea, cell cultures, tissue and organs as well as new medical drugs (Javaid and Haleem 2017). The capability of AM technology to produce complex high-resolution geometries that can be customised and personalised makes it very attractive for the production of medical implants. This technology has also the potential to decrease the production costs of medical devices (Schubert et al. 2014; Banks 2013).

3D printing medical devices market is expected to grow at a compound annual growth rate (CAGR) of 16.8% between 2017 and 2025, estimated to reach $2.77 billion by 2025 (Profshare 2018). The market growth of AM medical devices is mainly driven by technological advances, the possibility of customisation of a medical product, growing public and private funding and increased application in medical industry. There are more than 30 companies worldwide focussing their business on 3D printing for medical applications including 3D printers and materials, e.g. Stratasys, BASF, 3D Systems, FabRx Ltd., EOS GmbH and Cyfuse Biomedical K.K., among others.

AM is based on producing three-dimensional physical objects based on computer data normally using a layer-by-layer approach. AM, often termed 3D printing, removes the limitation in the possibility of fabricating a wide range of customised parts at a reasonable cost when compared to other manufacturing methods such as moulding, milling or casting. There are several AM technologies with promising applications for fabrication of medical implants including vat polymerisation, material extrusion, powder bed fusion, material jetting, binder jetting, sheet lamination and directed energy deposition (Ventola 2014).

Stereolithography is a form of vat polymerisation and is used to create solid objects from a photopolymeric liquid which cures under ultraviolet or visible laser light. Material extrusion-based methods such as fused deposition modelling (FDM) use molten thermoplastic filaments as a building material that are extruded layer by layer with a polymer material to support the main structure during printing which is removed after the printing process. Powder bed fusion-based methods such as laser sintering (LS) or selective laser melting (SLM) fabricate 3D structures using a high-power laser beam (Fig. 5.1) to melt and bind polymer and metallic powders, respectively, whereas material jetting-based technology creates objects through drop-on-demand deposition of polymers that cure at temperature or photopolymers that cure using light. Material jetting methods enable the use of multiple materials simultaneously. In binder jetting-based methods, a binding agent is deposited onto a thin layer of powder material. Consecutive layers bind to one another and form a solid object. Sheet lamination-based methods use sheets of material which are

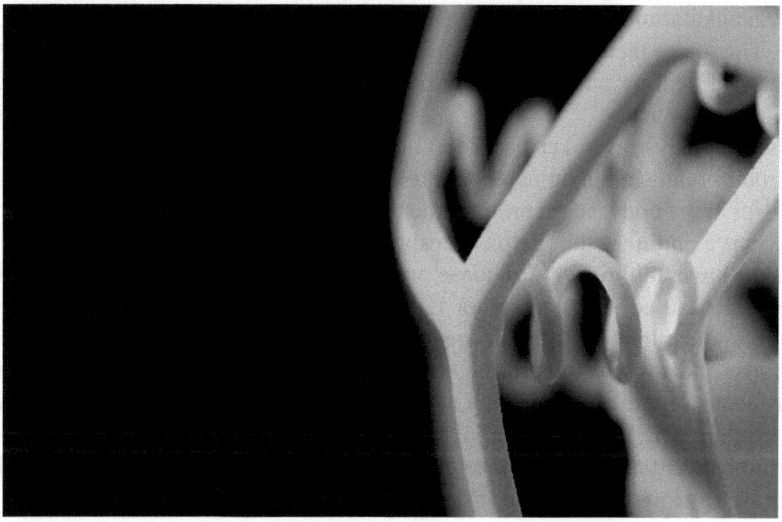

Fig. 5.1 Advanced functional geometry created via polymeric laser sintering

stacked and laminated to each other. In these approaches, unwanted parts are cut out layer by layer and removed once the part is fabricated. Directed energy deposition-based techniques produce parts by melting powder material or wire as it is fed into a laser or electron beam and are predominantly used for metal parts fabrication. The choice of the 3D printing technique for medical implants fabrication depends on material to be used, required complexity, resolution and final application of the product.

The number of medical implants produced with 3D printing has been reported with clinically effective results, and some are already available on the market. However, most of the products and technologies are still in an experimental phase. This chapter will focus on the application of 3D printing for the fabrication of biomedical implants such as ENT, dental products, artificial skin and cardiovascular stents in the research, clinical and industry stage.

5.2 Additive Manufacturing of Biomedical Implants

5.2.1 3D-Printed Skin Substitutes

Every year 11 million people worldwide suffer from skin injuries caused by burns, infections and surgery (Peck 2011). Although most patients receive medical attention, there is still a substantial number of morbidity and mortality occurrences due to the skin injuries. The 2016 global wound management market was estimated to reach $15 billion and predicted to be worth over $22 billion in 2024

(Vijayavenkataraman et al. 2016; MedMarketDiligence 2018) with a CAGR of 5.3%. The major factors driving the growth of the wound market are the usage of the products in healing and preventing infection and technological progress and new innovative products.

The skin plays several critical roles in the body and is regarded as one of the most important organs in the human body. It helps to regulate the temperature and moisture level in the body, and it protects the body from the outside environment. Having many nerve endings, skin helps to respond temperature, pressure, touch and pain. The skin has a multilayer structure consisting of the epidermal, dermal and hypodermis (also called subcutis) layers. Each of the layers contains various components and is responsible for a different function. The outermost layer (epidermis) controls water loss and consists of keratinocytes cells. The middle layer (dermis) contributes to physical strength and flexibility and supports the vasculature, lymphatic and nervous system, which is composed of extracellular matrix (ECM) and fibroblasts. The innermost layer (subcutis) helps to maintain body temperature and protect the bones and muscles from damage built from lipocytes and collagen.

5.2.1.1 Materials for Skin Substitutes

Over the last 25 years, the development of skin substitutes towards replacement of damaged skin and facilitation of wound healing has notably progressed. The first natural skin graft was introduced in 1871 by Reverdin. Since then, various techniques have been used involving autografts (grafting patient's own skin), allografts (grafting skin from another person) or xenografts (grafting skin from an animal) (Supp and Boyce 2005). However, a number of limitations associated with these types of skin substitutes, such as risk of infection, shortage of donors, slow speed of healing and risk of scar formation, ultimately impede skin regeneration (Metcalfe and Ferguson 2007; Shevchenko et al. 2010). For this reason, extensive research has taken place on finding skin substitutes for wound treatment other than natural skin grafts (Singh et al. 2016; Nathoo et al. 2014). Better understanding of a cell's behaviour and their interaction with the external environment has resulted in great achievements in the replacement of injured skin using acellular and cellular skin substitutes (Zhong et al. 2010; Catalano et al. 2013).

The acellular skin substitutes are composed of natural or synthetic polymers similar to host ECM and function as scaffolds to support the skin cells of a patient to form a new tissue (Ehrenreich and Ruszczak 2006). Compared to cellular matrices, acellular ones are easier to produce, less expensive and easier to store. Natural polymers are most often used for acellular skin substitutes that include collagen (Lee et al. 2001), elastin (Daamen et al. 2003) and chitosan (Ma et al. 2003). Synthetic polymers commonly used for the scaffolds include absorbable types, i.e. PCL, PLA and PEG (Vats et al. 2003; Almany and Seliktar 2005), and non-absorbable such as polyurethane or polytetrafluoroethylene (PTFE) (Zhong et al. 2010). There are commercially available acellular scaffolds, e.g. Alloderm®, DermACELL®, FlexHD® or Integra®, that have been proved to be effective scaffolds for wound healing, among others (Debels et al. 2015; Nyame et al. 2015). Although the advantages of synthetic polymers include relatively low price and desired mechanical properties, their compatibility still remains very low (Zhong et al. 2010; Chaudhari et al. 2016). Introducing viable cells into polymeric scaffolds (cellular scaffolds) opens the possibility to increase the compatibility of the materials. Some most common cells used with the cellular scaffolds include fibroblast, keratinocytes, melanocytes and stem cells (Shevchenko et al. 2010; Norouzi et al. 2015). Commercially available acellular scaffolds, e.g. Apligraf® and TransCyte®, have been described to work efficiently for the skin replacement therapy, among others (Debels et al. 2015).

5.2.1.2 Bioprinting of Skin Substitutes

Despite extensive research, currently no ideal skin substitute can be found available in the market. One reason is the difficulty in replicating a complex multilayer structure of natural skin. 3D printing is a promising technology to address this challenge. Bioprinting belongs to the AM technology spectrum and can precisely deposit living cells, factors and biopolymers and create 3D structures (Murphy and Atala 2014). 3D bioprinting brings advantages by offering the possibility to produce 3D structures with controlled geometry, tailored mechanical properties, controlled pore sizes and distribution which eventually can improve cell-cell interaction in a 3D environment (Gardien et al. 2014; Singh et al. 2016). The printer allows the printing of cells and creating multilayer structures with high precision. During 3D printing, the cells or aggregates are applied along with a hydrogel polymer which undergoes polymerisation under the influence of a UV light beam. This allows the placement of the cell in the desired place in the spatial structure. The biggest difficulty in bioprinting is the use of various types of cells and then the creation of a complete functional organ. Bioprinting generally requires the following steps: the first, collecting of autologous cells and culturing them and, the second, printing the cultured cells using a 3D printer. The second step can involve mixing the cultured cells with biomaterial or printing cells on the scaffold or printing cells on their own (Tarassoli et al. 2018).

Currently several research groups have investigated the application of 3D bioprinting technology for the fabrication of skin substitutes using in vivo and in vitro approaches (Vijayavenkataraman et al. 2016; Tarassoli et al. 2018).

The first researchers who deposited skin cells (keratinocytes and fibroblasts) in vitro and formed 3D structured tissue using 3D technology were Lee et al. (2009). In their work, they used a precursor of collagen to obtain hydrogel scaffolds which were printed and then fibroblast cells were subsequently printed on top, followed by a layer of keratinocytes. The printer consisted of four microvalves mounted onto three stages. The pneumatically driven valves connected to the syringes were dispensing the fluid (hydrogel precursor and cell suspension) in a precise computer-controlled position. Using this approach, the researchers were able to build a multilayer structure of skin. The results have shown that the morphology of the cells was maintained after printing. Additionally, it has been reported that the print resolution can control cell proliferation.

3D cell printing of in vivo skin has been underdevelopment as an appealing fabrication approach. Binder et al. (2010) printed a 'bioink' composed of fibroblasts, keratinocytes, fibrin and collagen directly onto a wound bed of an animal using movable inkjet printing system. The work demonstrated that two different types of cell skin could be printed in one process. The laser was used to scan the wound for depth and width before printing. After the printing process, complete epithelisation was observed after 8 weeks. Although it has been shown that the bioprinted skin formed skin tissue, it was highlighted that further investigation must be undertaken to make the process ready for using it for treatment of human wounds.

Lee et al. (2014b) designed a 3D structure for the maintenance of the morphology of keratinocytes and fibroblast cells using a solid freeform robotic system with attached dispensers for the ink. In their work, they have achieved a high density of printed keratinocytes and fibroblasts maintaining a high resolution and cell viability.

Koch et al. (2012) printed mixtures of keratinocytes and fibroblasts incorporated in collagen and successfully generated a real tissue with a layered composition (dermis and epidermis) using laser-assisted bioprinting (LaBP). Within the experimental set-up, they used two glass slides, the top one coated underneath with a 60 nm gold layer to absorb the laser and the bottom glass slide covered with a layer of the cell mixed with collagen. The hydrogel filled with cells was jetted with the help of the pressure produced by a laser-generated vapour bubble, and 3D cell pattern was generated using a layer-by-layer approach. The technique was proved to have a potential in further increase of the complexity of the printed structures and possibility to integrate vascular network in future skin models. LaBP technique has been further investigated by other researchers. Michael et al. (2013) used a mixture of fibroblasts and keratinocytes with hydrocortisone as a skin substitute source and printed onto open wound of nude mice using LaBP technique. The printed cells created a tissue comparable to natural skin, additionally containing fibroblasts produced by collagen. Their results showed that after less than 2 weeks, the printed skin was found to be attached to the nearby tissue of nude mice with a formed layer of epidermis.

A new method of extrusion for skin printing was developed by Cubo et al. (2016). They printed functional human skin which had an outer layer of the skin to protect a thicker layer that acted like a dermis and a layer made of fibroblast cells that in turn produced collagen and other proteins that gave the skin elasticity. Human cells and components were used as the ink which could generate its own human collagen. It was possible to print 100 cm of skin in less than 35 min. The product has been tested and successfully used in the treatment of burns and traumatic wounds. At present, artificial skin has passed the control phase in various European regulatory agencies.

In another work, presented by Skardal et al. (2012), a pressure-driven bioprinter has been used to print stem cells for skin formation with the fast wound healing effect. The ink was prepared by mixing fluid-derived and bone marrow-derived mesenchymal stem cells with fibrin-collagen gel and a cross-linking agent and then printed directly onto a skin wound of mice. The results indicated that the wound has been successfully closed and re-epithelialisation took place.

5.2.1.3 Electronic Skin

The idea of electronic skin takes us from the world of science fiction to medical and technological reality. Electronic skin is a thin, translucent material comparable to the human skin regarding its function and mechanical strength. Constant development in the field is fast and promising for the growth of a fully integrated and functional electronic skin very soon (Hammock et al. 2013). Today flexible electronics are used for many products, and conductive tracks can be easily printed on the flexible substrate (Fig. 5.2).

A self-healing skin has been developed by Zou et al. (2018) containing sensors for measuring pressure, temperature, humidity and airflow. Fabricated skin contained a thermosetting conductive silver nanoparticle, thanks to which it could regenerate itself even at room temperature after possible damage and it possessed sufficient mechanical strength, chemical stability and electrical conductivity. The healing of wounds formed on such skin occurred within 30 min under conditions in ambient conditions and only a few minutes at 60 °C. The new skin has been shown to imitate the functions of natural human skin. The invention has shown a promise to be used in both biomedical implants and minimally invasive surgery.

The application of 3D printing in production of electronic skin has been investigated by Zhu et al. (2018) who developed a method for printing flexible skin, equipped with a range of sensors collecting external stimuli. The multilayer structure

Fig. 5.2 Silver tracks deposited using material jetting

has been prepared using a new method that involves putting together direct ink writing, closed-loop feedback and computer vision technique. The whole process of printing wirelessly powered sensors on a human hand took only a few minutes at room temperature, maintaining a flexible structure and at the same time avoiding the risk of skin burns. The work of Zhu et al. finds real application in wearable devices, robotics and innovative implantology.

5.2.2 AM of Cardiovascular Stents

AM technology has been demonstrated to be also a promising technology for fabrication of cardiovascular stents (Fig. 5.3). The standard, efficient and safe procedure for the treatment of one of the most severe cardiovascular illnesses, that is, coronary artery disease, is stent implantation (Savage et al. 1994). Implanted stent keeps the artery open, thus enabling blood flow to the heart. However, placing the stent in the coronary artery may cause vessel injury which then triggers formation of neointimal tissue inside the stent and inflammatory response, consequently leading to restenosis (recurrent narrowing of the coronary artery). As a result, repeated procedures are required, which often are linked to reduced well-being of patient and increased healthcare costs (Kleinedler et al. 2012). To solve this problem, significant advances in stent development have taken place including drug-eluting stents, polymeric stents and stents fabrication methods (Ramadugu and Latha Alikatte 2016; Foerster et al. 2016).

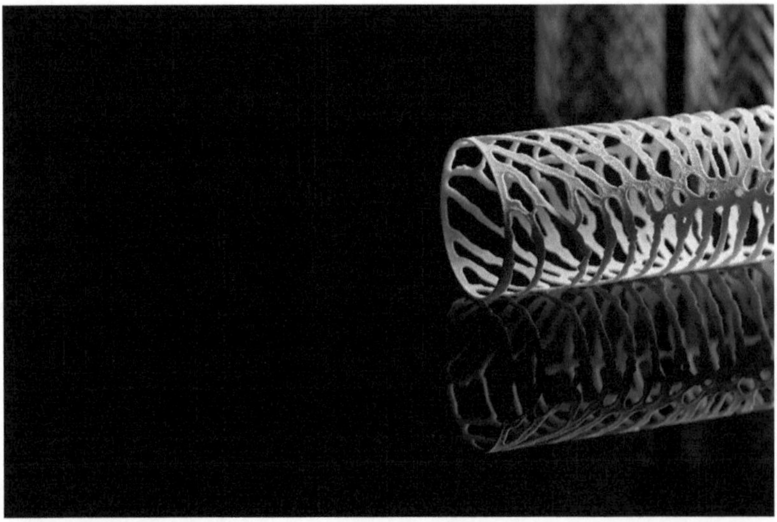

Fig. 5.3 3D-printed polymeric stent using selective laser sintering

5.2.2.1 Materials for Cardiovascular Stents

Development of metal-based drug-eluting stents reduced the risk of restenosis caused by traditional metal stents; however, the remaining struts in the vessel continue to pose a risk for inflammation (Scheinert et al. 2005). As a response to the above problem, new bioresorbable polymeric stents have been developed. These stents are made of biodegradable polymer which dissolves after 6–12 months leaving the vessel fully restored (Saito 2005). The most common material used for polymer-based stents is poly(L-lactide) (PLLA) which breaks down into absorbable lactic acid in the body over time. Other materials in development comprise polyglycolic acid, tyrosine, polycaprolactone, salicylic acid or polycarbonate (Ang et al. 2017). Currently there are few commercially available polymer-based stents including Absorb, Reva and ReZolve. The most common stent, Absorb, produced by Abbott, offers effective multicomponent design. The first layer of the Absorb stent is composed of poly(L-lactide), which serves as a base polymer, and it takes the form of a tube with zigzag hoops connected by bridges. The next layer is a mixture of drug, poly(D,L-lactide) (PDLLA), which serves as a drug-eluting layer, and then, on the top, additional features are included such as a pair of radiopaque platinum markers which aim to help in finding the stents during angiography and balloon delivery system.

5.2.2.2 3D Printing of Cardiovascular Stents

The most common methods for the fabrication of polymeric stents are laser machining and moulding processes. Laser machining techniques causing thermal effects has a risk of changing polymer properties, while the moulding process suffers from low resolution of produced stents (Martinez and Chaikof 2011). Additionally, the majority of stents are available off-the-shelf in various sizes and not always the size of the stent fit to the patient's vessel due to the geometry of the vessel. 3D printing not only addresses the mentioned problems but also opens the possibility of producing customised stents which minimises probability of post stent transplantation complications and offers possibility to control the strength by designing and printing the stent struts with different thickness. Using different design can also help to regulate degradation and drug loading of the stents.

A number of researchers have been investigating 3D printing techniques to improve stent fabrication. Flege et al. (2013) successfully fabricated a degradable coronary stent poly-L-lactic acid (PLLA) and PLLA-co-poly-e-caprolactone (PCL) using laser sintering technique (SLM) for the first time. A post-processing step of spray and dip-coating has been included in the process to smooth the stent surface and allow incorporation of drugs for the future.

van Lith et al. (2016) developed a new technique for the polymeric stent fabrication based on a bespoke micro-continuous liquid interface production system (microCLIP). By adding an oxygen-permeable membrane to the system, they advanced the traditional CLIP technique and provided high resolution and precision, which brought benefits to the fabrication of cardiovascular stents. Ink made of methacrylated polydiolcitrate and antioxidant were printed and polymerised under UV light. Printed stents have shown mechanical properties similar to those of metal stents.

Another 3D printing technique, which has been investigated for the stents fabrication, is extrusion. Park et al. (2015) using this technique successfully produced bioresorbable drug-eluting PCL stents. After printing, the stent was coated with the immunosuppressive drug followed by implantation into the artery and indicated no complication.

This research, for the first time, presented the method for fabrication of drug-eluting polymer biodegradable stent using 3D printing.

Cabrera et al. (2017) fabricated cardiovascular stent using FDM. The computer-designed stents were produced using commercially available co-polyester polymer which has been shown to biodegrade via hydrolysis. By optimising the process parameters, it was possible to adjust the stents geometry including stent diameter, width and thickness. Resulted mechanical properties were similar to that of nitinol stents.

5.2.3 3D-Printed ENT (Ear, Nose and Throat)

5.2.3.1 Ear

Otology is one of the first medical fields that explored 3D printing as a manufacturing method for their implants/devices. The hearing aid industry is a particularly interesting case as it has already transitioned its entire operations to using 3D printing (Sharma 2013). More than ten million hearing aid shells have already been manufactured using 3D printers. By 2006–2007, most of the hearing aid manufacturers had scaled up and implemented the technology across all their operations (Sandström 2015). Hearing aids can be classified as either behind-the-ear (BTE) or in-the-ear (ITE) products, the latter category being the more expensive one as it has to be customised for each patient. ITEs can in turn be classified as standard ITEs, in-the-canal (ITC) and completely-in-the-canal (CIC) products which are more or less invisible (Masters et al. 2006). Vat polymerisation is primarily used for ITE products as these need to be customised to each patient's ear.

In 1999, Siemens and Phonak started to explore different 3D printing technologies for the fabrication of hearing aids. SLA was rejected because at the time it was not able to offer a biocompatible material, while LS seemed more promising as it used nylon. In 2002 a biocompatible material developed by Dreve materials in Germany suitable for SLA became available, and the hearing aid company Widex started implementing this technology. The material was a resin composed of a monomeric dimethacrylate, a urethane methacrylate, a monomeric aliphatic dimethacrylate and the photoinitiator (Klare et al. 2004). In 2004, Phonak switched to use a vat polymerisation method utilising digital light processing, a UV curable technology by EnvisionTEC (EnvisionTEC n.d.). In the past, the process for manufacturing hearing aids consisted in more than nine steps which took more than a week of work. 3D printing has reduced the hearing aid manufacturing process to three steps: scanning, modelling and printing. This technology is able to mass-produce hundreds of thousands of custom-made products every year (Sonova n.d.).

The use of 3D printing for improving hearing is not limited to hearing aids. Surgeons are attracted to the technology and are utilising it for preclinical work and as tools to facilitate surgeries. One example is the 3D printed of ossicular prosthesis. Conductive hearing loss, which is the loss of hearing due to problems conducting sound waves, can be due to ossicular abnormalities including trauma, infection and congenital abnormalities. Surgical reconstruction to repair the ossicular chain is a well-established procedure but still with success rates of closing the air gap to less than 20 dB ranging from 55% to 75% and resulting in poor hearing (House and Teufert 2001; Mishiro et al. 2010). Long-term follow-up has shown that more than 40% of failure is attributed to prosthesis- or surgeon-related errors (Yung 2006). The University of Maryland School of Medicine investigated the use of 3D printing to manufacture the custom-made ossicular prosthesis; they evaluated how precise the 3D models were and how accurate they fitted in the space for which they were designed. The prosthesis was fabricated using SLA technology. The study showed that it is possible to manufacture custom-made middle ear prosthesis using routine CT imagining of the temporal bone, existing model software and desktop SLA printed. Additionally, the precision of the implants was demonstrated by showing that in four separate trials with different surgeons, they were all able to accurately match the correct prosthesis to its intended temporal bone during surgery (Hirsch et al. 2017).

Another application comprises the treatment of damaged tympanic membrane (TM). The TM function is to capture and transmit sound from the environment to the ossicular chain of the middle ear. It can be damaged due to chronic otitis media (COM) or traumatic perforation, resulting in hearing loss due to ineffective sound transmission (Strens et al. 2012). Researchers at the School of Harvard and Massachusetts Eye and Ear Infirmary used multi-material 3D extrusion printing to design and fabricate a biomimetic TM graft with the goal of reproducing specific structural features of the human TM (Kozin et al. 2016). The TM grafts were fabricated with polydimethylsiloxane (PDMS), flex-polylactic acid (PLA) and PCL materials followed by uniform infilling with a fibrin-collagen composite hydrogel. The study demonstrated the feasibility of creating TM grafts with acoustic properties that reflect sound-induced motion patterns of the human TM and mechanical properties that demonstrate increased resistance to deformation compared to temporalis fascia.

3D printing has also been used as an aid for microtia reconstruction. Microtia is a congenital disorder of the external ear that occurs in 1 of 5000 births. Children with microtia have malformed ears, or no outer ear at all. The standard auricular reconstruction method comprises harvesting rib cartilage and modelling it into an ear shape with the help of 2D templates (Baluch et al. 2014). Surgeons in the Seoul National University compared the use of traditional 2D dimensional templates with

3D-printed ear models during the reconstruction. The 3D models were designed based on the 3D laser scan of the patient's normal ear and fabricated using fused deposition modelling (FDM) and commercial PLA filament. Unlike using 2D templates, the 3D printing outcomes of each patient were fairly consistent and accurate, representing normal ear symmetry. The use of this technology showed that it is capable of reducing the operating time and effort (Jeon et al. 2016). A similar study was performed in the Hansjörg Wyss Department of Plastic Surgery in the New York School of Medicine where the same conclusions were withdrawn. The surgeons found that the operation was more efficient and precise when using 3D-printed ear models compared to the 2D templates (Witek et al. 2016; Flores et al. 2017).

A more advanced method for auricular reconstruction was used at the Plastic Surgery Hospital in the Chinese Academy of Science by combining 3D printing, moulding and tissue engineering. In this pilot clinical trial, CT scanning and 3D printing were employed to direct the fabrication of a biodegradable scaffold which replicated the exact auricular 3D structure symmetrical to the patient's healthy ear and possessed good mechanical properties. The ear scaffold used a 3D-printed mesh of polycaprolactone as an inner core which was wrapped with PGA unwoven fibres and coated with PLA. After the autologous chondrocytes derived from microtia cartilage were seeded onto the scaffold and in vitro cultured for 3 months, cartilage frameworks with patient-specific ear shapes were generated and then implanted to reconstruct auricles in five patients, with the longest follow-up time of 2.5 years. Satisfactory aesthetic outcome with mature cartilage formation was obtained in the first conducted case (Zhou et al. 2018).

Currently, several research groups are investigating the use of 3D bioprinting for auricular reconstruction. All of them are using the material extrusion method. One of the first examples is the work by Lee et al., where they extruded the biodegradable material polycaprolactone (PCL) and a dual cell-laden hydrogel to regenerate auricular cartilage and fat tissue using a system they developed called multi-head tissue/organ building system (MtoBS) (Shim et al. 2012). PCL was printed using a hot melt extrusion head, and polyethylene glycol was used as a sacrificial material to create pores within the structure. Afterwards, chondrocytes and adipocytes differentiated from adipose-derived stromal cells (ASCs) were encapsulated in an alginate hydrogel and printed into the framework to fabricate the cell-printed structure with an ear shape (Lee et al. 2014a). This was the first study that confirmed the feasibility of fabricating a cell-printed structure made from a framework and cell-laden hydrogels using 3D printing technology including a sacrificial layer process.

Similar work was the performed by Kang et al.; they developed a system called integrated tissue-organ printed (ITOP) to fabricate vascularised cellular constructs of clinically relevant size, shape and structural integrity. One of their tissue exemplars was auricular cartilage. They produced an ear shape implant by simultaneously printing a PCL mesh, a bioink containing the cells and Pluronic F-127 hydrogel as a sacrificial outer layer to support the 3D architecture. The bioink for delivering the cells was a composite hydrogel consisting of gelatin, fibrinogen, hyaluronic acid (HA) and glycerol mixed into culture medium. To determine whether the printed ear constructs would mature in vivo, they implanted them in the dorsal subcutaneous space of athymic mice and retrieved them 1 and 2 months after implantation. The results from the in vivo studies showed that the shape of the implants was well maintained, a substantial cartilage formation was achieved and maturation in vivo strengthened the tissue constructs (Kang et al. 2016). The 3D bioprinting technology has shown promising results, but there is need of further work in order to be able to implement it in a clinical set-up.

A different approach that has been investigated is the merge of biological and nanoelectronic functionalities via 3D printing to manufacture a bionic ear. Studies at Princeton and Johns Hopkins University used an extrusion 3D printing method to generate a bionic ear using a cell-seeded alginate hydrogel and an intertwined conducted polymer consisting of silver nanoparticles. The resulting ear exhibited enhanced auditory sensing for radio frequency reception, and complementary left and right ears can listen to stereo audio music. Additionally, after in vitro culture of the printed ear, formation of new cartilage was observed (Mannoor et al. 2013). This proof of concept approach demonstrates how the use of 3D printing can help the construction of hybrid implants that could not have been possible to manufacture with other techniques.

5.2.3.2 Nose

The use of 3D printing in rhinology has not reached the industrial manufacturing stage yet, but several cases where it has been applied in hospitals as individual case studies have been reported. One example is the 3D-printed nose prosthesis implanted in a boy that suffered extreme burn injuries (Placko 2016). A plastic surgeon at the New York Ear and Eye Infirmary of Mount Sinai in collaboration with Oxford Performance Materials designed and 3D printed a permanent nose prosthesis using SLS technology and a high-performance PEKK polymer formulation (OPM 2016). This surgery was a huge step forward in the use of 3D printing for personalised implants, but there is further research and preclinical work that has to be performed in order to comply with the medical devices regulators (FDA 2017) before it can be widely used.

3D printing has also been used to create templates for fabricating prosthesis for the closure of nasal septal perforations at Mayo Clinic (Altunay et al. 2016). The CT scans of 21 patients were used to design the 3D templates of the exact shape of the perforations. Then, the prostheses were casted using medical-grade silastic and fitted into the patients. These 3D-sized prostheses showed a 2.7-fold increased improvement in the retention compared with previous sizing methods. In order to be able to use 3D printing as a direct fabrication method for septal prosthesis, more biocompatible materials have to be developed and tested to be used in 3D printing processes. One of the materials biomaterials that has been investigated is PCL. Researchers in the ENT department at the Catholic University of Korea performed an in vivo tests of 3D-printed PCL-based implants for septoplasty and rhinoplasty applications (Park et al. 2017). The implants were fabricated using hot melt extrusion printing and surgically inserted as a septal extension graft sutured with alar cartilage for nasal reshaping. After 3 months, no definitive post-operative complications were observed, there were minimal inflammatory reactions and the implant's geometry stayed unchanged with no migration or extrusion experienced.

Fabrication of nasal stents for the treatment of cleft lip and palate is another possible application of 3D printing in rhinology. Pre-surgical orthopaedic treatments are often used as a means for guiding future surgical repair prior to the cleft lip repair (Mossey et al. 2009). Also, the early post-operative period and for months after cleft lip repair, patients wear standardised silicone stents that come in multiple sizes but require significant sculpting to fit the unique cleft deformity. Customised 3D-printed stent that matches the contours and unique features of each patient and permits modification and adjustments in size and shape has been fabricated using PLA and PCL filament with FDM (Mills et al. 2017). The feasibility study showed that the device can be designed at the first visit to create a part that could be worn intraorally to help shape the alveolus, lip and nose before surgical repair. Furthermore, the study showed that the stent could be coated with drugs to be worn post-surgically in order to aid repair.

3D bioprinting has also been investigated to be used as a facilitator of nasal reconstruction techniques. The gold standard graft used in rhinoplasty is autologous rib cartilage. Harvest of autologous rib cartilage requires a major additional surgical procedure which creates donor-site morbidity. Additionally, the procedure involves sculpting autologous cartilages to form a cartilage framework, which is complex, highly skill demanding and time-consuming (Cannady et al. 2009). The use of 3D bioprinting could tackle some of these problems, by creating a personalised implant that with time will also regenerate the cartilage tissue and remove the need for autologous grafts. Currently, the research on this area has been focussed on finding bioinks that allow the printing of encapsulated cartilage cells but also support the nose structure. A multi-material extrusion approach has been evaluated using PCL as a structural support material and a UV cross-linkable gelatin as a cell carrier (Ruiz-Cantu et al. 2016). The study showed that it was feasible to print both materials and that the encapsulated cells proliferated and secreted cartilage-relevant matrix during a 45-day in vitro culture (Fig. 5.4). Another bioink that has been studied for

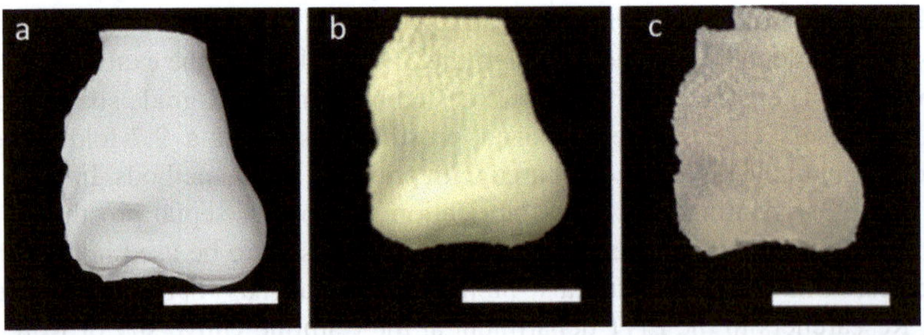

Fig. 5.4 Personalised 3D-printed biodegradable nose implant to aid nasal reconstruction after tumour removal. (**a**) CAD model, (**b**) polycaprolactone 3D-printed nose implant and (**c**) poly(lactic-co-glycolic acid) 3D-printed nose implant. Scale bar represents 3 cm

the same purpose is a composite formed by gellan gum, alginate and a clinical product called BioCartilage (Kesti et al. 2015). The final structure had good mechanical strength and the bioink supported the deposition of cartilage matrix proteins by the chondrocytes.

Given the early stage in the maturity of the 3D printing technology for implants fabrication, surgeons are also adopting it for patient education. In a clinical trial study, 50 candidate patients for endoscopic sinus surgery were given an explanation of their anatomy, disease state and treatment options using a 3D-printed education model as an aid. The models were designed using the CT scan of the patients and printed with an inkjet 3D printing technology called PolyJet. A statistically significant improvement in patients understanding of their surgical treatment was observed in the study, which, according to the authors, can lead to better patient compliance, improved medical outcomes and better patients' quality of life (Sander et al. 2017).

5.2.3.3 Throat

In the area of laryngology, tracheobronchomalacia (TBM) in newborns is the disease that 3D printing treatment has had most impact on. TBM is a disease that manifests with dynamic airway collapse and respiratory insufficiency which is difficult to treat (Carden et al. 2005). The first 3D-printed trachea splint implanted in an infant was reported by the University of Michigan (Zopf et al. 2013). The institutional review board of the University of Michigan consulted with the Food and Drug Administration and approved the use of the device under the emergency-use exemption, and written informed consent was provided by the patient's parents. The implant was fabricated using the PCL with LS technology. This material was selected because it has a long resorption time, which is very important in airway applications, because the implant needs to remain in place for at least 2 years before it is resorbed. Additionally, PCL is very ductile, so if it fails, it will not produce any particles that might puncture tissue (Jensen 2018). After transposition, sutures were placed around the malacic left bronchus and tied through interstices of the splint. After 1 month of the splint placement, CT analysis demonstrated an absolute increase in tracheal hydraulic diameter of 3.1 mm. One year after the surgery, no severe complications related to the splint were found (Morrison et al. 2017).

After the single exemption clinical case, Zopf et al. demonstrated the effectiveness of the splint in extending survival in a porcine model of severe TBM (Zopf et al. 2014). In all experimental animals after splint placement, there was resolution of symptoms and significantly longer survival than the control animals. Subsequently, they implanted the 3D-printed splint to another three children with severe TBM. Long-term data showed continued growth of primary airways and all patients had significant improvements in their respiratory symptoms (Morrison et al. 2015). Further work needs to be done to optimise safe design and manufacturing processes within regulatory mandates and clinical trials (Shieh and Jennings 2017).

Having demonstrated the advantages of the 3D-printed splints, researchers are now investigating the possibility of seeding cells in the devices to aid cartilage formation during implantation. An animal in vivo study was performed in order to attempt long segmental repair of trachea stenosis in rabbits. PCL scaffolds with a similar shape rather than the rabbit's entire trachea were 3D-printed and cultured with chondrocytes for 2 and 4 weeks before implantation to produce a tissue-engineered trachea (TET). In order to evaluate the feasibility of repairing the whole trachea, the rabbit's trachea was surgically replaced by the TET. The study showed that the TET that was cultured for 4 weeks had longer in vivo survival compared with the 2 week one, demonstrating that further in vitro maturation of the TET is needed in order to survive the in vivo environment.

5.2.4 Dental Applications

While the hearing aid market has been 3D printing custom products in large volumes for more than a decade, the dental industry seems to be on its way to go fully digital with its production (Siblani 2018). Digital dentistry and additive manufacturing are rapidly transforming the dental industry, and now 3D printing is used for a wide range of dental applications including dental and orthodontic models, direct crowns and bridges, dental aligners, night guards, surgical drill guides, flexible gingiva masks and denture bases (Dawood et al. 2015). The main companies that supply 3D printing services for these applications are 3D Systems, Stratasys Ltd., EnvisionTEC, Inc. DWS System and Invisalign (Bussiness Wire 2018). They are all developing their own printing methods and materials for the specific applications mentioned.

Using additive manufacturing methods for the fabrication of dental models is an approach that is increasingly been employed for surgical planning, simulation and orthodontics. Studies have compared the models manufactured with 3D printing techniques with those made with the traditional plaster method and found that the 3D-printed ones are clinically acceptable in terms of accuracy and reproducibility (Hazeveld et al. 2014). The advantages of using AM techniques instead of the traditional method comes from the ability to use digital models to fabricate them which increases efficiency. EnvisionTEC has developed its own materials called E-Model and E-Denstone, which are resins that can be photopolymerised and can be used in a large range of their DLP printers (EnvisionTEC 2017). Similarly, 3D Systems has developed two UV curable plastic polymers for the same application called VisiJet M3 Dentcast and Visijet M3 Pearlstone (3D Systems 2018).

The use of 3D printing for the preparation of long-term temporal crowns and bridges can potentially increase productivity in the clinic and allow for a more practical way of making provisional restorations. 3D printing of single-unit crowns may be done in as little as approximately 20 min. Therefore, it is not hard to imagine a clinical situation where the clinician could prepare a tooth, scan it, send it to a chairside 3D printer and proceed with other procedures in the same patient while the crown is being printed. The printed part would be easily separated from the supports and immediately cemented (Tahayeri et al. 2018).

Fabrication of dental aligners using 3D printing has completely transformed the orthodontics treatment. The company Invisalign® uses digital dentistry and SLA technology to manufacture around 220,000 aligners per day, almost 8 million per year (McCue 2017). Each set is customised to tackle patient-specific issues, a task that is easily overcome with 3D printing on the production line (Invisalign 2017). The material used for fabricating these devices can include a styrenic block copolymer, a silicone rubber, an elastomeric alloy, a thermoplastic elastomer (TPE), a thermoplastic vulcanizate (TPV) elastomer, a polyurethane elastomer, a block copolymer elastomer, a polyolefin blend elastomer and a thermoplastic co-polyester elastomer (Li and Chen 2012). Invisalign and traditional braces are two treatment modalities available for patients seeking orthodontic improvement. A study at Texas A&M University showed that there were no significant differences at any time point between the two methods, indicating that Invisalign® and traditional braces can have similar treatment results as well as settle similarly during 6 months of retention (Preston 2017). The main advantage of these dental aligners over conventional fixed appliances is that they are clear, and therefore more aesthetically acceptable, particularly for adults. Also, aligners are generally more comfortable to wear, and they are removable, allowing better oral hygiene (Boyd and Miller 2000).

In addition to using 3D printing for manufacturing dental devices, researchers and dental surgeons are also investigating its use for aiding periodontal repair. The first reported in human case of treatment of a large periodontal osseous defect with a 3D-printed bioresorbable patient-specific polymer scaffold and signalling growth factor was performed at the University of Michigan (Rasperini et al. 2015). In this study, a PCL scaffold containing 4% hydroxyapatite was printed using LS technology. The design consisted of perforations for fixation, an internal port for delivery of recombinant human platelet-derived growth factor BB (rhPDGF-BB) and pegs oriented perpendicularly to the root for periodontal ligament (PDL) formation. After a 12-month follow-up, the treated site remained intact, but the authors consider that optimisation of the growth factor release kinetics may further increase GF bioactivity in situ, and this could improve the outcome.

5.3 Summary

3D printing has been used to design unique items which promise to improve the way we create and build. As machine speed and product quality improve, 3D printers will continue to challenge conventional methods. Compared to conventional fabrication methods, 3D printing offers much faster and accurate process for medical device fabrication, additionally giving the possibility to personalise the produced parts for patients.

The application of 3D printing for fabrication of skin substitutes for the wound treatment has proved effective and accelerated the healing processes. The possibility of creating multiple complex geometries and using a wide range of biomaterials and cells has a great value for skin fabrication. Technological progression of 3D printing can improve wound healing further by creating larger patches of skin and cover larger areas of wounds with the hope of future use in clinical trials. Further developments will require new biomaterials, scaffold design optimisation and better understanding of cell functioning.

3D printing technology is also a developing and promising approach for polymer processing and fabrication of cardiovascular stents. Application of 3D printing helps in structure design procedure for the stents fabrication which ultimately influences the mechanical properties of the stents. There is still a need for materials which will be biocompatible with the right balance between mechanical properties and degradation rate.

The fields of ENT and dental were the first to implement 3D technology in their manufacturing methods; this technology has allowed them to manufacture hearing aids and dental aligners in customised mass production. Additional to these two medical devices already on the market, there are several 3D-printed implants such as a trachea, ear and nose replacements that have already been implanted in humans and showed promising results. Current work is undergoing on clinical trials to demonstrate the safety and efficacy of these implants in order to comply with all the regulations.

Although the application of 3D printing in biomedical area has reached the market for the applications mentioned above, the technology is still in the development process. Ongoing research work is looking promising towards the achievement of being used as a standard manufacturing method in both clinics and industry.

References

3D Systems. (2018). *Dental materials*. Rock Hill, SC: 3D Systems.

Almany, L., & Seliktar, D. (2005). Biosynthetic hydrogel scaffolds made from fibrinogen and polyethylene glycol for 3D cell cultures. *Biomaterials, 26*(15), 2467–2477. https://doi.org/10.1016/j.biomaterials.2004.06.047.

Altunay, Z. O., et al. (2016). Three-dimensional printing of large nasal septal perforations for optimal prosthetic closure. *American Journal of Rhinology and Allergy, 30*(4), 287–293. https://doi.org/10.2500/ajra.2016.30.4324.

Ang, H. Y., et al. (2017). Mechanical behavior of polymer-based vs. metallic-based bioresorbable stents. *Journal of Thoracic Disease, 9*(Suppl 9), S923–S934. https://doi.org/10.21037/jtd.2017.06.30.

Baluch, N., et al. (2014). Auricular reconstruction for microtia: A review of available methods. *Plastic Surgery, 22*(1), 39–43.

Banks, J. (2013). Adding value in additive manufacturing: Researchers in the United Kingdom and Europe look to 3D printing for customization. *IEEE Pulse, 4*(6), 22–26. https://doi.org/10.1109/MPUL.2013.2279617.

Binder, K. W., et al. (2010). In situ bioprinting of the skin for burns. *Journal of the American College of Surgeons, 211*(3), S76. https://doi.org/10.1016/j.jamcollsurg.2010.06.198.

Boyd, R. L., & Miller, R. J. (2000). The Invisalign system in adult orthodontics: Mild crowding and space closure cases. *Journal of Clinical Orthodontics, XXXIV*(4), 203–212.

Bussiness Wire. (2018). *Global dental 3D printing market 2018–2024—Drivers, opportunities, and trends*. San Francisco, CA: Bussiness Wire.

Cabrera, M. S., et al. (2017). Computationally designed 3D printed self-expandable polymer stents with biodegradation capacity for minimally invasive heart valve implantation. *3D Printing and Additive Manufacturing, 4*(1), 19–29. https://doi.org/10.1089/3dp.2016.0052.

Cannady, S. B., Cook, T. A., & Wax, M. K. (2009). The total nasal defect and reconstruction. *Facial Plastic Surgery Clinics of North America, 17*(2), 189–201. https://doi.org/10.1016/j.fsc.2009.01.002.

Carden, K. A., et al. (2005). Tracheomalacia and tracheobronchomalacia in children and adults. *Chest, 127*(3), 984–1005. https://doi.org/10.1378/chest.127.3.984.

Catalano, E., et al. (2013). Tissue-engineered skin substitutes: An overview. *Journal of Artificial Organs, 16*(4), 397–403. https://doi.org/10.1007/s10047-013-0734-0.

Chaudhari, A. A., et al. (2016). Future prospects for scaffolding methods and biomaterials in skin tissue engineering: A review. *International Journal of Molecular Sciences*. https://doi.org/10.3390/ijms17121974.

Cubo, N., et al. (2016). 3D bioprinting of functional human skin: Production and in vivo analysis. *Biofabrication, 9*(1), 015006.

Daamen, W., et al. (2003). Preparation and evaluation of molecularly-defined collagen–elastin–glycosaminoglycan scaffolds for tissue engineering. *Biomaterials, 24*(22), 4001–4009. https://doi.org/10.1016/S0142-9612(03)00273-4.

Dawood, A., et al. (2015). 3D printing in dentistry. *British Dental Journal, 219*(11), 521–529. https://doi.org/10.1038/sj.bdj.2015.914.

Debels, H., et al. (2015). Dermal matrices and bioengineered skin substitutes. *Plastic and Reconstructive Surgery Global Open, 3*(1), e284. https://doi.org/10.1097/GOX.0000000000000219.

Ehrenreich, M., & Ruszczak, Z. (2006). Update on tissue-engineered biological dressings. *Tissue Engineering, 12*(9), 2407–2424. https://doi.org/10.1089/ten.2006.12.2407.

EnvisionTEC. (2017). *Dental 3D printing materials | Clinically-approved | Industry-leading*. Dearborn, MI: EnvisionTEC.

EnvisionTEC. (n.d.). *Hearing aid | 3D printing materials*. Dearborn, MI: EnvisionTEC.

FDA. (2017). *Technical considerations for additive manufactured medical devices—Guidance for Industry and Food and Drug Administration Staff*. Silver Springer, MD: FDA.

Flege, C., et al. (2013). Development and characterization of a coronary polylactic acid stent prototype generated by selective laser melting. *Journal of Materials Science: Materials in Medicine, 24*(1), 241–255. https://doi.org/10.1007/s10856-012-4779-z.

Flores, R. L., et al. (2017). The technique for 3D printing patient-specific models for auricular reconstruction. *Journal of Cranio-Maxillofacial Surgery, 45*(6), 937–943. https://doi.org/10.1016/j.jcms.2017.03.022.

Foerster, A., et al. (2016). Stainless steel surface functionalization for immobilization of antibody fragments for cardiovascular applications. *Journal of Biomedical Materials Research Part A, 104*(4), 821–832. https://doi.org/10.1002/jbm.a.35616.

Gardien, K. L., Middelkoop, E., & Ulrich, M. M. (2014). Progress towards cell-based burn wound treatments. *Regenerative Medicine, 9*(2), 201–218. https://doi.org/10.2217/rme.13.97.

Hammock, M. L., et al. (2013). 25th anniversary article: The evolution of electronic skin (E-Skin): A brief history, design considerations, and recent progress. *Advanced Materials, 25*(42), 5997–6038. https://doi.org/10.1002/adma.201302240.

Hazeveld, A., Huddleston Slater, J. J. R., & Ren, Y. (2014). Accuracy and reproducibility of dental replica models reconstructed by different rapid prototyping techniques. *American Journal of Orthodontics and Dentofacial Orthopedics, 145*(1), 108–115. https://doi.org/10.1016/j.ajodo.2013.05.011.

Hirsch, J. D., Vincent, R. L., & Eisenman, D. J. (2017). Surgical reconstruction of the ossicular chain with custom 3D printed ossicular prosthesis. *3D Printing in Medicine, 3*(1), 7. https://doi.org/10.1186/s41205-017-0015-2.

House, J. W., & Teufert, K. B. (2001). Extrusion rates and hearing results in ossicular reconstruction. *Otolaryngology - Head and Neck Surgery, 125*(3), 135–141. https://doi.org/10.1067/mhn.2001.117163.

Invisalign. (2017). *Teeth straightening with clear, nearly invisible braces*. UK: Invisalign.

Javaid, M., & Haleem, A. (2017). Additive manufacturing applications in medical cases: A literature based review. *Alexandria Journal of Medicine*. https://doi.org/10.1016/j.ajme.2017.09.003.

Jensen, W. (2018). *Medical: University of Michigan—Customizing a biocompatible material for additively manufactured medical implants*. EOS.

Jeon, B., et al. (2016). Fabrication of three-dimensional scan-to-print ear model for microtia reconstruction. *Journal of Surgical Research, 206*(2), 490–497. https://doi.org/10.1016/j.jss.2016.08.004.

Kang, H.-W., et al. (2016). A 3D bioprinting system to produce human-scale tissue constructs with structural integrity. *Nature Biotechnology, 34*(3), 312–319. https://doi.org/10.1038/nbt.3413.

Kesti, M., et al. (2015). Bioprinting complex cartilaginous structures with clinically compliant biomaterials. *Advanced Functional Materials, 25*, 7406–7417. https://doi.org/10.1002/adfm.201503423.

Klare, M., et al. (2004). *Low-viscous, radiation curable formulation, particularly for the stereolithographical production of earpiece* (patent DE10328302A1).

Kleinedler, J. J., et al. (2012). Novel nanocomposite stent coating releasing resveratrol and quercetin reduces neointimal hyperplasia and promotes re-endothelialization. *Journal of Controlled Release, 159*(1), 27–33. https://doi.org/10.1016/j.jconrel.2012.01.008.

Koch, L., et al. (2012). Skin tissue generation by laser cell printing. *Biotechnology and Bioengineering, 109*(7), 1855–1863. https://doi.org/10.1002/bit.24455.

Kozin, E. D., et al. (2016). Design, fabrication, and in vitro testing of novel three-dimensionally printed tympanic membrane grafts. *Hearing Research, 340*, 191–203. https://doi.org/10.1016/j.heares.2016.03.005.

Lee, C. H., Singla, A., & Lee, Y. (2001). Biomedical applications of collagen. *International Journal of Pharmaceutics, 221*(1–2), 1–22. https://doi.org/10.1016/S0378-5173(01)00691-3.

Lee, W., et al. (2009). Multi-layered culture of human skin fibroblasts and keratinocytes through three-dimensional freeform fabrication. *Biomaterials, 30*(8), 1587–1595. https://doi.org/10.1016/j.biomaterials.2008.12.009.

Lee, J. S., et al. (2014a). 3D printing of composite tissue with complex shape applied to ear regeneration. *Biofabrication, 6*(2). https://doi.org/10.1088/1758-5082/6/2/024103.

Lee, V., et al. (2014b). Design and fabrication of human skin by three-dimensional bioprinting. *Tissue Engineering Part C: Methods, 20*(6), 473–484. https://doi.org/10.1089/ten.tec.2013.0335.

Li, C., & Chen, Y. (2012). *Multilayer dental appliances and related methods and systems* (patent US9655691B2).

Ma, L., et al. (2003). Collagen/chitosan porous scaffolds with improved biostability for skin tissue engineering. *Biomaterials, 24*(26), 4833–4841. https://doi.org/10.1016/S0142-9612(03)00374-0.

Mannoor, M. S., et al. (2013). 3D printed bionic ears. *Nano Letters, 13*(6), 2634–2639. https://doi.org/10.1021/nl4007744.

Martinez, A. W., & Chaikof, E. L. (2011). Microfabrication and nanotechnology in stent design. *Wiley Interdisciplinary Reviews: Nanomedicine and Nanobiotechnology, 3*(3), 256–268. https://doi.org/10.1002/wnan.123.

Masters, M., Velde, T., & McBagonluri, F. (2006). Rapid manufacturing in the hearing industry. In *Rapid manufacturing* (pp. 195–209). Chichester, UK: John Wiley & Sons, Ltd. https://doi.org/10.1002/0470033991.ch12.

McCue, T. (2017). 3D printing moves align technology toward $1.3 billion in sales. *Forbes*.

MedMarketDiligence, LLC. (2018). *Worldwide wound management, forecast to 2026: Established and emerging products, technologies and markets in the Americas, Europe, Asia/Pacific and rest of world*.

Metcalfe, A. D., & Ferguson, M. W. J. (2007). Tissue engineering of replacement skin: The crossroads of biomaterials, wound healing, embryonic development, stem cells and regeneration. *Journal of the Royal Society Interface, 4*(14), 413–437. https://doi.org/10.1098/rsif.2006.0179.
Michael, S., et al. (2013). Tissue engineered skin substitutes created by laser-assisted bioprinting form skin-like structures in the dorsal skin fold chamber in mice. *PLoS One, 8*(3), e5774. https://doi.org/10.1371/journal.pone.0057741.
Mills, D., et al. (2017). The use of 3D printing in the fabrication of nasal stents. *Inventions, 3*(1), 1. https://doi.org/10.3390/inventions3010001.
Mishiro, Y., et al. (2010). Prognostic factors of long-term outcomes after ossiculoplasty using multivariate analysis. *European Archives of Oto-Rhino-Laryngology, 267*(6), 861–865. https://doi.org/10.1007/s00405-009-1133-2.
Morrison, R. J., et al. (2015). Mitigation of tracheobronchomalacia with 3D-printed personalized medical devices in pediatric patients. *Science Translational Medicine, 7*(285), 285ra64. https://doi.org/10.1126/scitranslmed.3010825.
Morrison, R. J., et al. (2017). Treatment of severe acquired tracheomalacia with a patient-specific, 3D-printed, permanent tracheal splint. *JAMA Otolaryngology–Head & Neck Surgery, 143*(5), 523. https://doi.org/10.1001/jamaoto.2016.3932.
Mossey, P. A., et al. (2009). Cleft lip and palate. *Lancet (London, England), 374*(9703), 1773–1785. https://doi.org/10.1016/S0140-6736(09)60695-4.
Murphy, S. V., & Atala, A. (2014). 3D bioprinting of tissues and organs. *Nature Biotechnology, 32*(8), 773–785. https://doi.org/10.1038/nbt.2958.
Nathoo, R., Howe, N., & Cohen, G. (2014). Skin substitutes: An overview of the key players in wound management. *The Journal of Clinical and Aesthetic Dermatology, 7*(10), 44–48.
Norouzi, M., et al. (2015). Advances in skin regeneration: Application of electrospun scaffolds. *Advanced Healthcare Materials, 4*(8), 1114–1133. https://doi.org/10.1002/adhm.201500001.
Nyame, T. T., et al. (2015). Tissue-engineered skin substitutes. *Plastic and Reconstructive Surgery, 136*(6), 1379–1388. https://doi.org/10.1097/PRS.0000000000001748.
OPM. (2016). *OsteoFab® Technology*. South Windsor, CT: OPM.
Park, S. A., et al. (2015). In vivo evaluation and characterization of a bio-absorbable drug-coated stent fabricated using a 3D-printing system. *Materials Letters, 141*, 355–358. https://doi.org/10.1016/j.matlet.2014.11.119.
Park, S. H., et al. (2017). New application of three-dimensional printing biomaterial in nasal reconstruction. *The Laryngoscope, 127*(5), 1036–1043. https://doi.org/10.1002/lary.26400.
Peck, M. D. (2011). Epidemiology of burns throughout the world. Part I: Distribution and risk factors. *Burns, 37*(7), 1087–1100. https://doi.org/10.1016/j.burns.2011.06.005.
Placko, M. (2016). *New 3D printed nose implant is fully functional.*
Preston, K. A. (2017). *Treatment and post-treatment posterior occlusal changes in Invisalign® and traditional braces: A randomized controlled.* Retrieved from https://oaktrust.library.tamu.edu/handle/1969.1/161323.
Profshare. (2018). *3D printing medical devices market report by product type (external wearable devices, clinical study devices) by application/end user (hospital, medical center) by industry analysis, volume, share, growth, challenges, trends and forecast 2017–2025.* Retrieved from https://www.profsharemarketresearch.com/3d-printing-medical-devices-market-report/.
Ramadugu, P., & Latha Alikatte, K. (2016). A review on biodegradable and bioabsorbable stents for coronary artery disease. *Journal of Bioequivalence & Bioavailability, 08*(02), 64–67. https://doi.org/10.4172/jbb.1000269.
Rasperini, G., et al. (2015). 3D-printed bioresorbable scaffold for periodontal repair. *Journal of Dental Research, 94*(9_Suppl), 153S–157S. https://doi.org/10.1177/0022034515588303.
Ruiz-Cantu, L., et al. (2016). 3D bioprinting of mechanically strong chondrocyte-laden constructs for major nose reconstruction. In *16th Annual Meeting of the Tissue and Cell Engineering Society* (p. 39).
Saito, S. (2005). New horizon of bioabsorbable stent. *Catheterization and Cardiovascular Interventions, 66*(4), 595–596. https://doi.org/10.1002/ccd.20590.

Sander, I., et al. (2017). Patient education for endoscopic sinus surgery: Preliminary experience using 3D-printed clinical imaging data. *Journal of Functional Biomaterials, 8*(2), 13. https://doi.org/10.3390/jfb8020013.

Sandström, C. (2015). Adopting 3D printing for manufacturing—The case of the hearing aid industry. *Ratio Working Papers*, (262), 1–20.

Savage, M. P., et al. (1994). Long-term angiographic and clinical outcome after implantation of a balloon-expandable stent in the native coronary circulation. Palmaz-Schatz Stent Study Group. *Journal of the American College of Cardiology, 24*(5), 1207–1212.

Scheinert, D., et al. (2005). Prevalence and clinical impact of stent fractures after femoropopliteal stenting. *Journal of the American College of Cardiology, 45*(2), 312–315. https://doi.org/10.1016/j.jacc.2004.11.026.

Schubert, C., Van Langeveld, M. C., & Donoso, L. A. (2014). Innovations in 3D printing: A 3D overview from optics to organs. *British Journal of Ophthalmology, 98*(2), 159–161. https://doi.org/10.1136/bjophthalmol-2013-304446.

Sharma, R. (2013). The 3D printing revolution you have not heard about. *Forbes*.

Shevchenko, R. V., James, S. L., & James, S. E. (2010). A review of tissue-engineered skin bioconstructs available for skin reconstruction. *Journal of the Royal Society Interface, 7*(43), 229–258. https://doi.org/10.1098/rsif.2009.0403.

Shieh, H. F., & Jennings, R. W. (2017). Three-dimensional printing of external airway splints for tracheomalacia. *Journal of Thoracic Disease, 9*(3), 414–416. https://doi.org/10.21037/jtd.2017.02.53.

Shim, J.-H., et al. (2012). Bioprinting of a mechanically enhanced three-dimensional dual cell-laden construct for osteochondral tissue engineering using a multi-head tissue/organ building system. *Journal of Micromechanics and Microengineering, 22*(8), 085014. https://doi.org/10.1088/0960-1317/22/8/085014.

Siblani, A. (2018). The 3D printing race is on for 2018. *tctmagazine*, 4–7.

Singh, D., Singh, D., & Han, S. S. (2016). 3D printing of scaffold for cells delivery: Advances in skin tissue engineering. *Polymers, 8*(1), 1–17. https://doi.org/10.3390/polym8010019.

Skardal, A., et al. (2012). Bioprinted amniotic fluid-derived stem cells accelerate healing of large skin wounds. *Stem Cells Translational Medicine, 1*(11), 792–802. https://doi.org/10.5966/sctm.2012-0088.

Sonova. (n.d.). *3D printing technology for improved hearing*. Stäfa, Switzerland: Sonova.

Strens, D., et al. (2012). A pilot cost-of-illness study on long-term complications/sequelae of AOM. *B-ENT, 8*(3), 153–165.

Supp, D. M., & Boyce, S. T. (2005). Engineered skin substitutes: Practices and potentials. *Clinics in Dermatology, 23*(4), 403–412. https://doi.org/10.1016/j.clindermatol.2004.07.023.

Tahayeri, A., et al. (2018). 3D printed versus conventionally cured provisional crown and bridge dental materials. *Dental Materials, 34*(2), 192–200. https://doi.org/10.1016/J.DENTAL.2017.10.003.

Tarassoli, S. P., et al. (2018). Skin tissue engineering using 3D bioprinting: An evolving research field. *Journal of Plastic, Reconstructive and Aesthetic Surgery, 71*(5), 615–623. https://doi.org/10.1016/j.bjps.2017.12.006.

van Lith, R., et al. (2016). 3D-printing strong high-resolution antioxidant bioresorbable vascular stents. *Advanced Materials Technologies, 1*(9), 1–7. https://doi.org/10.1002/admt.201600138.

Vats, A., et al. (2003). Scaffolds and biomaterials for tissue engineering: A review of clinical applications. *Clinical Otolaryngology and Allied Sciences, 28*(3), 165–172.

Ventola, C. L. (2014). Medical applications for 3D printing: Current and projected uses. *P & T, 39*(10), 704–711. https://doi.org/10.1016/j.infsof.2008.09.005.

Vijayavenkataraman, S., Lu, W. F., & Fuh, J. Y. H. (2016). 3D bioprinting of skin: A state-of-the-art review on modelling, materials, and processes. *Biofabrication, 8*(3), 032001. https://doi.org/10.1088/1758-5090/8/3/032001.

Witek, L., et al. (2016). Patient-specific 3D models for autogenous ear reconstruction. *Plastic and Reconstructive Surgery - Global Open, 4*(10), e1093. https://doi.org/10.1097/GOX.0000000000001093.

Yung, M. (2006). Long-term results of ossiculoplasty: Reasons for surgical failure. *Otology & Neurotology, 27*(1), 20–26. https://doi.org/10.1097/01.mao.0000176173.94764.f5.

Zhong, S. P., Zhang, Y. Z., & Lim, C. T. (2010). Tissue scaffolds for skin wound healing and dermal reconstruction. *Wiley Interdisciplinary Reviews: Nanomedicine and Nanobiotechnology, 2*(5), 510–525. https://doi.org/10.1002/wnan.100.

Zhou, G., et al. (2018). In vitro regeneration of patient-specific ear-shaped cartilage and its first clinical application for auricular reconstruction. *EBioMedicine, 28*, 287–302. https://doi.org/10.1016/j.ebiom.2018.01.011.

Zhu, Z., et al. (2018). 3D printed functional and biological materials on moving freeform surfaces. *Advanced Materials, 30*(23), 1–8. https://doi.org/10.1002/adma.201707495.

Zopf, D. A., et al. (2013). Bioresorbable airway splint created with a three-dimensional printer. *New England Journal of Medicine, 368*(21), 2043–2045. https://doi.org/10.1056/NEJMc1301237.

Zopf, D. A., et al. (2014). Treatment of severe porcine tracheomalacia with a 3-dimensionally printed, bioresorbable, external airway splint. *JAMA Otolaryngology–Head & Neck Surgery, 140*(61), 66–71. https://doi.org/10.1001/jamaoto.2013.5644.Treatment.

Zou, Z., et al. (2018). Rehealable, fully recyclable, and malleable electronic skin enabled by dynamic covalent thermoset nanocomposite. *Science Advances, 4*(2), 1–9. https://doi.org/10.1126/sciadv.aaq0508.

Chapter 6
Orthopaedic 3D Printing in Orthopaedic Medicine

Margaret E. Brennan Fournet, Farah Alwani Azaman, Suzan Gunbay, Yuan Yuan Chen, and Declan M. Devine

Abstract Orthopaedic surgeries are commonly extremely challenging, and innovations are required to overcome a series of recognised difficulties and improve patient outcomes. Complications, in particular, the high occurrence of infections, often leads to prolonged patient pain, implant failure and loss of functions. 3D printing technology provides an ideal opportunity to integrate cutting edge technologies to address the major identified clinical problems directly. 3D-printed orthopaedic implants, designed to fit anatomical defects or malformations precisely, can resolve the current limited availability of appropriate well-fitting patient-specific implant parts. Integrated technologies can be included to target postoperative infections, which is an increasingly important concern, particularly considering population demographic and health profile changes. Improved osteointegration technologies can be incorporated to overcome inadequate tissue adherence to implants and bone on-growth. With increasing numbers of patients now outliving their implants, loosening difficulties and premature implant failure rates using conventional technologies are projected to increase dramatically. With these projections, the emergence of 3D printing is a welcome reprieve for patients and surgeons alike. 3D printing technologies hold the promise of significantly advancing current orthopaedic implant capabilities, delivering bespoke customised site-specific implants, which resolve fundamental clinical problems and achieve better results for patients.

This chapter explores the unprecedented solution offered by 3D-printed technologies for orthopaedic surgeries. The use of 3D printing for orthopaedic applications and advances, which may lead to future large-scale 3D-printed polymeric orthopaedic implant production, is discussed.

Keywords PEEK · Patient-specific implants · 3D printing · Surgical jigs

Margaret E. Brennan Fournet and Farah Alwani Azaman contributed equally to this work.

M. E. Brennan Fournet (✉) · F. A. Azaman · S. Gunbay · Y. Y. Chen · D. M. Devine
Athlone Institute of Technology, Co. Westmeath, Ireland
e-mail: mfournet@ait.ie

D. M. Devine (ed.), *Polymer-Based Additive Manufacturing*,
https://doi.org/10.1007/978-3-030-24532-0_6

6.1 Introduction

Orthopaedics is one of the first medical fields to adopt 3D printing technology. 3D-printed custom-created artificial bones are currently being utilised as replacements for missing or injured bone. Surgeons around the world are already implanting tens of thousands of 3D-printed replacement parts for hips, knees, ankles, sections of the pelvis, spine and skull. 3D-printed customised implants can go beyond the realm of where the patient does not fit the standard range of implant size. The inherent geometric freedom of 3D printing is highly advantageous for orthopaedic models and implants as it can fabricate the exact shape and size needed through the anatomical data obtained from medical scanning techniques using a range of suitable materials (Polo-Corrales et al. 2013; Thomas and Singh 2017). Considerable research is ongoing for the development of constructs that support bone formation. 3D-printed bone tissue engineering techniques are enabling precise spatial control of tissue scaffold microarchitecture, regulating porosity and pore size, shape and pore interconnectivity to promote osteointegration and bone tissue regeneration (Dehghani and Annabi 2011; Polo-Corrales et al. 2013). Several additive manufacturing (AM) techniques have been utilised to fabricate the jigs, scaffolds and implants for orthopaedic applications, and this is a promising approach to overcoming the limitations of current treatment options (Bose et al. 2016).

The use of 3D printing as patient-specific surgical guides as well as preoperative and intraoperative models from computed tomography (CT) data is already widely implemented (Honigmann et al. 2018). These 3D-printed operative models provide surgeons with an enhanced means of physical conceptualisation, which is especially beneficial in complex cases allowing surgeons to preselect and shape implants (Dehghani and Annabi 2011; Thomas and Singh 2017). Furthermore, 3D printing readily enables unprecedented precise tailoring to the subtle nuances of each patient's specific anatomy. 3D-printed designs improve the implant fit with the anatomy of the patient, ensuring good bone-to-implant contact and facilitating osteointegration and bony ingrowth (Turnbull et al. 2017). As groundbreaking 3D printing surgeries begin to make the headlines, such as the first ever sternum replacement in the United States, the outlook for 3D-printed orthopaedic technologies is buoyant.

6.2 Patient-Specific Implants

Surgeons need an accurate picture of bone defects to establish the optimal fit for grafts or implants. 3D-printed implants meet this need, as they are precisely translated from patient CT and/or magnetic resonance imaging (MRI) data. 3D-printed models of specific patient deformities facilitate surgeons in getting to know their patients' anatomy, before entering an operation theatre. The surgeon can study and plan the exact approach and even test procedures, including drilling and bone fixation plate bending, beforehand using a model, which delivers the feeling of the mechanical response of real bone. This technique increases the precision of the 3D detail of a patient's anatomy, compared with conventionally used radiographs, which often provide inadequate information on the extent of bone defects. Recently, there has been a rapid distribution of commercial 3D printers within hospital settings, and 3D-printed models are increasingly used in surgical planning for a variety of orthopaedic procedures. 3D-printed replica of various fractures and pathologic conditions including spine disorders, dysplasia of hips or bone tumours can be readily 3D printed for surgical planning, allowing the surgeon direct tactile and visual experience, which reduces surgical time and associated patients risks due to prolonged anaesthesia.

Accurate, high-resolution image acquisition is the most critical step in making a perfectly representative and well-fitting physical model. The high contrast of bone structure in comparison to the surrounding tissue lends to simple segmentation from the standardised Digital Imaging and Communication in Medicine (DICOM) images, in image post-processing. The process of converting DICOM images into a format suitable for 3D printers uses computer-aided design (CAD) software, which transforms the 3D images into a series of polygons. This data is translated into Standard Triangulate Language (STL) format, which is appropriate for 3D printers. This format defines individual sections of the 3D image as surfaces that enclose a specific region of space, enabling the 3D printers to add materials layer by layer to fabricate a precise replica model of the 3D image (Giannopoulos et al. 2015; Javaid and Haleem 2018; Partners et al. 2018). These steps can be listed as:

- A routine CT/MRI scan of the affected area, such as the pelvis, is first performed.
- The axial CT/MRI scan data is reconstructed into thin slices of the order of 0.75 mm at 0.5 mm intervals.
- The DICOM images are extracted for data reconstruction using post-processing software.
- CAD software transforms the images into a series of polygons and converts data into STL format, which is compatible with rapid prototyping.
- The STL file is then loaded on a commercial 3D printer.
- Colour may be added to the model by using the ZPR (*.zpr) format.
- 3D printers fabricate 3D physical model by adding material layer by layer. The materials used can be liquid, powder or sheet, comprising polymers, ceramics or metals.
- The bone-like models are thus printed and air-dried and fixed using different fixatives such as varnish or proprietary solutions.
- Printing time varies from 3 to 6 h depending on the vertical height of the model.

6.3 Orthopaedic Applications of 3D Printing

In the world of bone fixation, a broad range of materials have been developed to suit the numerous functions of orthopaedic implants. The material selection criteria can vary depending on both the application and the implant type. Here, advances in 3D printing of polymers for bone orthopaedic applications are discussed based on the medical indication.

6.3.1 Bone Fixation Devices

The human skeletal system is central to locomotion, respiration, protection of internal organs and supporting body weight. Long bones have additional essential load-bearing functions. There is a variation in the healing time of fractures according to different skeletal sites from 4–6 weeks for the upper limbs to 8–12 weeks for lower limbs, with almost all fractures taking about 3–4 months to heal completely. Many of the load-bearing implants utilised to treat fractures of the long bones of the arms and legs are fabricated from metallic materials, due to their high mechanical strength. The annual incidence rate of fractures of long bones is estimated to be 11.5 per 100,000 persons with 40% occurring in the lower limb. The incidence is highest in young men (higher energy fractures) and the elderly (osteoporotic fractures) (Francis and Sheski 2018). Open fractures of the tibial diaphysis are most common, with open femoral diaphyseal, distal femoral and proximal tibial fractures tend to occur in the most severely injured patients. Lower limb open fractures are more severe than open fractures of the upper limbs (Howe 2018). 3D-printed bone scaffolds hold strong potential for improved non-union outcomes in long bones. 3D-printed bone fixation plates which are patient-specific, precontoured or pre-bent can contribute to shortened surgery times, reduced patient pain and improved outcomes (Dobbe et al. 2013).

PEEK is an aromatic semicrystalline thermoplastic polymer with favourable mechanical properties and provides an alternative to conventional metallic or ceramic orthopaedic implants. PEEK has a modulus of elasticity of 3–4 GPa, melting temperature of 334 °C and glass transition temperature of 145 °C. While ketone and ether functional groups within the structure of PEEK make it very chemically and thermally stable, PEEK, like many other thermoplastic polymers, is hydrophobic which can lead to the formation of fibrous tissue capsules around the implant. This property is advantageous in areas where tissue adhesion is undesirable such as the dura in the skull, muscles around the orbits, etc. However, in cases where osteointegration is required to prevent implant loosening and dead spots where bacteria can create biofilms, it may be necessary to enhance the bioactivity of the polymer. Many strategies have been tested with particular reference to PEEK as it is becoming increasingly popular as a load-bearing implantable material for orthopaedics (Panayotov et al. 2016a).

The biomechanical properties and bioactivity of PEEK can be improved through surface modification and composite formation (Ma and Tang 2014). Surface modification approaches include plasma treatment and surface coating as listed in Table 6.1. Plasma ionised gases can modify PEEK surface chemical and physical properties to deliver enhanced osteointegrative features. PEEK has been treated with oxygen plasma (O_2) (Inagaki et al. 1997), argon plasma (Ar) (Zhang et al. 2011), methane and oxygen (CH_4/O_2) (Awaja et al. 2010), ammonia plasma (NH_4) (Comyn et al. 1996), oxygen and ammonia (O_2/NH_4) (Waser-Althaus et al. 2014), nitrogen and oxygen plasma (N_2/O_2) (Occhiello et al. 1992), hydrogen (H_2) (Welle et al. 2014) and accelerated neutral atom beam (ANAB) (Khoury et al. 2013). Various bioactive materials have been successfully coated onto the surface of PEEK to improve bioactivity, including hydroxyapatite (HA) (Wu et al. 2009), silica (Stawarczyk et al. 2013), titanium (Ti) (Han et al. 2010), gold (Yao et al. 2007), titanium dioxide (TiO_2) (Kizuki and Matsushita 2015), diamond-like carbon (DLC) (Wang et al. 2010) and calcium phosphate (Ha et al. 1997).

Carbon fibre-reinforced polyether ether ketone (CFR-PEEK) has been proposed as an alternative to metallic implants. These polymeric implants have the added advantage of enabling bone healing to be monitored in vivo using conventional X-ray without obstruction (Niemeyer et al. 2010). CFR-PEEK have an elastic modulus close to that of cortical bone (18 GPa) and are potential alternatives to titanium, titanium alloys, Cr-Co-Mo alloys and biological ceramics in orthopaedic surgery (Panayotov et al. 2016b). IcoTec Medical AG has launched a range of products

Table 6.1 Approaches to adding osteointegrative properties to PEEK

PEEK surface modification		PEEK composites
Plasma treatment	Oxygen plasma (O_2)	PEEK/carbon fibre
	Argon plasma (Ar)	PEEK/glass fibre
	Methane and oxygen (CH_4/O_2)	PEEK/hydroxyapatite (HA)
	Ammonia plasma (NH_4)	PEEK/β-tricalcium-phosphate (β-TCP)
	Oxygen and ammonia (O_2/NH_4)	PEEK/calcium silicate (CS)
	Nitrogen and oxygen plasma (N_2/O_2)	PEEK/magnesium (Mg)
	Hydrogen (H_2)	PEEK/nano-silicon dioxide (nano-SiO_2)
	Accelerated neutral atom beam (ANAB)	PEEK/nano-HA
Coating	Hydroxyapatite (HA)	PEEK/nano-titanium oxide (TiO_2)
	Silica	PEEK/carbon nanofibre
	Titanium (Ti)	PEEK/nano-fluorohydroxyapatite (nano-FHA). Carbon fibre-reinforced PEEK
	Gold	
	Titanium dioxide (TiO_2)	
	Diamond-like carbon (DLC)	
	Calcium phosphate	

under the registered trademark Black Amour®. These implants are based on carbon fibre-reinforced, and there are several reports of these successful use in orthopaedic applications. PEEK Invibio™ Biomaterial Solutions is the leading manufacturer of medical-grade PEEK with a product selection including PEEK (PEEK-OPTIMA Natural), PEEK/HA composite (PEEK-OPTIMA HA Enhanced material), PEEK/short carbon fibre (PEEK-OPTIMA Reinforced) and PEEK/continuous carbon fibre (PEEK-OPTIMA Ultra-Reinforced). Other materials and fillers, which can be incorporated into PEEK composites, include β-tricalcium-phosphate (β-TCP) (Petrovic et al. 2006), calcium silicate (CS) (Kim et al. 2009), magnesium (Mg) (Jung et al. 2014), nano-silicon dioxide (nano-SiO_2) (Tourani et al. 2013), nano-titanium oxide (TiO_2) (Wu et al. 2012), carbon nanofibre (Chan et al. 2016) and nano-fluorohydroxyapatite (nano-FHA) (Wang et al. 2014). While many of these implants are manufactured by conventional means, modern 3D printing machines, which are capable of handling continuous fibres and polymer filaments such as the MarkForged®, are capable of 3D printing personalised implants.

While less popular, glass fibre-reinforced PEEK (GFR-PEEK) has also been reported with similar elastic modulus to that of cortical bone (Lin et al. 1997; Wang et al. 2010).

Importantly, 3D printing technologies are making progress in the fabrication of PEEK polymer-based implants. 3D printing of PEEK, using fused filament fabrication (FFF), can overcome the high melting point challenges and allow construction of almost any complex design geometry. The resulting high-precision implants, which cannot be manufactured using other technologies, can offer better fatigue strength, good fracture toughness and radiolucency, overcoming the imaging constraints of metallic implants.

Postoperative infections are an increasing concern in orthopaedics with infection rates reported between 1% and 5% for patients receiving devices (Rochford et al. 2016). Additionally, up to 28% of revision surgeries report patient infections. Bacteria readily attach to implant surfaces, proliferating and forming biofilms, inducing inflammation and further medical complications (Veerachamy et al. 2014; Blanchette and Wenke 2018; Khatoon et al. 2018). Research on the impact of infection on artificial hips and knees, related to 3D-printed PEEK implants, has been conducted on developing surface-modified 3D-printed PEEK with anti-infection properties. Anti-infection 3D-printed PEEK orthopaedic implants have the potential to rapidly advance orthopaedic treatments, reduce healthcare costs and improve patient recovery and outcomes (Wang et al. 2017).

Traditional methods of mitigating microbial activity on the surface of orthopaedic implants involved incorporation of antibiotics into the bulk implant material as well as fabrication of surface coatings composed of materials with antimicrobial properties. However, considering the prevalence of antibiotic resistance the search for non-coating and non-pharmaceutical means of reducing implant-associated infection is of great importance. One current alternative to traditional methods of making implant surface antimicrobial has been to reinforce the surface of orthopaedic materials such as PEEK with nanosized physical features. Generally, this can be achieved utilising simple template-moulding methods, in which a material with a unique structure is used as a template to imprint its structure onto another article (Hong et al. 2009). A study by Kelleher et al. (2016) used a nano-topographical pattern closely replicating the naturally hydrophobic and antimicrobial structure of the cicada insect wing which is composed of randomly interspaced nanopillars. Subsequently, the research group compared levels of bacteria culture proliferation on commercial orthopaedic PEEK and the nanosurface-modified PEEK. The results showed a 40% and 50% reduction in levels of *Staphylococcus epidermidis* and *Pseudomonas aeruginosa*, respectively, after 5 days on the nanosurface-modified PEEK compared to that of the commercial PEEK (Wang et al. 2017). This is primarily due to the physically destructive effect of the nanopillars on the membrane of bacterium which is generally not seen with mammalian cells due to their larger and more robust nature (Bagherifard et al. 2015). 3D printing can potentially be used to produce nano-topography on the surface of medical devices directly. This approach could be advantageous as it would simplify the regularity approval pathway as no active pharmaceuticals are used.

It is also reported that surface wettability and associated surface energy of the implant material can ultimately lead to promising antibacterial properties and desirable cellular functions. Liu (2017) reported a direct linear correlation between surface energy and surface roughness and that surface energy plays a significant role in determining cell and bacteria functionality. Moreover, the incorporation of nano-textured silicon nitride onto the surface of orthopaedic implants is being investigated and has been shown to decrease bacterial growth and encourage the formation of bone simultaneously. It is suggested that silicon nitride is textured in a way that attracts osteoblasts and repels microbes (Ishikawa et al. 2017). Recently, implants comprised entirely of silicon nitride have been fabricated using a new emerging AM technique Robocasting (Koslow 2016).

Another area of interest in the field of bone fixation is that of fully biodegradable bone fixation devices which appear to be great alternatives for patients. These implants have several advantages over traditional materials due to their biodegradation once their function is fulfilled. These advantages include no need for implant removal, minimal risk of implant-related complications and early functional rehabilitation. Fully biodegradable bone fixation devices are often made of biodegradable polymers such as polylactic acid (PLA), polyglycolic acid (PGA), poly(lactic-co-glycolic acid) (PLGA), polydioxanone (PDO), polysulphate and polycarbonate. PLA is a relatively strong biodegradable polymer and has been used intensively in medical applications. With L- and D-enantiomeric forms of lactic acid, PLA appears as PLLA with only the L-isomer, poly(D-lactic acid) (PDLA) with just the D-isomer, poly(D,L-lactic acid) (PDLLA) with half L-isomer and half D-isomer and PLA with a majority of the L-isomer and a small amount of the D-isomer. These polymers are frequently melt processed by extrusion which is fundamentally the technology used in FFF. Thus, customisation of implants from these polymers is relatively straight forward.

Biodegradable bone fixation devices which biodegrade in an osteotransductive manner have many advantages over non-biodegradable counterparts. Loosening of permanent implants is a severe complication requiring subsequent surgical removal of these implants. Removal of permanent implants in the spine due to instability or dislocation may leave a deficit leading to instability of the spine. Therefore, biodegradable bone fixation devices offer an ideal option for spine fixation. Indeed, biodegradable bone fixation devices are superior to traditional metallic devices in terms of biocompatibility and functional recovery. Research has been conducted to compare tissue reactions to titanium and biodegradable screws upon postoperative irradiation in rats. In this study, biodegradable screws induced a lower inflammatory response compared to titanium screws which led the researchers to hypothesise that biodegradable screws may be safer reconstructive devices for patients undergoing radiotherapy. Several studies have shown the rate of infection and plate removal of biodegradable plates to be significantly lower when biodegradable plates are utilised instead of titanium plates.

The benefits of 3D-printed biodegradable bone fixation devices outweigh those of traditional bone fixation devices. Polymer implants do not interfere with imaging and allow a clear view of fracture sites on radiographs. 3D printing of patient-specific implants eliminates the need to shape the implant during surgery, and bending instruments, such as pliers, are not needed to place biodegradable plates against bone surfaces (Ellis and Walker 1996). There is no need for a procedure to remove them, which reduces costs, and there is evidence that biodegradable bone fixation devices promote better bone healing than traditional plates and screws. However, 3D-printed biodegradable implants also present drawbacks which include the need to pre-drill and tap holes for biodegradable screws. This can lead to increased operative time, as well as more difficult implant placement due to concerns with heat generation during drilling and polymer debris, which has been reported to cause inflammatory foreign-body responses.

6.3.2 Craniomaxillofacial

Craniomaxillofacial (CMF) plastic and reconstructive surgeons specialise in the soft tissue and skeletal structure of the skull and face. The face defines and expresses our identity and individuality and is enabled by the incredibly complex structure of the facial skeleton. Reconstruction of craniomaxillofacial defects and deformities, which arise as a result of congenital defects, cancers and injuries, requires consideration of both functional and aesthetic objectives. Reconstruction aims to restore volume, shape, bone continuity and symmetry of the bone skeleton. Oral maxillofacial soft and hard tissues enable functions such as mimics, mastication, swallowing and articulation. Restoration of multifaceted maxillofacial geometry such as orbital wall contours involves mastery of complicated and specific 3D anatomy and can also present significant clinical risks. In the case of orbital reconstruction, the reproduction of the contralateral unaffected orbit may result in clinical failures such as diplopia, enophthalmos and blindness, for even the most experienced surgeons (Chul et al. 2017). Implant fit in this region is critical as ill-fitting implants can be visible under the skin or can cause discomfort or irritation to overlying tissues.

Hence, 3D printing of biodegradable implants is highly advantageous in this region. Biodegradable polymers have already been successfully utilised in this field, and it has been reported that maxillofacial fractures which were treated with biodegradable Inion CPS® reported favourable healing and no local tissue immune reaction. Biodegradable bone fixation devices have adequate shear resistance but less load resistance and stiffness compared with titanium, which helps to avoid stress shielding and promotes healing. Additionally, since the bones of the CMF region are not significantly weight bearing studies have shown that fractures of the mandibular body stabilised using Inion CPS® show sufficient mechanical strength and fixation stability. The use of biodegradable bone fixation devices is a promising technology, and a reasonable degree of healing can be achieved in patients of all ages, as well as mandible fractures in early childhood (Gaba et al. 2014; Ali et al. 2016).

Significant advances in craniofacial reconstruction have resulted from the development of rigid biomaterials and medical modelling based on 3D printing over the last 15 years. Accurate but nonsterile 3D-printed models of a patient's maxillofacial skeleton can readily be produced using CT data. Widely used 3D-printed prototype models provide aids for the production of surgical implants (Mohamed et al. 2014), surgical planning improvements (Wang and Yeung 2017), service as orientation aids during surgery (Thomas and Singh 2017), enhanced diagnostic quality (Polo-Corrales et al. 2013), assisted preoperative simulation (Dehghani and Annabi 2011; achieving a patient's consent prior to surgery (Bose et al. 2016) and resection templates for surgeons (Turnbull et al. 2017), as well as providing an educational tool for medical students (Choi and Kim 2015).

A facial prosthetic fitting is much quicker than full reconstructive surgery, offering patients a low-risk and high-impact treatment option to address underlying facial defects. Conventionally, facial prosthetics are handmade and require several labour-intensive manufacturing phases and numerous patient consultations. Production is, therefore, a time-consuming, costly and highly subjective process, being highly dependent on the skills of the anaplastologist (Peart et al. 2017). High-resolution 3D printing has been demonstrated to deliver superior auricular prostheses, which provide patients with a realistic look while streamlining the current production methodology. Accurate topology of the patient's uncompromised anatomy from CT scan data is used to print a 3D-mirrored template to use in the casting process for prosthesis production or for the direct 3D printing of a prosthetic such as an eye or ear for patients who have lost or were born without them. While, currently, 3D-printed prosthetics do not restore function, these prosthetics can be seamlessly incorporated into the patient's face providing a highly realistic outcome (Mohammed et al. 2018).

A recent study developed a computer algorithm to create a 3D nose model based on 2D photographs, which is similar in appearance to the patient's actual anatomy (Isaiah et al. 2018). 3D-printed high-impact polystyrene nasal prosthetics were created for five volunteers. Surveyed clinical evaluators indicated that they would consider using this technology to create a temporary prosthesis instead of referring to an anaplastologist. 3D-printed patient-specific implants provide a personalised approach to reconstructive and aesthetic maxillofacial surgery. Maxillofacial patient-specific implants can now be designed using preoperative imaging data for input into computer-aided design (CAD)/computer-aided manufacturing for fabrication using 3D printing. This approach increases precision and decreases or eliminates the need for intraoperative modification of implants (Owusu and Boahene 2015). 3D-printed patient-specific implants have been successfully used in mandible reconstruction (Rana et al. 2017), cranioplasty (Zegers et al. 2017) and orbital reconstruction (Stoor et al. 2014). Furthermore, customised implants have reduced implant-related complications, improved surgical accuracy, increased surgical speed, improved quality of life, decreased postoperative pain and provide aesthetically good results.

In a study evaluating 104 patients with immediate orbital defect reconstructions and 23 post-traumatic orbital deformity reconstructions using mirror-imaged rapid prototype skull models and a pre-moulded synthetic scaffold, all reconstructions were successful without immediate postoperative complications (Park et al. 2015). 3D simulation was performed using Mimics 3D software (Materialise NV, Inc., Leuven, Belgium). The unaffected orbit was reflected onto the contralateral side by mirroring techniques, and then rapid prototype skull models (AMK, Inc., Goyang-si, Gyeonggi-do, Korea) representing the individual model of the uninjured state were generated using a 3D printer (ProJet 660 Pro; 3D Systems, Inc., Rock Hill, SC). Synthetic scaffolds were anatomically pre-moulded using the skull model as a guide and inserted into the orbital defect. 3D-printed CMF devices and patient-specific implants are being found to outperform their conventional comparators consistently. The availability of inexpensive compact desktop 3D printers, high-performance biocompatible materials and excellent patient outcomes is now paving the way for surgeons to access on-site production of medically certified 3D-printed patient-specific implants in hospital settings.

6.3.3 3D Printing for the Repair of Pelvis Fractures

Ageing populations are now requiring more orthopaedic surgeries, with a doubling in annual joint replacement operations reported in the United States between 2000 and 2010. 3D orthopaedic printing technologies are also advancing at a considerable rate which is benefiting complicated sections of the skeleton such as the pelvis. In 2011, the first 3D-printed pelvic implant was carried out on a patient suffering from chondrosarcoma, who required half of his pelvis replaced due to the bone cancer. 3D-printed-mirrored pelvic implant surgery is a further approach, which has been made available by 3D printing. In this technique, surgery teams use a 3D model of the healthy hemipelvis to precontour the plates using the assumption that the healthy hemipelvis was anatomically symmetric to the fractured half (Upex et al. 2017). Using DICOM files, OsiriX™ software was used to generate a 3D volumetric reconstruction and surface rendering of the fractured and healthy hemipelvis ready for export as STL files. The healthy hemipelvis was mirrored (reversed horizontally) and 3D printed along with the fractured hemipelvis using a 3D printer

(Ultimaker™, Geldermalsen, Netherlands) using polylactic acid (PLA) (0.15 mm layer thickness, 100 mm/s extrusion speed, 20% filling).

The feasibility of 3D printing technology in pelvic endoprostheses and artificial hip joints for patients suffering from pelvic tumours has been demonstrated. Between September 2013 and December 2015, 25 out of 35 patients that underwent complete pelvic reconstruction surgery at the hospital survived without significant complications. The research team is supporting 3D printing as a viable and practical route for the precise matching of implants and patient's host bone (Liang et al. 2017). Further research also supports the use of 3D-printed pelvic prostheses for reconstruction of the bony defect after pelvic tumour resection, stating its safety, the lack of additional complications and the good short-term functional results experienced by patients.

The application of 3D printing has been reported for improved surgical outcomes for often devastating both-column acetabular fractures, which are breaks in the socket portion of the ball and hip joint where the fracture plane travels vertically through the innominate bone, dividing it anteriorly and posteriorly into at least three fragments. 3D patient-specific printed models of the pelvis facilitate preoperative planning, simulation of the fracture reduction procedure and precontouring of the fixation plates, contributing to good medium-term functional outcomes.

6.3.4 3D Printing in Bone Tissue Engineering

Conventionally, non-union fractures are treated using autografts, which meets the required properties in terms of osteoconduction, osteoinduction and osteogenesis, as it contains both osteogenic cells and osteoconductive mineralised extracellular matrix material that promotes cell growth and proliferation (Polo-Corrales et al. 2013). Nevertheless, autografts are restricted by the size of the graft that can be harvested and carry a risk of donor morbidity including infection and ongoing pain following surgery (Turnbull et al. 2017). The alternative, known as allografts where bone tissue is harvested from cadaveric and living sources, are also widely adopted as this approach can eliminate the donor-site morbidity. However, allograft drawbacks include the potential risk of disease transmission and immune rejection (Polo-Corrales et al. 2013), in addition to also lacking a cellular component to aid tissue regeneration (Turnbull et al. 2017). Orthopaedic fixations using plates and screws as well as bone cement and substitutes are a mainstay approach to aiding bone regeneration. These approaches have considerable limitations including bone thinning due to stress shielding, loosening and failure over time, resulting in the need for revision surgeries. Non-invasive therapies such as ultrasound treatment have shown promise for aiding bone healing by delivering ultrasound waves with low-intensity pulses to accelerate the healing of non-union fractures of long bones in suitable adult patient (Higgins et al. 2014).

Porous 3D bone tissue engineering scaffolds are currently the subject of intense research due to their recognised potential for osteoconduction, osteoinduction and osteogenesis. 3D printing scaffolds with spatially controlled scaffold microarchitecture can be fabricated using stereolithography, fused deposition remodelling and selective laser sintering. New emergent technologies have been used to 3D bioprint scaffolds. These technologies include inkjet bioprinting, laser-assisted bioprinting, microvalve bioprinting and extrusion bioprinting (Turnbull et al. 2017). Examples of the use of polymer-based 3D printing in these applications include incorporating 3D printing, electrospinning and a physical punching process to create a composite polycaprolactone/alginate construct with nanofibrous content and improved mechanical strength (Kim et al. 2009). This method showed a significant enhancement of cell viability within 7 days, with evidence of a high increase of osteoblast formation at 14 days and considerably increased water absorption due to the improved hydrophilicity contributed by the scaffold alginate content. Porosity can be added to the composite polymer-ceramic scaffolds using techniques including 3D printing and electrospinning which are capable of depositing material within gaps between fibres, thereby facilitating interconnectivity. The scaffold micro- and macroporosity can also be modulated by using a combination of porogens with 3D printing techniques during the fabrication process. Cheng et al. (2013) describe combining polydopamine (PDA)-coated/HA precipitate in developing polycaprolactone scaffolds using stereolithography technology, fabricating 3D structures through spatially controlling the solidification of liquid photo-polymerisable resin. Injectable hydrogels have also attracted significant attention within bone and cartilage tissue engineering applications due to its potential advantages in the ability to perform minimally invasive injections and the capacity to mould hydrogels in situ to match irregular patient defects.

6.3.5 3D-Printed Fixtures and Jigs for Surgical Applications

The primary function of customised surgical guides/instruments (CSGs/CSI) is to guide, direct and fix locations for the use of surgical tools including drilling, cutting and sawing tools (Fig. 6.1). Hananouchi and Saito (2010) described CSGs as being made up of two working components: a guide (Mohamed et al. 2014), i.e. cylinders, holes and slots for tool guidance, and a base part (Wang and Yeung 2017) for fixation of the guide directly onto the patient's anatomy. CSGs are a by-product of virtual surgery planning, a preoperative technique that emerged out of advancements in medical imaging technologies and image processing software. Virtual surgery planning utilises personalised, highly accurate CAD models of patient anatomy to improve procedure and outcomes of complex surgeries (Oliveira et al. 2008; Azari and Nikzad 2009; Logan et al. 2013; Shen et al. 2017). CSGs have also been referred to in the literature as patient-specific instruments, surgical jigs, fixtures, blocks, templates and stents (Oliveira et al. 2008; Dahake et al. 2016), wherein all of the above refer to tools or aids that are specifically designed to improve pre-surgical planning and/or improve operative efficiency.

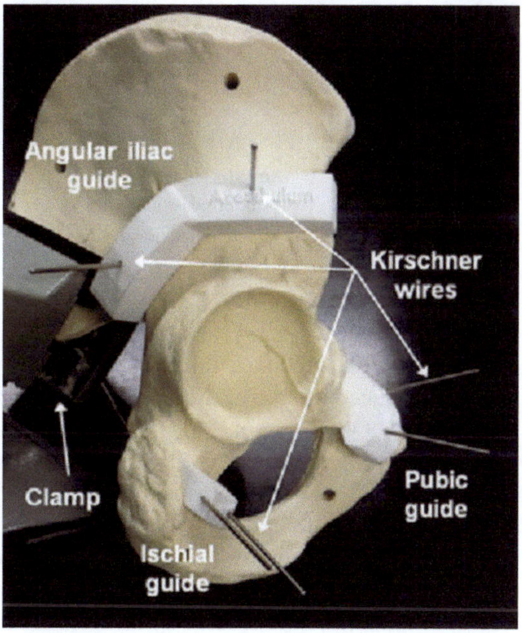

Fig. 6.1 Clamping of the pelvic bone model during the cuts and pinning of the patient-specific instruments with Kirschner wires

Selection of a rapid prototyping fabrication technique for printing surgical jigs and fixtures is dependent on the availability of the technology as well as cost and time. The most common types of medical rapid prototyping technologies reported on are stereolithography, FFF, extrusion-based bioprinting, selective laser sintering as well as material jetting techniques (Tappa and Jammalamadaka 2018). Post-processing of 3D-printed CSGs prior to use in surgery involves removal of printing support structures which, depending on application, can affect the matching of the base with the patient's anatomical surface with the part as well as a suitable method of sterilisation which depends on the material used to produce the part (Winder and Bibb 2005). While the use of biocompatible metallic materials fulfils many of these aspects, there are some medical grade polymers currently available for use in medical rapid prototyping technology applications where quality, strength and durability are demonstrated (Wang et al. 2014; Chan et al. 2016). Research into improving medical-grade polymers for 3D printing applications remains ongoing. Current commercially available medical-grade polymers for 3D printing applications are summarised in Table 6.2 (Dahake et al. 2016; Ligon et al. 2017). The actual fitting of 3D-printed surgical jigs and fixtures during surgery is generally carried out using traditional tools such as monocortical and titanium screws, Kirschner wires and Steinman pins (Fu et al. 2013; Hanasono and Skoracki 2013). 3D-printed surgical jigs and fixtures are unique in that they allow for the production of complex and intricate geometries as well as compact and accurate fit on the surface of the anatomy. In some cases, 3D-printed CSGs are designed to fit snugly onto the patient's anatomy without the need for screws and fixtures. Incorporation of CSGs into

Table 6.2 Commercially available 3D printing techniques and their associated materials used in surgical guide applications

Commercial technique	Process	Material
SLA (1987)	Photo-curing	Acrylate
SLS (1991)	Sintering of polymer powders	Polyamide
FDM (1991)	Melt extrusion	ABS, PEEK, polyurethane foams
MJ (1998)	Inkjet/poly-jet printing	ABS, PMMA, acryl, polyamide
Robocasting (2016)	Paste extrusion AM	Silicon nitride

Source: Negi et al. (2014), Dahake et al. (2016), Koslow (2016)

modern surgical practices through the medium of virtual surgery planning has reported reduced operative timeline as well as a higher level of surgical precision and accuracy (Van Brussel et al. 1997; Iii et al. 2012; Fantini et al. 2013; Polley and Figueroa 2013; Cartiaux et al. 2014; Berrone and Tos 2015; Khan et al. 2013).

There is a growing consensus that the use of custom jigs improves orthopaedic reconstruction, particularly in orthopaedic oncology. There is evidence that techniques utilising custom jigs result in more accurate fitting between a custom implant and host bone compared to the traditional freehand method which requires mitigation of alignment errors created by reliance on anatomic landmarks for the positioning of the guide or fixture. There are a wide variety of areas for which custom surgical jigs and fixtures have been applied, including dentistry (Ganz 2005; Ito 2006; Kim et al. 2007; Sohmura et al. 2010; Kashani et al. 2012; Di Giacomo et al. 2014; Paek et al. 2014), craniomaxillofacial surgery (Poukens et al. 2003; Bibb et al. 2009; Schepers et al. 2012), total hip and knee arthroplasty (Ng et al. 2012; Kerens et al. 2013), corrective osteotomy (Dobbe et al. 2013, 2014; Holzapfel et al. 2014), bone tumour resection and allograft reconstruction (Bellanova et al. 2013). Other applications have been reported including orbital implant placement (Hirao et al. 2014) and even neurosurgery as guides for performing brain biopsies, brain tumour resections, custom implantation systems for deep brain stimulation electrodes (Ding et al. 2013) as well as placement of ventricular shunts (Konrad et al. 2011; Huang and Zhao 2016).

Osteosarcoma resection surgery is a procedure that involves initial removal of cancerous tissue within the confines of an oncologically safe margin followed by defect reconstruction utilising a customised prosthesis. The overall objective is to treat the disease while maintaining a high level of functionality for the patient. The procedure is complicated and utilises large computer navigation systems that require additional operational personnel in the theatre (Wong et al. 2012; Zhang et al. 2015). A 2012 study by Wong et al. (2012) attempted to simplify the procedure for femur osteosarcoma resection surgery and reconstruction via a custom prosthesis. A 3D CAD preoperative model of the affected femur enabled the generation of a custom surgical cutting jig by FDM 3D printing. This jig could be placed in a desired orientation that matched that of the custom prosthesis precisely. Ceramic femoral bone models were also 3D printed to test initial positioning of surgical jigs. The method was first validated on cadaveric femurs and took the operating surgeon 1 min in total to set the location of the jig and a further 3 min to perform the bone resections via cutting slits incorporated into the jig compared to a reported 24–28 min for the traditional navigation-guided approach. The method was then tested on a 46-year-old man diagnosed with low-grade osteosarcoma of the left distal femoral metaphysis. Reported dimensional differences between the achieved and planned bone resection were less than 1 mm and were comparable to that of an experienced surgeon with a navigation system. Overall, this study showed that intraoperative incorporation of virtual surgery planning and custom surgical jigs may facilitate and expedite bone resection along a desired orientation, resulting in an accurate prosthesis reconstruction with enormous reductions in procedure time.

Cartiaux et al. (2014) reported on the surgical accuracy of 24 surgeons in performing a pelvic tumour resection on a synthetic 3D pelvic bone model compared to the traditional freehand technique. Generally, in real-time surgery, pelvic bone tumour resections are difficult due to the complex geometry of the pelvis, limited visibility and restricted working space in the pelvic region. The 3D model was imaged from a patient using a CT scanner. The set of images was reconstructed in 3D and resection of a simulated periacetabular tumour was defined with four target planes (ischium, pubis, anterior ilium and posterior ilium) with a 10-mm desired safe margin. Patient-specific instruments for bone-cutting were also designed based on the fitting of the morphology of the model, and the resultant STL file was sent away externally to be rapidly prototyped from polyamide material using FDM 3D printing technology with a dimensional tolerance of 0.2 mm. The instruments were equipped with 2-mm-diameter holes to be pinned on the pelvic bone using Kirschner wires. It was noted that more accurate mapping of bone beforehand is necessary to ensure better contact points. Overall, the planned surgical margin for the exercise was given as 1 cm. While the achieved, surgical margins were subject to statistical variation, the standard deviation of the 1 cm surgical margins was decreased by use of the synthetic bone model, compared traditional variation technology methods.

Khan et al. (2013) showed that CAD/CAM-produced surgical jigs improve the accuracy of primary bone tumour resection, enabling a surgeon to reproduce a given preoperative plan reliably and consistently. Jigs were fabricated using FDM 3D printing and ABS-based resin. Surgery was performed on six matched pairs of cadaveric femurs with a joint-sparing, hemimetaphyseal wide resection precisely outlined on each femur. The resection was performed on each pair using the standard manual technique on one and the custom jig-assisted technique on the other. Superimposition of preoperative and post-resection images enabled quantitative analysis of resection accuracy. The mean maximum deviation from the preoperative plan was 9.0 mm for the manual group and 2.0 mm for the custom-jig group, demonstrating the increased accuracy achievable for 3D-printed custom jigs. Building on this, using a 3D model of patient's bone constructed from CT/MRI scans, an algorithm was developed, to determine a minimal surface area that would provide for stable contact between a customised jig and a bone (Zhang et al. 2015). Once the accurate point of contact was established, the jig was fabricated by means of FDM. The increased accuracy achievable with the custom jigs will continue to allow for new surgical advances in reconstructive orthopaedic surgery. Validation and broader application through further studies will facilitate the acceptance of 3D-printed custom jigs as surgical tools.

6.4 Summary

The geometric freedom of 3D printing and its rapid individual production capabilities at relatively low cost provide significant benefits to the medical community. Amongst the different faculties of medicine, orthopaedics has been the first discipline to embrace the technology fully. 3D printing has demonstrated its potential to revolutionise a range of orthopaedic medicine aspects, including (1) eliminating the traditional time-consuming and labour-intensive method of prosthesis manufacture; (2) enabling precise prosthetic implants which are individually tailored to the patients anatomy; (3) removing the need for inter-operative implant manipulation, currently widely required to achieve suitable implant fitting; and (4) facilitating preoperative planning and preparation for surgical procedures. Furthermore, 3D printing is highly amenable to the integration of additional technologies which address significant concerns including prevention of postoperative infections and enhancement of tissue implant adherence. Rising geriatric populations throughout the world are requiring increasing numbers of orthopaedic surgeries and more complex orthopaedic interventions. 3D printing supports the advancement of orthopaedic technologies, enabling the rapid development and refinement of prosthetic manufacturing, the generation of innovative, intricate skeletal replacement parts and the progression of surgical techniques.

A broad range of polymers have been developed for 3D medical printing of polymers and composite materials demonstrated as suitable for orthopaedic implants. 3D-printed microarchitecture using these materials has been shown to exhibit physical properties which are similar or approach those of natural bone, making 3D printing a viable solution for low cost, accurate and rapid production of bespoke implants. The use of osteoconductive and anti-infective materials has been demonstrated to improve patient outcomes. Continued 3D orthopaedic implant printing research collaboration in conjunction with 3D-printed tissue engineering and regenerative medicine promises to facilitate the development of further healthcare advancements. Given the rapid uptake of 3D printing within orthopaedics, and the continuing strong research advancements in materials, science and related medical disciplines, a positive future is assured for 3D printing technologies within the broader medical field. The many benefits compared to traditional orthopaedic methods underline the pivotal role 3D printing is set to play in the future of orthopaedic medicine.

Acknowledgements This work was supported in part by financial support from Athlone Institute of Technology under the Presidents Seed Fund and Enterprise Ireland funding under the Technology Gateway Programme, grant number TG-2017-0114, and the Commercialisation Fund Programme, grant number CF-2016-0600-P, co-funded by the European Regional Development Fund. Dr. Conor Coffey is thanked for advice and discussion.

References

Ali, A. A. A., et al. (2016). Use of biodegradable plates and screws in the treatment of pediatric facial bone fractures. *Egyptian Journal of Oral and Maxillofacial Surgery, 7*(3), 86–93. https://doi.org/10.1097/01.omx.0000496444.08336.3d.

Awaja, F., et al. (2010). Enhanced autohesive bonding of polyetheretherketone (PEEK) for biomedical applications using a methane/oxygen plasma treatment. *Plasma Processes and Polymers, 7*, 1010–1021. https://doi.org/10.1002/ppap.201000072.

Azari, A., & Nikzad, S. (2009). The evolution of rapid prototyping in dentistry: A review. *Rapid Prototyping Journal, 15*(3), 216–225. https://doi.org/10.1108/13552540910961946.

Bagherifard, S., et al. (2015). The influence of nanostructured features on bacterial adhesion and bone cell functions on severely shot peened 316L stainless steel. *Biomaterials, 73*, 185–197. https://doi.org/10.1016/J.BIOMATERIALS.2015.09.019.

Bellanova, L., Paul, L., & Docquier, P. (2013). Surgical guides (patient-specific instruments) for pediatric tibial bone sarcoma resection and allograft reconstruction. *Sarcoma, 2013*(4), 787653. https://doi.org/10.1155/2013/787653.

Berrone, G. S. M., & Tos, B. B. P. (2015). Step-by-step surgical technique for mandibular reconstruction with fibular free flap: Application of digital technology in virtual surgical planning. *European Archives of Oto-Rhino-Laryngology, 272*(6), 1491–1501. https://doi.org/10.1007/s00405-014-3078-3.

Bibb, R., et al. (2009). Rapid manufacture of custom-fitting surgical guides. *Rapid Prototyping Journal, 15*, 346–354. https://doi.org/10.1108/13552540910993879.

Blanchette, K. A., & Wenke, J. C. (2018). Current therapies in treatment and prevention of fracture wound biofilms: Why a multifaceted approach is essential for resolving persistent infections. *Journal of Bone and Joint Infection, 3*(2), 50–67. https://doi.org/10.7150/jbji.23423.

Bose, S., Tarafder, S., & Bandyopadhay, A. (2016). Effect of chemistry on osteogenesis and angiogenesis towards bone tissue engineering using 3D printed scaffolds. *Biomedical Engineering Society, 45*, 261. https://doi.org/10.1007/s10439-016-1646-y.

Cartiaux, O. L. C., et al. (2014). Improved accuracy with 3D planning and patient-specific instruments during simulated pelvic bone tumor surgery. *Annals of Biomedical Engineering, 42*(1), 205–213. https://doi.org/10.1007/s10439-013-0890-7.

Chan, K. W., et al. (2016). Preparation of polyetheretherketone composites with nanohydroxyapatite rods and carbon nanofibers having high strength, good biocompatibility and excellent thermal stability. *RSC Advances, 6*, 19417–19429. https://doi.org/10.1039/C5RA22134J.

Cheng, Y. L., Chen, Y. W., Wang, K., & Shie, M. Y. (2013). Enhanced adhesion and differentiation of human mesenchymal stem cell inside apatite poly(dopamine)-coated poly(ε-caprolactone) scaffolds by stereolithography. *Journals of Materials Chemistry B, 38*, 1–3. https://doi.org/10.1039/C6TB01377E.

Choi, J. W., & Kim, N. (2015). Clinical application of three-dimensional printing technology in craniofacial plastic surgery. *Archives of Plastic Surgery, 42*, 267–277. https://doi.org/10.5999/aps.2015.42.3.267.

Chul, Y., et al. (2017). The accuracy of patient specific implant prebented with 3D-printed rapid prototype model for orbital wall reconstruction. *Journal of Cranio-Maxillofacial Surgery, 45*(6), 928–936. https://doi.org/10.1016/j.jcms.2017.03.010.

Comyn, J., Mascia, L., Xiao, G., & Parker, B. M. (1996). Plasma-treatment of polyetheretherketone (PEEK) for adhesive bonding. *International Journal of Adhesion and Adhesives, 16*(2), 97–104.

Dahake, S. W., et al. (2016). Applications of medical rapid prototyping assisted customized surgical guides in complex surgeries. *Rapid Prototyping Journal, 22*(6), 934–946. https://doi.org/10.1108/RPJ-02-2015-0021.

Dehghani, F., & Annabi, N. (2011). Engineering porous scaffolds using gas-based techniques. *Current Opinion in Biotechnology, 22*(5), 661–666. https://doi.org/10.1016/j.copbio.2011.04.005.

Di Giacomo, G., et al. (2014). Computer-designed selective laser sintering surgical guide and immediate loading dental implants with definitive prosthesis in edentulous patient: A preliminary method. *European Journal of Dentistry, 8*(1), 100–106. https://doi.org/10.4103/1305-7456.126257.

Ding, H., et al. (2013). Computer-aided resection and endoprosthesis design for the management of malignant bone tumors around the knee: Outcomes of 12 cases. *BMC Musculoskeletal Disorders, 14*(331), 1–9. https://doi.org/10.1186/1471-2474/14/331.

Dobbe, J. G. G., et al. (2013). Patient-tailored plate for bone fixation and accurate 3D positioning in corrective osteotomy. *Medical and Biological Engineering, 51*, 19–27. https://doi.org/10.1007/s11517-012-0959-8.

Dobbe, J. G. G., et al. (2014). Patient-specific distal radius locking plate for fixation and accurate 3D positioning in corrective osteotomy. *Strategies in Trauma and Limb Reconstruction, 9*(3), 179–183. https://doi.org/10.1007/s11751-014-0203-1.

Ellis, E., & Walker, L. R. (1996). Treatment of mandibular angle fractures using one noncompression miniplate. *Journal of Oral and Maxillofacial Surgery, 54*(7), 864–871.

Fantini, M., De Crescenzio, F., & Ciocca, L. (2013). Design and rapid manufacturing of anatomical prosthesis for facial rehabilitation. *International Journal on Interactive Design and Manufacturing, 7*(1), 51–62. https://doi.org/10.1007/s12008-012-0159-7.

Francis, D., & Sheski, M. (2018). *Medical thoracoscopy: Equipment, procedure, and complications* (pp. 17–19). Official reprint from UpToDate® www.uptodate.com ©2018 UpToDate, Inc. and/or its affiliates. All Rights Reserved. https://doi.org/10.1007/s11920-014-0463-y.

Fu, M., et al. (2013). Construction and accuracy assessment of patient-specific biocompatible drill template for cervical anterior transpedicular screw (ATPS) insertion: An in vitro study. *PLoS One, 8*(1), e53580. https://doi.org/10.1371/journal.pone.0053580.

Gaba, S., et al. (2014). To evaluate the efficacy of biodegradable plating system for fixation of maxillofacial fractures: A prospective study. *National Journal of Maxillofacial Surgery, 4*(2), 167. https://doi.org/10.4103/0975-5950.127645.

Ganz, S. D. (2005). Presurgical planning with CT-derived fabrication of surgical guides. *Journal of Oral and Maxillofacial Surgeons, 63*, 59–71. https://doi.org/10.1016/j.joms.2005.05.156.

Giannopoulos, A. A., et al. (2015). Medical 3D printing for the radiologist. *Radiographics, 35*, 1965–1988. https://doi.org/10.1148/rg.2015140320.

Ha, S. W., et al. (1997). Surface activation of polyetheretherketone (PEEK) and formation of calcium phosphate coatings by precipitation. *Journal of Materials Science: Materials in Medicine, 8*(11), 683–690. https://doi.org/10.1023/A:1018535923173.

Han, C., et al. (2010). The electron beam deposition of titanium on polyetheretherketone (PEEK) and the resulting enhanced biological properties. *Biomaterials, 31*(13), 3465–3470. https://doi.org/10.1016/j.biomaterials.2009.12.030.

Hananouchi, T., & Saito, M. (2010). Tailor-made surgical guide reduces incidence of outliers of cup placement. *Clinical Orthopaedics and Related Research, 468*, 1088–1095. https://doi.org/10.1007/s11999-009-0994-4.

Hanasono, M. M., & Skoracki, R. J. (2013). Computer-assisted design and rapid prototype modeling in microvascular mandible reconstruction. *Laryngoscope, 123*(3), 597–604. https://doi.org/10.1002/lary.23717.

Higgins, A., Glover, M., & Yang, Y. (2014). EXOGEN ultrasound bone healing system for long bone fractures with non-union or delayed healing: A NICE medical technology guidance. *Applied Health Economics and Health Policy, 12*, 477–484. https://doi.org/10.1007/s40258-014-0117-6.

Hirao, M., et al. (2014). Computer assisted planning and custom-made surgical guide for malunited pronation deformity after first metatarsophalangeal joint arthrodesis in rheumatoid arthritis: A case report. *Computer Aided Surgery, 19*(1–3), 13. https://doi.org/10.3109/10929088.2014.885992.

Holzapfel, B. M., et al. (2014). Customised osteotomy guides and endoprosthetic reconstruction for periacetabular tumours. *International Orthopaedics*. https://doi.org/10.1007/s00264-014-2314-1.

Hong, S.-H., Hwang, J., & Lee, H. (2009). Replication of cicada wing's nano-patterns by hot embossing and UV nanoimprinting. *Nanotechnology, 20*(38), 385303. https://doi.org/10.1088/0957-4484/20/38/385303.

Honigmann, P., et al. (2018). Patient-specific surgical implants made of 3D printed PEEK: Material, technology, and scope of surgical application. *BioMed Research International.* https://doi.org/10.1155/2018/4520636.

Howe, A. S. (2018). *General principles of fracture management: Early and late complications* UpToDate. Retrieved from https://www.uptodate.com/contents/general-principles-of-fracture-management-early-and-late-complications.

Huang, Y., & Zhao, L. (2016). Virtual surgical planning and 3D printing in prosthetic orbital reconstruction with percutaneous implants: A technical case report. *International Medical Case Reports Journal, 9*, 341–345. https://doi.org/10.2147/IMCRJ.S118139.

Iii, P. D., Rannar, L. E., & Hirsch, J. M. (2012). Imaging, virtual planning, design, and production of patient-specific implants and clinical validation in craniomaxillofacial surgery. *Craniomaxillofacial Trauma and Reconstruction, 5*(3), 137–143. https://doi.org/10.1055/s-0032-1313357.

Inagaki, N., et al. (1997). Surface modification of poly(aryl ether ether ketone) film. *Journal of Applied Polymer Science, 68*(6), 3–5. https://doi.org/10.1002/(SICI)1097-4628(19980411)68:2<271::AID-APP9>3.0.CO;2-N.

Isaiah, A., Greywoode, J. D., & Vakharia, K. T. (2018). Assessment of a novel computer algorithm for printing a 3-dimensional nasal prosthetic. *JAMA Otolaryngology - Head and Neck Surgery, 144*, 557–563. https://doi.org/10.1001/jamaoto.2018.0360.

Ishikawa, M., et al. (2017). Surface topography of silicon nitride affects antimicrobial and osseointegrative properties of tibial implants in a murine model. *Journal of Biomedical Materials Research Part A, 105*(12), 3413–3421. https://doi.org/10.1002/jbm.a.36189.

Ito, T. (2006). A virtual environment approach for planning implant placement. *International Journal of Advanced Manufacturing Technology, 30*, 974–978. https://doi.org/10.1007/s00170-005-0071-4.

Javaid, M., & Haleem, A. (2018). Additive manufacturing applications in orthopaedics: A review. *Journal of Clinical Orthopaedics and Trauma, 9*, 6–10. https://doi.org/10.1016/j.jcot.2018.04.008.

Jung, H., et al. (2014). Polyetheretherketone/magnesium composite selectively coated with hydroxyapatite for enhanced in vitro bio-corrosion resistance and biocompatibility. *Materials Letters, 116*, 20–22. https://doi.org/10.1016/j.matlet.2013.10.062.

Kashani, J., Kadir, M. R. A.. and Arabshahi, Z. (2012). *A surgical drill guide for pedicle screw placement*, IEEE International Conference on Biomedical Engineering. https://doi.org/10.1109/ICoBE.2012.6179026.

Kelleher, S. M., et al. (2016). Cicada wing surface topography: An investigation into the bactericidal properties of nanostructural features. *ACS Applied Materials and Interfaces, 8*(24), 14966–14974. https://doi.org/10.1021/acsami.5b08309.

Kerens, B., et al. (2013). Patient-specific guide for revision of medial unicondylar knee arthroplasty to total knee arthroplasty: Beneficial first results of a new operating technique performed on 10 patients. *Acta Orthopaedica, 84*(2), 165–169. https://doi.org/10.3109/17453674.2013.785908.

Khan, F. A., et al. (2013). Computer-generated custom jigs improve accuracy of wide resection of bone tumors. *Clinical Orthopaedics and Related Research, 471*, 2007–2016. https://doi.org/10.1007/s11999-012-2769-6.

Khatoon, Z., et al. (2018). Bacterial biofilm formation on implantable devices and approaches to its treatment and prevention. *Heliyon, 4*(12), e01067. https://doi.org/10.1016/j.heliyon.2018.e01067.

Khoury, J., et al. (2013). Neutral atom beam technique enhances bioactivity of PEEK. *Nuclear Instruments and Methods in Physics Research B, 307*, 630–634. https://doi.org/10.1016/j.nimb.2012.11.087.

Kim, S., et al. (2007). Surgical positioning of orthodontic mini-implants with guides fabricated on models replicated with cone-beam computed tomography. *American Journal of Orthodontics and Dentofacial Orthopedics, 131*(4), 82–89. https://doi.org/10.1016/j.ajodo.2006.01.027.

Kim, I. Y., et al. (2009). Bioactive composites consisting of PEEK and calcium silicate powders. *Journal of Biomaterials Applications, 24*(2), 105–118. https://doi.org/10.1177/0885328208094557.

Kizuki, T., & Matsushita, T. (2015). Apatite-forming PEEK with TiO2 surface layer coating. *Journal of Materials Science: Materials in Medicine, 26*(41), 1–9. https://doi.org/10.1007/s10856-014-5359-1.

Konrad, P. E., et al. (2011). Customized, miniature rapid-prototype stereotactic frames for use in deep brain stimulator surgery: Initial clinical methodology and experience from 263 patients from 2002 to 2008. *Stereotactic and Functional Neurosurgery, 89*(1), 34–41. https://doi.org/10.1159/000322276.

Koslow, T. (2016). *Silicon nitride biomaterial gets 3D print treatment for the first time ever*. 3D printing Industry.com.

Liang, H., Ji, T., Zhang, Y., Wang, Y., & Guo, W. (2017). Reconstruction with 3D printed pelvic endoprostheses after resection of a pelvic tumour. *Bone and Joint Journal, 9*(22), 5–6.

Ligon, S. C., et al. (2017). Polymers for 3D printing and customized additive manufacturing. *Chemical Reviews, 117*(15), 10212–10290. https://doi.org/10.1021/acs.chemrev.7b00074.

Lin, T. W., et al. (1997). Glass peek composite promotes proliferation and osteocalcin production of human osteoblastic cells. *Journal of Biomedical Materials Research, 36*(2), 137–144.

Liu, L. (2017). *Fabrication of nanostructures on implantable biomaterials for biocompatibility enhancement and infection resistance*. Dissertation Presented at the Department of Chemical Engineering in partial fulfillment of the requirements for the degree.

Logan, H., et al. (2013). Exploratory benchtop study evaluating the use of surgical design and simulation in fibula free flap mandibular reconstruction. *Journal of Otolaryngology - Head and Neck Surgery, 42*(42), 1–9.

Ma, R., & Tang, T. (2014). Current strategies to improve the bioactivity of PEEK. *International Journal of Molecular Sciences, 15*, 5426–5445. https://doi.org/10.3390/ijms15045426.

Mohamed, K. R., Beherei, H. H., & El-Rashidy, Z. M. (2014). In vitro study of nano-hydroxyapatite/chitosan-gelatin composites for bio-applications. *Journal of Advanced Research, 5*(2), 201–208. https://doi.org/10.1016/j.jare.2013.02.004.

Mohammed, M. I., et al. (2018). Augmented patient-specific facial prosthesis production using medical imaging modelling and 3D printing technologies for improved patient outcomes. *Virtual and Physical Prototyping, 13*(3), 164–176. https://doi.org/10.1080/17452759.2018.1446122.

Negi, S., Dhiman, S., & Sharma, R. K. (2014). Basics and applications of rapid prototyping medical models. *Rapid Prototyping Journal, 20*(3), 256–267. https://doi.org/10.1108/RPJ-07-2012-0065.

Ng, V. Y., et al. (2012). Improved accuracy of alignment with patient-specific positioning guides compared with manual instrumentation in TKA. *Clinical Orthopaedics and Related Research, 470*, 99–107. https://doi.org/10.1007/s11999-011-1996-6.

Niemeyer, P., et al. (2010). Xenogenic transplantation of human mesenchymal stem cells in a critical size defect of the sheep tibia for bone regeneration. *Tissue Engineering Part A, 16*(1), 33–43. https://doi.org/10.1089/ten.tea.2009.0190.

Occhiello, E., Guerrini, G. L., & Garbassi, F. (1992). Adhesion properties of plasma-treated carbon/PEEK composites. *Composites, 23*(3), 193–200. https://doi.org/10.1016/0010-4361(92)90440-6.

Oliveira, M., et al. (2008). 3-D biomodelling technology for maxillofacial reconstruction. *Materials Science and Engineering C, 28*(8), 1347–1351. https://doi.org/10.1016/j.msec.2008.02.007.

Owusu, J. A., & Boahene, K. (2015). Update of patient-specific maxillofacial implant. *Current Opinion in Otolaryngology and Head and Neck Surgery, 23*(4), 261–264. https://doi.org/10.1097/MOO.0000000000000175.

Paek, J., et al. (2014). Virtually fabricated guide for placement of the C-tube miniplate. *American Journal of Orthodontics and Dentofacial Orthopedics, 145*(5), 694–702. https://doi.org/10.1016/j.ajodo.2013.02.036.

Panayotov, I. V., Orti, V., Cuisiner, F., & Yachouh, J. (2016a). Polyetheretherketone (PEEK) for medical applications. *Clinical Applications of Biomaterials, 27*(118), 1–11. https://doi.org/10.1007/s10856-016-5731-4.

Panayotov, I. V., et al. (2016b). Polyetheretherketone (PEEK) for medical applications. *Journal of Materials Science: Materials in Medicine, 27*(7). https://doi.org/10.1007/s10856-016-5731-4.

Park, S. W., Choi, J. W., & Koh, K. S. (2015). Mirror-imaged rapid prototype skull model and pre-molded synthetic scaffold to achieve optimal orbital cavity reconstruction. *Journal of Oral Maxillofacial Surgery, 73*, 1540–1553. https://doi.org/10.1016/j.joms.2015.03.025.

Partners, O. S., et al. (2018). American surgical team successfully completes first 3D printed titanium and polymer sternum implant procedure in the United States. *Anatomics*, 1–8.

Peart, G., et al. (2017). Advanced auricular prosthesis development by 3D modelling and multimaterial printing. *KnE Engineering, 2*(2), 37. https://doi.org/10.18502/keg.v2i2.593.

Petrovic, L., et al. (2006). Effect of b TCP filled polyetheretherketone on osteoblast cell proliferation in vitro. *Journal of Biomedical Science, 13*, 41–46. https://doi.org/10.1007/s11373-005-9032-z.

Polley, J. W., & Figueroa, A. A. (2013). Orthognathic positioning system: Intraoperative system to transfer virtual surgical plan to operating field during orthognathic surgery. *Journal of Oral Maxillofacial Surgery, 71*, 1–10. https://doi.org/10.1016/j.joms.2012.11.004.

Polo-Corrales, L., Latorre-Esteves, M., & Ramirez-Vick, J. (2013). Scaffold design for bone regeneration. *Journal of Nanoscience and Nanotechnology, 31*(9), 1713–1723. https://doi.org/10.1109/TMI.2012.2196707.Separate.

Poukens, J., Haex, J., & Riediger, D. (2003). The use of rapid prototyping in the preoperative planning of distraction osteogenesis of the cranio-maxillofacial skeleton. *Computer Aided Surgery, 8*(3), 145–154. https://doi.org/10.3109/10929080309146049.

Rana, M., et al. (2017). Increasing the accuracy of mandibular reconstruction with free fibula flaps using functionalized selective laser-melted patient-specific implants. *Journal of Cranio-Maxillo-Facial Surgery, 45*, 1212. https://doi.org/10.1016/j.jcms.2017.04.003.

Rochford, E. T. J., et al. (2016). Monitoring immune responses in a mouse model of fracture fixation with and without Staphylococcus aureus osteomyelitis. *Bone, 83*, 82. https://doi.org/10.1016/j.bone.2015.10.014.

Schepers, R. H., et al. (2012). Full 3-D digital planning of implant-supported bridges in secondary mandibular reconstruction with prefabricated fibula free flaps: Case study. *Head and Neck Oncology, 4*(2), 2–7.

Shen, S., et al. (2017). Freeform fabrication of tissue-simulating phantom for potential use of surgical planning in conjoined twins separation surgery. *Scientific Reports, 7*, 11048. https://doi.org/10.1038/s41598-017-08579-6.

Sohmura, T., Kumazawa, Y., & Cam, C. A. D. (2010). Original computer aided support system for safe and accurate implant placement. *Japanese Dental Science Review, 46*(2), 150–158. https://doi.org/10.1016/j.jdsr.2010.01.002.

Stawarczyk, B., et al. (2013). Polyetheretherketone—A suitable material for fixed dental prostheses? *Journal of Biomedical Materials Research Part B, 101*, 1209–1216. https://doi.org/10.1002/jbm.b.32932.

Stoor, P., et al. (2014). Rapid prototyped patient specific guiding implants for orbital wall defects. *Journal of Cranio-Maxillo-Facial Surgery, 42*, 1644. https://doi.org/10.1016/j.jcms.2014.05.006.

Tappa, K., & Jammalamadaka, U. (2018). Novel biomaterials used in medical 3D printing techniques. *Journal of Functional Biomaterials*. https://doi.org/10.3390/jfb9010017.

Thomas, D., & Singh, D. (2017). 3D printing in surgery—The evolving paradigm-shift in surgical implants on demand. *International Journal of Surgery, 42*, 58–59. https://doi.org/10.1016/j.ijsu.2017.04.027.

Tourani, H., et al. (2013). Effects of fibers and nanoparticles reinforcements on the mechanical and biological properties of hybrid composite polyetheretherketone/short carbon fiber/nano-SiO2. *Polymer Composites, 34*, 1961–1969. https://doi.org/10.1002/pc.22603.

Turnbull, G., et al. (2017). 3D bioactive composite scaffolds for bone tissue engineering. *Bioactive Materials, 3*, 278. https://doi.org/10.1016/j.bioactmat.2017.10.001.

Upex, P., Jouffroy, P., & Riouallon, G. (2017). Application of 3D printing for treating fractures of both columns of the acetabulum: Benefit of pre-contouring plates on the mirrored healthy pelvis. *Orthopaedics and Traumatology: Surgery and Research, 103*(3), 331–334. https://doi.org/10.1016/j.otsr.2016.11.021.

Van Brussel, K., et al. (1997). Medical image-based design of an individualized surgical guide for pedicle screw insertion. In *Engineering in Medicine and Biology Society, 1996. Bridging disciplines for biomedicine. Proceedings of the 18th Annual International Conference of the IEEE* (Vol. 1, pp. 225–226). https://doi.org/10.1109/IEMBS.1996.656927.

Veerachamy, S., et al. (2014). Bacterial adherence and biofilm formation on medical implants: A review. *Proceedings of the Institution of Mechanical Engineers, Part H: Journal of Engineering in Medicine, 228*(10), 1083–1099. https://doi.org/10.1177/0954411914556137.

Wang, W., & Yeung, K. W. K. (2017). Bioactive materials bone grafts and biomaterials substitutes for bone defect repair: A review. *Bioactive Materials, 2*(4), 224–247. https://doi.org/10.1016/j.bioactmat.2017.05.007.

Wang, H., et al. (2010). Mechanical and biological characteristics of diamond-like carbon coated poly aryl-ether-ether-ketone. *Biomaterials, 31*(32), 8181–8187. https://doi.org/10.1016/j.biomaterials.2010.07.054.

Wang, L., et al. (2014). Polyetheretherketone/nano-fluorohydroxyapatite composite with antimicrobial activity and osseointegration properties. *Biomaterials, 35*, 6758–6775. https://doi.org/10.1016/j.biomaterials.2014.04.085.

Wang, M., Bhardwaj, G., & Webster, T. J. (2017). Antibacterial properties of PEKK for orthopedic applications. *International Journal of Nanomedicine, 12*, 6471–6476. https://doi.org/10.2147/IJN.S134983.

Waser-Althaus, J., et al. (2014). Differentiation of human mesenchymal stem cells on plasma-treated polyetheretherketone. *Journal of Materials Science: Materials in Medicine, 25*(2), 515–525. https://doi.org/10.1007/s10856-013-5072-5.

Welle, A., Weigel, S., & Bulut, Ö. D. (2014). Patterning of polymeric cell culture substrates. *Methods in Cell Biology, 119*, 35–53. https://doi.org/10.1016/B978-0-12-416742-1.00003-2.

Winder, J., & Bibb, R. (2005). Medical rapid prototyping technologies: State of the art and current limitations for application in oral and maxillofacial surgery. *Journal of Oral and Maxillofacial Surgery, 63*(7), 1006–1015. https://doi.org/10.1016/j.joms.2005.03.016.

Wong, K. C., et al. (2012). Use of a patient-specific CAD/CAM surgical jig in extremity bone tumor resection and custom prosthetic reconstruction. *Computer Aided Surgery, 17*(6), 284–293. https://doi.org/10.3109/10929088.2012.725771.

Wu, G. M., Hsiao, W. D., & Kung, S. F. (2009). Investigation of hydroxyapatite coated polyether ether ketone composites by gas plasma sprays. *Surface and Coatings Technology, 203*(17–18), 2755–2758. https://doi.org/10.1016/j.surfcoat.2009.02.115.

Wu, X., et al. (2012). Nano-TiO2/PEEK bioactive composite as a bone substitute material: In vitro and in vivo studies. *International Journal of Nanomedicine, 7*, 1215–1225. https://doi.org/10.2147/IJN.S28101.

Yao, C., Storey, D., & Webster, T. J. (2007). Nanostructured metal coatings on polymers increase osteoblast attachment. *International Journal of Nanomedicine, 2*(3), 487–492.

Zegers, T., et al. (2017). The therapeutic effect of patient-specific implants in cranioplasty. *Journal of Cranio-Maxillofacial Surgery, 45*(1), 82–86. https://doi.org/10.1016/j.jcms.2016.10.016.

Zhang, S., et al. (2011). Autohesion of plasma treated semi-crystalline PEEK: Comparative study of argon, nitrogen and oxygen treatments. *Colloids and Surfaces A: Physicochemical and Engineering Aspects, 374*(1–3), 88–95. https://doi.org/10.1016/j.colsurfa.2010.11.013.

Zhang, X., Chan, K-C., Wang, C. C. L., Wong, K-C. and Kumt, S. M., (2015). *Computing stable contact interface for customized surgical jigs,* IEEE International Conference on Robotics and Automation.

Chapter 7
Customised Interventions Utilising Additive Manufacturing

Evert A. Fuenmayor, Andrew V. Healy, Maurice Dalton, and Ian Major

Abstract Additive manufacturing is being utilised in biomedical applications for drug delivery and soft tissue engineering as a key enabling technology for precision medicine. Unlike conventional manufacturing techniques, additive manufacturing can fabricate customised interventions with relative ease. In drug delivery, other drivers for growth in the technology is the production of complex dosage forms, incorporating multiple drugs and materials, and the pursuit of on-demand manufacturing. For tissue engineering applications, additive manufacturing is creating 3D scaffolds to aid or replace the native tissue with morphology and internal architecture which mimic the final shape and structure of the defective tissue. The chapter details the utilisation of additive manufacturing in the production of pharmaceutical dosage forms, spanning simple flat-faced oral tablets to more complex drug-delivery systems that provide for the independent release of multiple drug compounds. For tissue engineering applications, the chapter describes the production of scaffolds for various soft tissue scaffolds currently under development by research teams worldwide.

Keywords Drug delivery · Tissue engineering · Fused-filament fabrication · Precision medicine

7.1 Introduction

Precision medicine is an emerging field in the healthcare sector, and its aim is to tailor preventative and therapeutic strategies for the individual or a population subgroup. This strategy takes the patients' unique physiology, biochemistry and lifestyle into account in order to fabricate the medical intervention, whether a drug delivery system or tissue engineering scaffold. This methodology has already been exploited in certain biomedical sectors such as dentistry, patient-specific

E.A. Fuenmayor · A.V. Healy · M. Dalton · I. Major (✉)
Materials Research Institute, Athlone Institute of Technology, Athlone, Ireland
e-mail: imajor@ait.ie

D. M. Devine (ed.), *Polymer-Based Additive Manufacturing*,
https://doi.org/10.1007/978-3-030-24532-0_7

tissue-organs and prosthetics (Chia and Wu 2015; Blake et al. 2016). It remains a challenge to fabricate customised interventions via conventional manufacturing techniques. It has been argued that these limitations could be attributed to lack of suitable biomanufacturing approaches capable of generating geometrically programmable architectures augmented with biomimetic physical, biochemical and cellular components (Blake et al. 2016). This chapter explores the utilisation of additive manufacturing in biomedical applications for drug delivery and soft tissue engineering, two areas that can provide the most impact for personalised medicine (Cengiz et al. 2016).

Drugs are agents administered for the diagnosis, mitigation, treatment, cure or prevention of diseases in humans or in other animals (FDA 2013). Drugs are a cornerstone in modern medicine, used for full array of purposes, from treating common infections to stopping the proliferation of cancer. Modern surgery would be virtually impossible without the implementation of anaesthetics, analgesics and antibiotics (Allen et al. 2015). Drug compounds are derived from plant and animal sources, from by-products of microbial growth or through chemical synthesis and molecular modification (Mahato and Narang 2012). The method of administration for such drugs plays a role equally important to the effect of the drug. Miraculous drugs are useless if administered orally and do not bypass the gastrointestinal (GI) barrier into the bloodstream or are unstable in their suspension media (Mitra et al. 2013). Thus, it is imperative to devise the most suitable strategy to enhance or exploit the capabilities of therapeutics through drug delivery. Drug delivery system is a broad term which engulfs strategies, materials, dosage methodologies and frequency, manufacturing processes, routes of administration and instruments used in the administration of drugs (Wilson and Crowley 2011). The development of a drug delivery system focuses on therapeutic accuracy and efficacy. This objective is driven by three recurrent elements found in the literature: drug, formulation and route of administration (Aulton 2001; Wilson and Crowley 2011). The first two factors are controllable by technological means. Drugs are characterised by biochemical and physical properties, attributes that determine which pharmacological effects drug molecules will exert and how they will exert them as well dose response and duration of action, among others. Formulations are known for playing a crucial role in the ability to enhance (or compromise) the effectiveness of the active pharmaceutical ingredient (API). Drug delivery is then an engineering problem, creating a system capable of hitting or maintaining a desired plasma/drug concentration ratio, above the minimum effect concentration and below the minimum toxic concentration.

The intrinsic characteristics that differentiate additive manufacturing from traditional pharmaceutical manufacturing processes are the main drive for its exponential growth in recent years (Lim et al. 2018). Among these traits, we found that readily product complexity, personalisation and on-demand manufacturing are three of the main features that additive manufacturing offers to pharmaceutic and formulation scientists (Norman et al. 2017). The capacity to produce objects with complex geometries (Olszewski et al. 2014; Goyanes et al. 2015b; Genina et al. 2016), dosage forms containing multiple active pharmaceutical ingredients (API) (Khaled et al. 2015; Park 2015; Goyanes et al. 2015a) as well prosthetics and implants (Shor et al. 2009; Eckardt et al. 2011; Seyednejad et al. 2012; Mannoor et al. 2013) are within the achievable applications for fused-filament fabrication (FFF). As for personalisation, there is a trend among academia to stress FFF as the manufacturing revolution to allow for the feasible and replicable treatment of medical conditions in a patient-tailored fashion (Alomari et al. 2015). These claims, although sensational, are not fully far-fetched when considering the results obtained so far by researchers (Alomari et al. 2015; Goyanes et al. 2015a; Poulin et al. 2015; Choonara et al. 2016; Jonathan and Karim 2016; Melocchi et al. 2016) and predicted projections estimated 3D-printed drug products in the market in a minimum of 10 years from now due to regulatory setbacks (Shanler and Basiliere 2017). Examples of on-demand manufacturing using 3D printing applications seem to suggest that FFF is the best option due to its cost-effectiveness and availability (Lim et al. 2016), and there is a hint that in the future this technology could be adapted by pharmacies worldwide to dispense personalised treatments on demand (Lim et al. 2018).

The global market for tissue engineering was estimated at US$ 23 billion in 2015 and is expected to register a CAGR of about 23% between 2016 and 2022, reaching approximately US$ 94.2 billion by 2022 (Ye et al. 2018). Tissue regeneration is the restoration or replacement of defective organs or tissues. In order to aid in the regeneration process, 3D scaffolds are commonly used to aid or replace the native tissue. High porosity is a mandatory feature when fabricating scaffolds for providing three-dimensional growth in cell regeneration. The morphology and internal architecture of the scaffold need to mimic the final shape and structure of the defective tissue. Scaffolds must satisfy specific mechanical properties such as pore size, distribution and interconnectivity (Ceretti et al. 2017). The ability to design and fabricate 3D biomedical devices is essential in tissue engineering. When fabricating scaffolds, the design of the architecture on the macro, micro and nano level is vital for structural nutrient transport and cellular matrix interaction (Chia and Wu 2015). Unfortunately, conventional methods used to generate 3D scaffolds such as freeze drying, gas foaming, fibre bonding and moulding can only control the bulk properties of the scaffold but are limited in the control of internal architecture and topology of a scaffold (Zhu et al. 2016). None of these conventional techniques have allowed researchers to build scaffolds with a complete interconnected pore network with large interconnected channels (Zein et al. 2002). Widely exploited technologies such as electrospinning are still being utilised to fabricate scaffolds either using small diameter tubular scaffolds or two-dimensional meshes. Unfortunately, the electrospinning apparatus is limited when the design must be patient specific. The need for a more customisable method such as additive manufacturing is essential for patient-tailored scaffolds (Rabionet et al. 2018). 3D printing has been used to fabricate a wide arrange of scaffolds for various tissue engineering applications including bone, cartilage, tendons, ligaments and nerve regeneration.

7.2 Pharmaceutical Applications of Additive Manufacturing

In 2014, Goyanes led a team of researchers (Goyanes et al. 2014) in what is believed to be the first documented utilisation of FFF in the fabrication of drug-loaded dosage forms. The main aim of their research was to impregnate fluorescein as a model active pharmaceutical ingredient (API) into commercially available polyvinyl alcohol (PVA) extruder filament and produce tablets that may be suitable with regard to personalised medicine and individual patient doses. A final drug-loading concentration of 0.29% w/w was achieved by placing the extruded filament in an ethanolic drug solution in which the drug passively diffused into the polymeric matrix and became trapped after the drying phase. However, it was found that the printed tablets demonstrated 0.28% w/w drug content, indicating that the thermal printing process, 220 °C, did not impart any drug degradation, fluorescein melting temperature (T_m) being 320 °C. However, it should be noted that this process may not be suitable for utilisation when fabricating dosage forms where thermally liable drugs need to be incorporated as it may result in degradation of the drug (Pietrzak et al. 2015). The authors, Goyanes et al. (2014), utilised the desktop software to manipulate the percentage infill of the tablets, (0%, 10%, 25%, 50%, 90% or 100%), and it was found that by doing so the weight and volume of the dosage forms could be altered. The results from conducting dissolution studies indicated that the release profiles were found to be dependent on the infill percentage with faster drug release being detected at a lower infill percentage, 10% infill complete drug release in 6 h in comparison to complete release in 90% infill only observed after a duration of 20 h. The authors concluded that although the drug loading via passive diffusion was low, 0.28% w/w printed dosage form, it was possible to detect by dissolution studies demonstrating the proof-of-concept 3DP affords in drug delivery with infill percentage playing a role in the modulation of drug release.

Goyanes et al. have continued to expand on this initial work and utilised both hot-melt extrusion (HME) and FFF to fabricate dosage forms with different geometries that would not be possible using conventional methods available such as powder compaction. In the study, Goyanes and co-authors (2015b) used commercially available, pelletised, PVA filament and incorporated paracetamol into the matrix using HME to form melt-blended filament with a suitable diameter, 1.75 mm, for the FFF process. They then printed tablets of five different geometries: cube, pyramid, cylinder, sphere and torus, some of which were complex and intricate, demonstrating that 3DP offers an advantageous route of production for dosage forms of novel geometries. The authors evaluated what role the geometric shape imparted on drug release. Although 5% drug w/w was incorporated, the actual drug loading of the filament was found to be 3.95% w/w, owing to the API adhering to both the walls of the container during the mixing process and the barrel during the extrusion process, with the drug loading being 3.78% w/w in the printed tablets, indicating that little degradation occurred during the printing process with similar results found in the previous study (Goyanes et al. 2014). It was established from dissolution testing that the printed geometry influenced drug release. The studies found that when the surface area was kept constant, the release rates varied significantly. It was noted the drug release occurred in the following order, with the fastest first, pyramid > torus > cube > sphere and cylinder with 90% release observed in the printed pyramid dosage form in less than 2 h, while it took almost 12 h for the same release rate to be observed in the sphere and cylinder dosage forms. However, when the authors evaluated the release profiles when the surface area/volume ratio was kept constant, drug release occurred in a different fashion. It was observed that drug release was fastest in the sphere and cube dosage forms and the slowest release rate was seen in the pyramid printed tablet. The 90% release rate values found that most geometries exhibited values between 2 and 3 h with the exception of the pyramid geometry being significantly longer with a slower release rate of 9 h. Furthermore, in the study drug release from the printed dosage forms was assessed when the tablet weight was kept constant, and it was found that the release rates were comparable to one another. This work demonstrates that it was possible to utilise HME to incorporate drugs into suitable FFF filament with the resultant filament printing tablets of intricate geometries, thus illustrating the potential benefits 3DP presents in the manufacturing of pharmaceutics. In addition to these findings, the authors concluded that drug modulation was dependent on surface area/volume ratio showing that the geometrical shape influenced its release, which highlights that the use of 3DP technology may allow for the controlled release of API when the shape of the dosage form is considered.

In 2015, Skowyra and researchers (2015) conducted a similar study to that of Goyanes et al. (2014) study in which they incorporated drug (the steroid prednisolone) into commercially available PVA filament by soaking the filament in a saturated drug solution and evaluating the release profiles from FFF printed tablets. Through manipulation of the printer's software, the authors could also control and vary the printed dosage with a good correlation between the target dosage and achieved dosage. The authors stated that the technology held the potential to control a patient's dosage by means of simple software manipulation. In vitro release studies of the printed tablets indicated that similar release profiles were observed irrespective of tablet weight. The time taken for 80% drug release for dosage forms of 2 and 3 mg took place after a duration of 12 h, whereas the other dosages, 4, 5, 7.5 and 10 mg, only reached the same release after 18 h which was longer than what was required for complete release from the 2 and 3 mg printed tablets, 16 h, with the quicker release rate being attributed to the tablets' larger surface area/mass ratio, allowing for the drug to dissipate from the already eroding PVA matrix. The research presented clearly demonstrated that FFF could be utilised in the fabrication of tuneable drug delivery systems via software manipulation with an excellent accuracy of dosage forms produced leading to personalised patient-specific dosing, e.g. paediatric patients versus geriatric patients.

A team of researchers led by Pietrzak (Pietrzak et al. 2015) also conducted research utilising both HME and FFF to fabricate 3DP tablets. The authors evaluated the release rate of theophylline from the polymer matrices, using a realistic drug loading of 50%, aiming to explore the feasibility of developing immediate and extended release dosage forms of an easy-to-swallow caplet design. Further, the team built on a previous publication (Skowyra et al. 2015) in which they demonstrated the linear relationship that exists between the volume of a tablet and the mass of a 3DP tablet. They fabricated 3DP tablets of varying therapeutic dosages: 60, 120, 200, 250 and 300 mg, with Eudragit RL being the polymeric matrix and theophylline the model drug. Actual drug content was in the range of 91–96%, thus demonstrating that FFF may be functionally suitable to print a dosage form of desired strength by simple manipulation of computer software. In addition to this in vitro drug release from the Eudragit RL, FFF tablets were found to extend over a duration of 16 h. In order to verify that FFF can be employed in the production of both immediate and extended dosage forms, the team of research workers examined the release of theophylline when encapsulated within commonly used polymers for the aforementioned dosage forms. Eudragit E and hydroxypropyl cellulose were selected for the immediate release application with Eudragit RS and RL chosen for the extended release application. With respect to the immediate release formulation, it was observed that the FFF process imparted a difference in the release profiles of the extruded filament, non-printed when compared to a printed dosage form with the FFF tablet drug release taking longer. It is plausible that this increase in drug release can be attributed to the thermal printing process, allowing further drug encapsulation within the polymeric matrices. Additionally, when the researchers aimed to evaluate theophylline drug release in an extended FFF tablet, they combined Eudragit RL with Eudragit RS with Eudragit RS shown in the literature to slow down drug release from dosage forms. The in vitro theophylline release profiles demonstrated that when the ratio of Eudragit RL:RS was 1:1, there was a significant decline in the release of drug when compared to the Eudragit RL matrix alone in both the non-printed filament and the FFF tablets as well as the FFF process delaying the release of the active. The work presented by Pietrzak et al. (2015) demonstrates the feasibility and adaptability that FFF technology offers in the development of both immediate and extended release dosage forms which may be beneficial depending on the patient to receive the therapeutic dose.

The possibility of fabricating modified release oral tablets using enteric polymers was evaluated by Goyanes et al. (2017). FFF was coupled with HME for the fabrication of paracetamol-loaded hypromellose acetate succinate (HPMCAS) filaments. Up to the publication of this work, most approaches to achieve enteric drug release consisted in coating 3D-printed tablets to achieve protection against the acidic condition of the stomach (Goyanes et al. 2015c) or dual printing of a capsule containing highly hydrophilic drug-loaded core protected by an outer shell made of enteric materials (Okwuosa et al. 2016b). However, the researchers managed to fabricate a modified release tablet that could be tailored to discharge its drug content at different places in the gastrointestinal tract by modifying the grade of HPMCAS used to produce the filament or by changing the percentage infill used to produce the tablets. Not only that, but it was possible to fabricate filament with a drug loading of 50% without compromising the printability of the material which has been one of the limitations of FFF for drug delivery applications (Fuenmayor et al. 2018).

Most research related to the development of FFF as a pharmaceutical production method revolves around oral dosage forms, and although they are the most used method of administration of API (Jones 2016), this is not without its limitations. Oral drug absorption is dominated by several parameters which have compelled formulation scientist into inventing strategies to improve the bioavailability of active compounds (Davis et al. 2005). These strategies rely on polymer bioadhesive system, swelling and expanding systems, high-density systems and floating systems (Pawar et al. 2012). Domperidone (DOM) is a weak base drug with a high solubility in acidic conditions, making the perfect drug candidate for a stomach floating system, and it was used by a team of researchers from Shanghai, China, to develop a 3D-printed tablet for intragastric floating delivery system (Chai et al. 2017). DOM is prescribed for three-times-a-day dosage due to its short half-time and its poor solubility in neutral and alkaline conditions (Reddymasu et al. 2007), motivating the development of a prolonged release system for the upper GI tract where the drug's bioavailability peaks. Chai et al. (2017) proposed developing a low-density tablet by enhancing buoyancy via a lower infill percentage 3D printing setting and different number of outer shells.

Modifying the inner structures of a dosage form allows for the creation of tablets with controlled drug release characteristics as demonstrated by Li et al. (2017). This study used HME for the incorporation of the diabetes drug glipizide into PLA filaments in two different drug loadings (2.5% and 5%) for the manufacture of a cylindrical tablet with well-defined inner and outer structure. The design consisted of a shell structure engulfing a solid inner compartment, and three different combinations were fabricated, drug outer/inner blank (a), inner drug/outer blank (b) and both structures drug loaded (c). Filaments with low and high drug loading were used for the creation of outer and inner parts of the tablets, respectively, which produced three distinct drug release curves. While (a) released 90% of its drug after 120 min, (b) had a lag time of 90 min before any drug was dissolved in the media, while (c) had a sustained release over a period of 5 h, fitting a Korsmeyer-Peppas release kinetic model, demonstrating how the intrinsic ability of FFF for complex geometry fabrication can be exploited to modify drug release with little-to-no modification of batches and manufacturing equipment.

Hybrid manufacturing techniques often aim to merge the advantages of different processes into a single processing step. 3DP tablet (printlets) infused with deflazacort-loaded nanocapsules is one of the examples of these hybrid techniques being implemented and was developed by in a collaborative effort between University College London (UK) and the Federal University of Rio Grande do Sul (Brazil) (Beck et al. 2017). Twin-screw HME was used to adapt PCL and Eudragit® RL100 for FFF applications, and the effects of using mannitol as a channelling agent were also evaluated. The drug loading was achieved by soaking the printlets in a suspension of deflazacort-loaded nanocapsules for 24 h. The swelling characteristics of the matrix polymers and the presence of mannitol used in this study had an influence on the percentage of drug loaded into the printed devices. Overall, this technique offered a higher presence of API in the finalised product than other approaches proposed before for the loading of drug-filled nanoparticles into tablets (Friedrich et al. 2010a, b).

The possibilities of manufacturing drug dosage forms with tailored and well-differentiated properties are one of the advantages of additive manufacturing. There is evidence supporting that reducing dosing frequency will in return increase patient compliance (Paes et al. 1997), and FFF is being investigated to produce tablets that combine multiple therapeutic agents in a single dosage form (Goyanes et al. 2015a; Khaled et al. 2015; Gioumouxouzis et al. 2018). In the work undertaken by researchers led by Gioumouxouzis (Gioumouxouzis et al. 2017), two antidiabetic drugs, metformin and glimepiride, were incorporated into a dosage form with two distinct drug release kinetics. The drugs are usually co-prescribed, but their dosage varies (once daily for glimepiride and 2–3 times for metformin (Hwang et al. 2013)), so a single dosage combining sustained and immediate release kinetics is desired. Combining two polymer matrices known to provide this drug release profiles and a handful of plasticisers to improve the printability of the polymers, the team devised a bilayer tablet that would eradicate the need for multiple daily dosage for these drugs and at the same time attempted to offer a mechanical profile for the affinity of materials for FFF applications. Polyvinyl alcohol and Eudragit™ RL PO were selected as the polymer carriers for the immediate and sustained release of the drugs, respectively. Several combinations of these polymers with plasticisers where characterised using an instrumented indentation test to determine their elastic and hardness values (Mansour and Tzetzis 2013; Tzetzis et al. 2013) with the intention of determining what values for these properties a material should display for FFF applications. Among the main findings, there was that a material should possess an elastic modulus between 3 and 4 GPa in combination with an indentation hardness of 50 N/mm^2 for reliable FFF manufacture. An equation was also developed to determine the amount of drug per tablet using the known mass and density of printed parts using a 100% infill setting; this equation defines the amount of drug contained by a 3D-printed layer based on the total height of each. They successfully managed to produce a bilayer tablet with the desired amount of drug and release profiles for the treatment of type 2 diabetes in a single daily dosage form.

Efforts to fully understand the mechanical and rheological properties that qualify a material for FFF are of the utmost importance if researchers want to break through the material selection threshold limiting the reach of extruded-based 3D printing. This is why a team of researchers recently publish their findings on a detailed investigation/review on the material considerations for the fabrication of solid dosage forms (Fuenmayor et al. 2018). Based on the literature available up to date, the researchers developed an action plan to fully characterised PCL and Kollidon VA 64™ blends in order to create a range for mechanical and rheological properties where materials are considered ideal for FFF applications. In addition, the researchers manufactured caffeine-loaded oral dosage forms using a range of blends via both direct compression and FFF processes and compared how the manufacturing method had an effect on the drug release. The team also produced an extensive guide to solving the most common issues encountered while adapting pharmaceutical-grade polymers for FFF processes. Using common polymer characterisation techniques and polymer processing concepts, this publication provided a full guideline for improving the FFF printability of a material.

The geometry complexity achievable using additive manufacturing processes was demonstrated by a group of researchers from Switzerland who managed to manufacture personalised drug-loaded mouthguards for the targeted API delivery to the buccal cavity (Liang et al. 2018). The team managed to create a tailored mouthpiece based on maxilla scans of six participants and adapted CAD models to improve comfort and fit of the pieces that would deliver model of anti-inflammatory drug clobetasol propionate over a 10-day period in vitro. For the human study, the drug was replaced by the food-grade flavour vanillic acid. Participants were asked to wear the device over a 2-h period when water and food were restricted, and saliva samples were taken every half an hour. The mouthguard successfully delivered model active ingredients to the buccal cavity over an extended usage period of 3 weeks for 2 h daily, displaying the possibility of adapting FFF for the fabrication of personalised wearable drug delivery dosage forms.

Compromised populations, children, for example, represent a challenge for therapeutic strategies that would guarantee compliance: right drug dosage, high variability among intended patients and taste masking (Strickley et al. 2008). The most common administration route for this group is not oral but liquid drug dosage forms which have many disadvantages that could be resolved by using solid oral dosage forms (Yin et al. 2010; Mennella et al. 2015). Starmix™ is a famous brand of chewable candy that was used as a base for manufacturing chewable indomethacin (IND)-loaded 3D-printed tablets by a group of researchers from the University of Greenwich in the aims to solve some of these common issues (Scoutaris et al. 2018). Using HME, the research team adapted hypromellose acetate succinate (HPMCAS) for the FFF process and loaded it with IND. There were five different designs, and all remarkably achieved similar drug release in vitro, and all participants reported a significant reduction in bitterness when comparing FFF samples to the simple drug and polymer combination.

The recent advancement in FFF for drug delivery has also helped researchers understand better the limitations of using this fabrication method, including the hindered capacity to produce immediate release tablets (Okwuosa et al. 2016a; Sadia et al. 2016). This is a consequence of the large amount of polymer necessary to produce a material capable of being used as feedstock for FFF (40%) and the erosion and diffusion mechanism dominating drug release from 3D-printed products (Goyanes et al. 2014; Skowyra et al. 2015). A research group in the UK managed to control the surface area/volume and surface area/mass ratio of FFF printed tablets to accelerate the release of low solubility/low permeability drug hydrochlorothiazide (Sadia et al. 2018). Their solution consisted in creating channels in the long and short axis of the capsule-shaped tablet (caplet), increasing the available surface area and reducing the resistance to aqueous media flushing through the drug dosage form. They determined that including this orifice on the longer axis of the caplet increases the drug release by reducing the friction area for the flow of media as well as increasing the surface area/volume ratio. The result was a tablet that qualifies as an immediate release drug dosage form according to the United States Pharmacopeia (USP).

7.3 Soft Tissue Engineering Applications of Additive Manufacturing

3D printing for soft tissue engineering applications has mostly centred around techniques that can use soft hydrogels during fabrication since these materials share many biofunctional similarities to physiological tissues (Scott et al. 2018). Since FFF is a hot-melt extrusion-based process, hydrogels are unsuitable, and appropriate thermoplastic materials are more limited in number. However, 3D-printed biodegradable implants also present drawbacks which include the need to pre-drill and tap holes for biodegradable screws. This can lead to increased operative time, as well as more difficult implant placement due to concerns with heat generation during drilling and polymer debris, which has been reported to cause inflammatory foreign-body responses (Pok et al. 2013). PCL has a glass transition temperature at −60 °C and a melt point ranging from 59 to 64 °C (depending on the crystalline nature of PCL). The low Tg of PCL allows melt processing at lower temperatures. Due to a slow in vivo degradation rates (2–4 years), other resorbable polymers, such as polylactides and polyglycolides, are increasing in popularity due to faster in vivo degradation times (2–4 months). However, due to superior rheological and viscoelastic properties over many other resorbable polymers, the material has been exploited in FFF printing over many years (Woodruff and Hutmacher 2010). Researchers have previously demonstrated the use of FFF to fabricate thermoplastic PCL into a functional 3D scaffold for tissue engineering applications (Yeong et al. 2004). Lam et al. (2008, 2009) investigated the degradation rate of PCL processed via FFF into porous scaffolds both in vitro and in vivo. Zein et al. fabricated a PCL honeycomb-pattern scaffold with interconnected and controllable porosity via FFF. According to Zein et al., honeycomb PCL scaffolds were fabricated by the deposition of layers of directionally aligned microfilaments utilising computer-controlled extrusion, and deposition process, displaying a channel sizes of 160–700 × 260–370 mm and porosity 48–77%. Mechanical testing was performed to examine the mechanical deformation of porous PCL scaffolds via compression testing, indicating the stress–strain behaviour of the scaffolds was found to be dictated by the scaffolds' porosity, regardless of the lay-down pattern and channel size (Zein et al. 2002). Hutmacher et al. (2001a) demonstrated that using FFF allows for design and fabrication of highly reproducible bioresorbable 3D scaffolds with a fully interconnected pore network. Similar to Zein et al., Hutmacher and research team fabricated a honeycomb-like PCL scaffold with different lay-down patterns (0/60/120° and 0/72/144/36/108° lay-down), displaying porosity of 61 ± 1% size and falling size within the range of 360 × 430 × 620 mm. Mechanical analysis was done in terms of compressive stiffness and yield strength. Scaffolds with 0/60/120° lay-down pattern showed better mechanical properties in comparison with 0/72/1444/36/108°, that is, the lay-down patterns of 0/60/120° and 0/72/1444/36/108° effected the mechanical properties of the scaffolds. Cell culture analysis was performed using human fibroblasts and periosteal cells via light, environmental scanning electron and confocal laser microscopy as well as immunohistochemistry, demonstrating the cells can proliferate, differentiate and produce a cellular tissue in the 3D polycaprolactone matrix (Hutmacher et al. 2001b). Cao et al. designed and fabricated porous PCL scaffolds via FFF to successfully coculture osteogenic and chondrogenic cells on PCL scaffolds. Cao and research team seeded and precultured the osteogenic cells in one-half of the scaffold. Chondrogenic cells were later seeded into the other half, and cell lines were cultured for over 50 days. Both cell types successfully proliferated, migrated and linked in their scaffold compartments to achieve an integrated interface. Cao et al. indicated that seeded PCL scaffold has the potential for osteochondral defect repair (Cao et al. 2003; Bose et al. 2018).

Kolan et al. investigated the fabrication of a 3D porous scaffolds with PCL/13-93B3 bioactive borate glass and Pluronic F-127 hydrogel for tissue engineering applications via FFF. Kolan describes the printing of PCL/13-93B3 composite as a scaffolding material and Pluronic F-127 hydrogel as a cell suspension medium. Weight loss analysis indicated the presence of the F-127 hydrogels after 7 days. Scaffold porosity with the presence of macro- and micropores dictated the glass dissolution speed. In addition, hydroxyapatite formation layers were detected on the surface showing scaffold's bioactivity. According to Kolan et al. (2017), novel fabrication approach to produce porous scaffolds shows the high potential of extrusion-based 3D printing for tissue engineering applications.

Wang et al. reported the use of precision extrusion deposition modelling similar to FFF; the only difference is that it extrudes scaffolding material in granule form (Wang et al. 2004). Wang et al. utilised this novel method to fabricate PCL scaffolds with a controlled pore size of 250 mm for tissue engineering applications. The scaffold compression modulus was in the range between 150 and 200 MPa. Biocompatibility was also evident via alamarBlue staining and cytofluorimetry using cardiomyoblasts, fibroblasts and smooth muscle cells (Wang et al. 2004). Other researchers, including Jeon et al., manufactured a multiphasic scaffold composed of an alginate hydrogel surmounting a biphasic PCL scaffold via FFF. Jeon et al. fabricated $100 \times 100 \times 6\ mm^3$ medical-grade PCL scaffolds with a 100% interconnectivity, 70% porosity and 08/908 lay-down pattern for in vivo osteochondral defects. Bovine osteochondral plugs with 4 mm diameter osteochondral defect was fixed with multiphasic osteochondral grafts consisting of chondrocyte-seeded alginate gels and osteoblast-seeded onto PCL scaffolds. Seeded PCL scaffolds were subcutaneously implanted into athymic nude rats for up to 12 weeks to allow for screening of osteochondral repair potential of multiphasic construct (Jeon et al. 2014). Korpela et al. examined the use of FFF to evaluate the printability of porous biodegrade material including PCL/BAG composite and PLC copolymer scaffold structures. Korpela further emphasises that the mechanical properties of the materials were depended on porosity level and structural geometry. In addition, fibroblast proliferation was higher in polylactide in comparison with PCL or PCL/BAG composite via optical microscope and SEM microscopy. Korpela's research reports the expanded application of FFF for tissue engineering applications (Korpela et al. 2013).

The other major polyester for soft tissue applications has been PLA. PLA is the most studied aliphatic polyester and demonstrated potential in biomedical applications such as resorbable surgical sutures, meshes or in drug delivery (Boccaccini and Maquet 2003). PLA is a semicrystalline thermoplastic aliphatic polyester that has fragile characteristics including high elastic modulus and low elongation at break and as a result is more preferred in bone engineering. This thermoplastic polymer possesses unique characteristics such as high strength and biocompatibility, and the monomeric units within the polymer can be synthesised via nontoxic renewable feedstock as well as can be naturally derived (Scaffaro et al. 2016). Implantable PLA hydrolyses into intermediate α-hydroxy acid and is broken down metabolically via the tricarboxylic acid cycle and excreted. Researchers have explored PLA using FFF from composite to scaffold fabrication. Drummer et al. fabricated a thermoplastic composite via FFF composed of PLA and tricalcium phosphate for tissue engineering and maxillofacial surgery. Drummer evaluated the suitability of PLA for the processing with FFF via thermal, mechanical and microscopic analysis (Drummer et al. 2012; Singh et al. 2017). Ravi et al. (2017) employed the use of FFF for rapid fabrication of medical-grade PLA scaffolds with different lay-down patterns. Zeltinger et al. investigated the influence of pore size and porosity on cell adhesion and proliferation of disc-shaped PLA scaffolds (Singh et al. 2017). Chhaya et al. fabricated poly(D,L)-lactide-based scaffolds via FFF from models generated by laser scanning of patients. The breast-shaped scaffold was incubated for 4 weeks in a rotating bioreactor, to encourage seeding with umbilical cord perivascular cells. The seed scaffold was then human umbilical vein endothelial cells and implanted subcutaneously into athymic nude rats for 24 weeks. The research team demonstrated personalised engineered breast contrast as a facile approach towards engineering clinically relevant volumes of adipose tissue for breast reconstruction (Chhaya et al. 2015). Unfortunately, for the fabrication of scaffolds, PLA-based materials are limited due to poor thermal stability during the melt processing as ester linkages tend to degrade into smaller fractions (Poh et al. 2016).

Polyhydroxyalkanoate (PHA) is a biodegradable, biocompatible aliphatic polyester. PHA can be divided into two groups according to the carbon chain length of the monomer, at the 3–5-carbon chain showing high degree of stiffness and brittleness and 50–80% crystallinity which is an unfavourable characteristic for soft tissue engineering and FFF fabrication. Fortunately, researchers have copolymerised poly(3-hydroxybutyrate) (PHB) with hydroxyvalerate (HV) to lower melt temperature and decrease PHB stiffness and mechanical strength, making it an ideal candidate for biomedical applications, in particular as scaffolds and implants (Chen and Wang 2002; Chen and Wu 2005; Chiulan et al. 2017). Other copolymers have also been investigated using FFF for tissue engineering applications. Poly(lactide-co-glycolide) (PLGA) has also been exploited for FFF scaffold fabrication applications; however, due to its high glass transition (46–60 °C), melt processing is challenging at lower temperatures (Zein et al. 2002; Felgueiras and Amorim 2017). Yen et al. (2009) fabricated highly porous PLGA scaffolds for tissue engineering application via FFF. Kim and Cho (2009) fabricated PCL/PLGA scaffolds for tissue engineering applications with an interconnected and highly porous (69.6%) network. Lee et al. discuss the utilisation of 3D printing technology to create porous PLGA scaffold for tissue engineering applications (Lee et al. 2005), particularly conduits for peripheral nerve.

7.4 Conclusion

Future medicine will be smarter, targeted and precise. Precision medicine is a goal to provide therapies that are specific to the genetic make-up of the individual patient or geometry of injury site and replaces the current 'one-size-fits-all' approach of standard treatments. To meet these ambitious goals, there must be an evolution in the fabrication of medicine and other biomedical interventions. The impact of additive manufacturing in medicine is only beginning to be understood by academics and practitioners. Several 3D printing processes, particularly FFF, have enabled the production of complex drug delivery systems and biomimetic soft tissue scaffolds. Due to current limitations, the technology has yet to find a foothold in modern clinical practice. Nevertheless, due to its enormous potential and the increasing importance of precision medicine, 3D printing will continue to be utilised in the biomedical academic research and clinical development, and with the advent of new resins and techniques, the growth of additive manufacturing in customised interventions is on course to grow exponentially.

References

Allen, L. V., Popovich, N. G., & Ansel, H. (2015). Ansel's pharmaceutical dosage forms and drug delivery systems. *Journal of Chemical Information and Modeling*. https://doi.org/10.1017/CBO9781107415324.004.

Alomari, M., et al. (2015). Personalised dosing: Printing a dose of one's own medicine. *International Journal of Pharmaceutics, 494*(2), 568–577. https://doi.org/10.1016/j.ijpharm.2014.12.006.

Aulton, M. E. (2001). *Pharmaceutics: The science of dosage form design*. Edinburgh: Churchill Livingstone. https://doi.org/10.1017/CBO9781107415324.004.

Beck, R. C. R., et al. (2017). 3D printed tablets loaded with polymeric nanocapsules: An innovative approach to produce customized drug delivery systems. *International Journal of Pharmaceutics*. https://doi.org/10.1016/j.ijpharm.2017.05.074.

Blake, N. J., et al. (2016). HHS public access. *Advanced Functional Materials, 25*(39), 6205–6217. https://doi.org/10.1002/adfm.201501760.3D.

Boccaccini, A. R., & Maquet, V. (2003). Bioresorbable and bioactive polymer/Bioglass® composites with tailored pore structure for tissue engineering applications. *Composites Science and Technology, 63*, 2417–2429. https://doi.org/10.1016/S0266-3538(03)00275-6.

Bose, S., et al. (2018). Additive manufacturing of biomaterials. *Progress in Materials Science, 93*, 45–111. https://doi.org/10.1016/j.pmatsci.2017.08.003.

Cao, T., Ho, K.-H., & Teoh, S.-H. (2003). Scaffold design and *in vitro* study of osteochondral coculture in a three-dimensional porous polycaprolactone scaffold fabricated by fused deposition modeling. *Tissue Engineering, 9*(Suppl 1), 103–112. https://doi.org/10.1089/10763270360697012.

Cengiz, I. F., et al. (2016). Building the basis for patient-specific meniscal scaffolds: From human knee MRI to fabrication of 3D printed scaffolds. *Bioprinting, 1–2*, 1–10. https://doi.org/10.1016/j.bprint.2016.05.001.

Ceretti, E., et al. (2017). Multi-layered scaffolds production via fused deposition modeling (FDM) using an open source 3D printer: Process parameters optimization for dimensional accuracy and design reproducibility. *Procedia CIRP, 65*, 13–18. https://doi.org/10.1016/j.procir.2017.04.042.

Chai, X., et al. (2017). Fused deposition modeling (FDM) 3D printed tablets for intragastric floating delivery of domperidone. *Scientific Reports, 7*(1), 2829. https://doi.org/10.1038/s41598-017-03097-x.

Chen, L. J., & Wang, M. (2002). Production and evaluation of biodegradable composites based on PHB-PHV copolymer. *Biomaterials, 23*(13), 2631–2639. https://doi.org/10.1016/S0142-9612(01)00394-5.

Chen, G. Q., & Wu, Q. (2005). The application of polyhydroxyalkanoates as tissue engineering materials. *Biomaterials, 26*(33), 6565–6578. https://doi.org/10.1016/j.biomaterials.2005.04.036.

Chhaya, M. P., et al. (2015). Sustained regeneration of high-volume adipose tissue for breast reconstruction using computer aided design and biomanufacturing. *Biomaterials, 52*(1), 551–560. https://doi.org/10.1016/j.biomaterials.2015.01.025.

Chia, H. N., & Wu, B. M. (2015). Recent advances in 3D printing of biomaterials. *Journal of Biological Engineering, 9*(1), 1–14. https://doi.org/10.1186/s13036-015-0001-4.

Chiulan, I., et al. (2017). Recent advances in 3D printing of aliphatic polyesters. *Bioengineering, 5*(1), 2. https://doi.org/10.3390/bioengineering5010002.

Choonara, Y. E., et al. (2016). 3D-printing and the effect on medical costs: A new era? *Expert Review of Pharmacoeconomics & Outcomes Research, 16*(1), 23–32. https://doi.org/10.1586/14737167.2016.1138860.

Davis, S. S., et al. (2005). Formulation strategies for absorption windows. *Drug Discovery Today, 10*(4), 249–257.

Drummer, D., Cifuentes-Cuéllar, S., & Rietzel, D. (2012). Suitability of PLA/TCP for fused deposition modeling. *Rapid Prototyping Journal, 18*(6), 500–507. https://doi.org/10.1108/13552541211272045.

Eckardt, M. A., et al. (2011). Not all hemangiomas are benign—Epithelioid hemangioendothelioma: An aggressive vascular lesion in children. *Pediatric Hematology and Oncology, 28*, 622–624. https://doi.org/10.3109/08880018.2011.584609.

FDA. (2013). Federal food, drug, and cosmetic act. In *Pharmacy law: Desk reference* (pp. 1–692). Boca Raton, FL: CRC Press. https://doi.org/10.1002/9780471462422.eoct415.

Felgueiras, H. P., & Amorim, M. T. P. (2017). Functionalization of electrospun polymeric wound dressings with antimicrobial peptides. *Colloids and Surfaces B: Biointerfaces, 156*, 133–148. https://doi.org/10.1016/j.colsurfb.2017.05.001.

Friedrich, R. B., et al. (2010a). Drying polymeric drug-loaded nanocapsules: The wet granulation process as a promising approach. *Journal of Nanoscience and Nanotechnology, 10*(1), 616–621. https://doi.org/10.1166/jnn.2010.1732.

Friedrich, R. B., et al. (2010b). Tablets containing drug-loaded polymeric nanocapsules: An innovative platform. *Journal of Nanoscience and Nanotechnology, 10*(9), 5885–5888. https://doi.org/10.1166/jnn.2010.2464.

Fuenmayor, E., et al. (2018). Material considerations for fused-filament fabrication of solid dosage forms. *Pharmaceutics, 10*(2), 44. https://doi.org/10.3390/pharmaceutics10020044.

Genina, N., et al. (2016). Ethylene vinyl acetate (EVA) as a new drug carrier for 3D printed medical drug delivery devices. *European Journal of Pharmaceutical Sciences, 90*, 53–63. https://doi.org/10.1016/j.ejps.2015.11.005.

Gioumouxouzis, C. I., et al. (2017). 3D printed oral solid dosage forms containing hydrochlorothiazide for controlled drug delivery. *Journal of Drug Delivery Science and Technology, 40*, 164–171. https://doi.org/10.1016/j.jddst.2017.06.008.

Gioumouxouzis, C. I., et al. (2018). A 3D printed bilayer oral solid dosage form combining metformin for prolonged and glimepiride for immediate drug delivery. *European Journal of Pharmaceutical Sciences, 120*(2017), 40–52. https://doi.org/10.1016/j.ejps.2018.04.020.

Goyanes, A., et al. (2014). Fused-filament 3D printing (3DP) for fabrication of tablets. *International Journal of Pharmaceutics, 476*(1–2), 88–92. https://doi.org/10.1016/j.ijpharm.2014.09.044.

Goyanes, A., Wang, J., et al. (2015a). 3D printing of medicines: Engineering novel oral devices with unique design and drug release characteristics. *Molecular Pharmaceutics, 12*(11), 4077–4084. https://doi.org/10.1021/acs.molpharmaceut.5b00510.

Goyanes, A., Robles Martinez, P., et al. (2015b). Effect of geometry on drug release from 3D printed tablets. *International Journal of Pharmaceutics*. https://doi.org/10.1016/j.ijpharm.2015.04.069.

Goyanes, A., Chang, H., et al. (2015c). Fabrication of controlled-release budesonide tablets via desktop (FDM) 3D printing. *International Journal of Pharmaceutics, 496*(2), 414–420. https://doi.org/10.1016/j.ijpharm.2015.10.039.

Goyanes, A., et al. (2017). Development of modified release 3D printed tablets (printlets) with pharmaceutical excipients using additive manufacturing. *International Journal of Pharmaceutics, 527*(1–2), 21–30. https://doi.org/10.1016/j.ijpharm.2017.05.021.

Hutmacher, D. W., et al. (2001a). Mechanical properties and cell cultural response of polycaprolactone scaffolds designed and fabricated via fused deposition modeling. *Journal of Biomedical Materials Research, 55*(2), 203–216. https://doi.org/10.1002/1097-4636(200105)55:2<203::AID-JBM1007>3.0.CO;2-7.

Hutmacher, D. W., Schantz, T., Zein, I., Ng, K. W., Teoh, S. H., & Tan, K. C. (2001b). Mechanical properties and cell cultural response of polycalrolactone scaffolds designed and fabricated via fused deposition modelling. *Journal of Biomedial Materials Research, 55*, 203–216.

Hwang, Y. C., et al. (2013). Efficacy and safety of glimepiride/metformin sustained release once daily vs. glimepiride/metformin twice daily in patients with type 2 diabetes. *International Journal of Clinical Practice, 67*(3), 236–243. https://doi.org/10.1111/ijcp.12071.

Jeon, J. E., et al. (2014). Multiphasic construct studied in an ectopic osteochondral defect model. *Journal of the Royal Society Interface, 11*(95), 20140184–20140184. https://doi.org/10.1098/rsif.2014.0184.

Jonathan, G., & Karim, A. (2016). 3D printing in pharmaceutics: A new tool for designing customized drug delivery systems. *International Journal of Pharmaceutics, 499*(1–2), 376–394. https://doi.org/10.1016/j.ijpharm.2015.12.071.

Jones, D. S. (2016). *FASTtrack: Pharmaceutics—Dosage form and design, antimicrobial agents and chemotherapy*. Cambridge: Cambridge University Press. https://doi.org/10.1128/AAC.03728-14.

Khaled, S. A., et al. (2015). 3D printing of five-in-one dose combination polypill with defined immediate and sustained release profiles. *Journal of Controlled Release, 217*, 308–314. https://doi.org/10.1016/j.jconrel.2015.09.028.

Kim, J. Y., & Cho, D. W. (2009). Blended PCL/PLGA scaffold fabrication using multi-head deposition system. *Microelectronic Engineering, 86*(4–6), 1447–1450. https://doi.org/10.1016/j.mee.2008.11.026.

Kolan, K., et al. (2017). Solvent based 3D printing of biopolymer/bioactive glass composite and hydrogel for tissue engineering applications. *Procedia CIRP, 65*, 38–43. https://doi.org/10.1016/j.procir.2017.04.022.

Korpela, J., et al. (2013). Biodegradable and bioactive porous scaffold structures prepared using fused deposition modeling. *Journal of Biomedical Materials Research Part B: Applied Biomaterials, 101*(4), 610–619. https://doi.org/10.1002/jbm.b.32863.

Lam, C. X., et al. (2008). Dynamics of in vitro polymer degradation of polycaprolactone-based scaffolds: Accelerated versus simulated physiological conditions. *Biomedical Materials, 3*(3), 34108. https://doi.org/10.1088/1748-6041/3/3/034108.
Lam, C. X. F., et al. (2009). Evaluation of polycaprolactone scaffold degradation for 6 months in vitro and in vivo. *Journal of Biomedical Materials Research Part A, 90*(3), 906–919. https://doi.org/10.1002/jbm.a.32052.
Lee, M., Dunn, J. C. Y., & Wu, B. M. (2005). Scaffold fabrication by indirect three-dimensional printing. *Biomaterials, 26*(20), 4281–4289. https://doi.org/10.1016/j.biomaterials.2004.10.040.
Li, Q., et al. (2017). Preparation and investigation of controlled-release glipizide novel oral device with three-dimensional printing. *International Journal of Pharmaceutics, 525*(1), 5–11. https://doi.org/10.1016/j.ijpharm.2017.03.066.
Liang, K., et al. (2018). 3D printing of a wearable personalized oral delivery device: A first-in-human study. *Science Advances, 4*(5), 1–12. https://doi.org/10.1126/sciadv.aat2544.
Lim, S. H., et al. (2016). Three-dimensional printing of carbamazepine sustained-release scaffold. *Journal of Pharmaceutical Sciences, 105*(7), 2155–2163. https://doi.org/10.1016/j.xphs.2016.04.031.
Lim, S. H., et al. (2018). 3D printed drug delivery and testing systems—A passing fad or the future? *Advanced Drug Delivery Reviews*. https://doi.org/10.1016/j.addr.2018.05.006.
Mahato, R. I., & Narang, A. S. (2012). *Pharmaceutical dosage forms and drug delivery* (2nd ed.). Boca Raton, FL: CRC Press.
Mannoor, M. S., et al. (2013). 3D printed bionic ears. *Nano Letters, 13*(6), 2634–2639. https://doi.org/10.1021/nl4007744.
Mansour, G., & Tzetzis, D. (2013). Nanomechanical characterization of hybrid multiwall carbon nanotube and fumed silica epoxy nanocomposites. *Polymer - Plastics Technology and Engineering, 52*(10), 1054–1062. https://doi.org/10.1080/03602559.2013.769581.
Melocchi, A., et al. (2016). Hot-melt extruded filaments based on pharmaceutical grade polymers for 3D printing by fused deposition modeling. *International Journal of Pharmaceutics, 509*(1–2), 255–263. https://doi.org/10.1016/j.ijpharm.2016.05.036.
Mennella, J. A., et al. (2015). Children's perceptions about medicines: Individual differences and taste. *BMC Pediatrics, 15*(1), 1–6. https://doi.org/10.1186/s12887-015-0447-z.
Mitra, A. K., Lee, C. H., & Cheng, K. (2013). Advance drug delivery. *Journal of Chemical Information and Modeling*. https://doi.org/10.1017/CBO9781107415324.004.
Norman, J., et al. (2017). A new chapter in pharmaceutical manufacturing: 3D-printed drug products. *Advanced Drug Delivery Reviews, 108*, 39–50. https://doi.org/10.1016/j.addr.2016.03.001.
Okwuosa, T. C., Stefaniak, D., et al. (2016a). A lower temperature FDM 3D printing for the manufacture of patient-specific immediate release tablets. *Pharmaceutical Research, 33*(11), 2704–2712. https://doi.org/10.1007/s11095-016-1995-0.
Okwuosa, T. C., Pereira, B. C., et al. (2016b). Fabricating a shell-core delayed release tablet using dual FDM 3D printing for patient-centred therapy. *Pharmaceutical Research, 1*, 1–11. https://doi.org/10.1007/s11095-016-2073-3.
Olszewski, R., Szymor, P., & Kozakiewicz, M. (2014). Accuracy of three-dimensional, paper-based models generated using a low-cost, three-dimensional printer. *Journal of Cranio-Maxillofacial Surgery, 42*(8), 1847–1852. https://doi.org/10.1016/j.jcms.2014.07.002.
Paes, A. H. P., Barker, A., & Soe-Agnie, C. J. (1997). Impact of dosage frequency on patient compliance. *Diabetes Care, 20*(10), 1512–1517. https://doi.org/10.2337/diacare.20.10.1512.
Park, K. (2015). 3D printing of 5-drug polypill. *Journal of Controlled Release, 217*, 352. https://doi.org/10.1016/j.jconrel.2015.10.014.
Pawar, V. K., et al. (2012). Industrial perspective of gastroretentive drug delivery systems: Physicochemical, biopharmaceutical, technological and regulatory consideration. *Expert Opinion on Drug Delivery, 9*(5), 551–565.
Pietrzak, K., Isreb, A., & Alhnan, M. A. (2015). A flexible-dose dispenser for immediate and extended release 3D printed tablets. *European Journal of Pharmaceutics and Biopharmaceutics, 96*, 380–387. https://doi.org/10.1016/j.ejpb.2015.07.027.

Poh, P. S. P., et al. (2016). Polylactides in additive biomanufacturing. *Advanced Drug Delivery Reviews, 107*, 228–246. https://doi.org/10.1016/j.addr.2016.07.006.
Pok, S., et al. (2013). A multilayered scaffold of a chitosan and gelatin hydrogel supported by a PCL core for cardiac tissue engineering. *Acta Materialia, 9*(3), 5630–5642. https://doi.org/10.1016/j.actbio.2012.10.032.
Poulin, E., et al. (2015). Towards real-time 3D ultrasound planning and personalized 3D printing for breast HDR brachytherapy treatment. *Radiotherapy and Oncology, 114*(3), 335–338. https://doi.org/10.1016/j.radonc.2015.02.007.
Rabionet, M., et al. (2018). 3D-printed tubular scaffolds for vascular tissue engineering. *Procedia CIRP, 68*(4), 352–357. https://doi.org/10.1016/J.PROCIR.2017.12.094.
Ravi, P., Shiakolas, P. S., & Welch, T. R. (2017). Poly-L-lactic acid: Pellets to fiber to fused filament fabricated scaffolds, and scaffold weight loss study. *Additive Manufacturing, 16*, 167–176. https://doi.org/10.1016/j.addma.2017.06.002.
Reddymasu, S. C., Soykan, I., & Mccallum, R. W. (2007). Domperidone: Review of pharmacology and clinical applications in gastroenterology. *The American Journal of Gastroenterology, 102*, 2036–2045. https://doi.org/10.1111/j.1572-0241.2007.01255.x.
Sadia, M., et al. (2016). Adaptation of pharmaceutical excipients to FDM 3D printing for the fabrication of patient-tailored immediate release tablets. *International Journal of Pharmaceutics, 513*(1–2), 659–668. https://doi.org/10.1016/j.ijpharm.2016.09.050.
Sadia, M., et al. (2018). Channelled tablets: An innovative approach to accelerating drug release from 3D printed tablets. *Journal of Controlled Release, 269*(9), 355–363. https://doi.org/10.1016/j.jconrel.2017.11.022.
Scaffaro, R., Lopresti, F., Botta, L., Rigogliuso, S., & Ghersi, G. (2016).Preparation of three-layered porous PLA/PEG scaffold: relationship between morphology, mechanical behavior and cell permeability. *Journal of the Mechanical Behavior of Biomedical Materials, 54*, 8–20. https://doi.org/10.1016/j.jmbbm.2015.08.033
Scott, S., et al. (2018). Polymeric 3D printed structures for soft-tissue engineering. *Journal of Applied Polymer Science, 135*(24), 46279. https://doi.org/10.1002/app.46279.
Scoutaris, N., Ross, S. A., & Douroumis, D. (2018). 3D printed "Starmix" drug loaded dosage forms for paediatric applications. *Pharmaceutical Research, 35*(2), 1–11. https://doi.org/10.1007/s11095-017-2284-2.
Seyednejad, H., et al. (2012). In vivo biocompatibility and biodegradation of 3D-printed porous scaffolds based on a hydroxyl-functionalized poly(ε-caprolactone). *Biomaterials, 33*(17), 4309–4318. https://doi.org/10.1016/j.biomaterials.2012.03.002.
Shanler, M., & Basiliere, P. (2017). *Hype cycle for 3D printing*. Stamford, CT: Gartner Inc..
Shor, L., et al. (2009). Precision extruding deposition (PED) fabrication of polycaprolactone (PCL) scaffolds for bone tissue engineering. *Biofabrication, 1*(1), 015003. https://doi.org/10.1088/1758-5082/1/1/015003.
Singh, S., Ramakrishna, S., & Singh, R. (2017). Material issues in additive manufacturing: A review. *Journal of Manufacturing Processes, 25*, 185–200. https://doi.org/10.1016/j.jmapro.2016.11.006.
Skowyra, J., Pietrzak, K., & Alhnan, M. A. (2015). Fabrication of extended-release patient-tailored prednisolone tablets via fused deposition modelling (FDM) 3D printing. *European Journal of Pharmaceutical Sciences, 68*, 11–17. https://doi.org/10.1016/j.ejps.2014.11.009.
Strickley, R. G., et al. (2008). Pediatric drugs—A review of commercially available oral formulations. *Journal of Pharmaceutical Sciences, 97*(5), 1731–1774. https://doi.org/10.1002/jps.21101.
Tzetzis, D., et al. (2013). Nanoindentation measurements of fumed silica epoxy reinforced nanocomposites. *Journal of Reinforced Plastics and Composites, 32*(3), 163–173. https://doi.org/10.1177/0731684412463978.
Wang, F., Shor, L., Darling, A., Khalil, S., Sun, W., Güçeri, S., & Lau, A. (2004). Precision extruding deposition and characterization of cellular poly-ϵ-caprolactone tissue scaffolds. *Rapid Prototyping Journal, 10*(1), 42–49. https://doi.org/10.1108/13552540410512525

Wilson, C. G., & Crowley, P. J. (2011). *Controlled release in oral drug delivery*. Boston, MA: Springer. https://doi.org/10.1007/978-1-4614-1004-1.

Woodruff, M. A., & Hutmacher, D. W. (2010). The return of a forgotten polymer—Polycaprolactone in the 21st century. *Progress in Polymer Science, 35*(10), 1217–1256. https://doi.org/10.1016/j.progpolymsci.2010.04.002.

Ye, H., et al. (2018). Polyester elastomers for soft tissue engineering. *Chemical Society Reviews, 47*(12), 4545–4580. https://doi.org/10.1039/c8cs00161h.

Yeong, W. Y., et al. (2004). Rapid prototyping in tissue engineering: Challenges and potential. *Trends in Biotechnology, 22*(12), 643–652. https://doi.org/10.1016/j.tibtech.2004.10.004.

Yen, H. J., Tseng, C. S., Hsu, S. H., & Tsai, C. L. (2009). Evaluation of chondrocyte growth in the highly porous scaffolds made by fused deposition manufacturing (FDM) filled with type II collagen. *Biomedical microdevices, 11*(3), 615–624. https://doi.org/10.1007/s10544-008-9271-7

Yin, H. S., et al. (2010). Parents' medication administration errors. *Archives of Pediatrics & Adolescent Medicine, 164*(2), 181–186. https://doi.org/10.1001/archpediatrics.2009.269.

Zein, I., et al. (2002). Fused deposition modeling of novel scaffold architectures for tissue engineering applications. *Biomaterials, 23*(4), 1169–1185. https://doi.org/10.1016/S0142-9612(01)00232-0.

Zhu, W., et al. (2016). 3D printing of functional biomaterials for tissue engineering. *Current Opinion in Biotechnology, 40*, 103–112. https://doi.org/10.1016/j.copbio.2016.03.014.

Chapter 8
3D Bioprinting Hardware

Ross Burdis and Daniel J. Kelly

Abstract Tissue engineering aims to restore complete functionality to damaged or diseased organs by implanting cell-free or cell-laden scaffolds or in vitro engineered tissues. Accurately recapitulating the complexity of native tissues and organs remains a challenge using traditional tissue engineering strategies, and hence clinical translation of such biological implants has been limited. Inadequate spatial control over the distribution of cells and bioactive cues has been identified as a significant challenge that could be overcome using additive manufacturing technologies. 3D bioprinting is an emerging technology used to biofabricate complex, spatially inhomogeneous cell-laden structures as a means to engineer tissues and organs. This chapter will discuss the current state-of-the-art 3D bioprinting technologies. The working principles of microextrusion, droplet-based, laser-assisted and lithography bioprinting methodologies are discussed. Additionally, the benefits and challenges associated with each technique are provided to help better inform the choice of 3D bioprinting hardware. Finally, the future direction and putative areas of application of the 3D bioprinting platforms are addressed within this chapter.

Keywords 3D bioprinting · Biofabrication · Microextrusion · Drop-on-demand · Laser-assisted bioprinting · Lithography

R. Burdis
Trinity Centre for Bioengineering, Trinity Biomedical Sciences Institute, Trinity College Dublin, Dublin, Ireland

Department of Mechanical and Manufacturing Engineering, School of Engineering, Trinity College Dublin, Dublin, Ireland

D. J. Kelly (✉)
Trinity Centre for Bioengineering, Trinity Biomedical Sciences Institute, Trinity College Dublin, Dublin, Ireland

Department of Mechanical and Manufacturing Engineering, School of Engineering, Trinity College Dublin, Dublin, Ireland

Department of Anatomy, Royal College of Surgeons in Ireland, Dublin, Ireland

Advanced Materials and Bioengineering Research Centre (AMBER), Royal College of Surgeons in Ireland and Trinity College Dublin, Dublin, Ireland
e-mail: kellyd9@tcd.ie

D. M. Devine (ed.), *Polymer-Based Additive Manufacturing*,
https://doi.org/10.1007/978-3-030-24532-0_8

8.1 Introduction

Additive manufacturing is a rapidly expanding field. It is now challenging the long-standing dogma of subtractive manufacturing as the principle method of fabrication in a number of fields. At present, a multitude of additive manufacturing approaches exists. Generally, three-dimensional (3D) structures formed via additive manufacturing involve drawing a two-dimensional (2D) raster scan of a pattern from a material to fabricate each layer. Sequential deposition of layers builds the part/construct in a bottom-up fashion. In order to convert a whole part into a printable sequence of lines, the object is first sliced into 2D sections. However, recent advances have taken 3D bioprinting beyond the confines of a 2D print substrate. The BIOASSEMBLYBOT® (Advanced solutions—Life Sciences) operates printheads attached to a 6-axis robotic arm permitting 3D contour printing alongside additive 3D printing. Generally, to facilitate the incorporation of viable cells, the entire bioprinting procedure takes place under sterile conditions.

The capacity to deposit material with high spatial accuracy and reproducibility via 3D printing became an attractive tool for tissue engineering and regenerative medicine applications. This interest percolated throughout various fields and ultimately the term '*bioprinting*' was borne, wherein scaffolds, cells and exogenous biological cues are printed with the intention of generating de novo tissues for different applications, from drug discovery and preclinical testing to regenerative medicine. Bioprinting is also suitable for the fabrication of complex, anatomically accurate geometries using medical images, and as such, it offers the capacity to engineer patient-specific biological implants. Although generally successful, traditional tissue engineering approaches—reliant on the adhesion of cells to the surface of scaffolds with a stochastic architecture—have been restricted to small-sized defects due to several limitations. Bioprinting aims at mitigating these disadvantages via incorporation of the cells within the construct during fabrication. The simultaneous fabrication of the structural and cellular components improves seeding efficiency and distribution homogeneity throughout the construct. Additionally, the spatial resolution achievable allows the creation of user-defined cellular microenvironments by the direct structuring of multiple biological materials. This digitally controlled fabrication offers a tighter reign over multi-scaled, multicomponent biological constructs.

8.2 Microextrusion

Collectively, microextrusion systems operate via pressure. This approach, alternatively branded pressure-assisted bioprinting (PAB) (Sklare et al. 2015; Li et al. 2016), is the deposition of a continuous material filament building up a 3D construct in a layer-by-layer fashion. The material, driven by the pressure, travels through a needle or nozzle onto a solid substrate. Accessory hardware is combined with the standard extrusion system depending on the materials' printing requirements. There have been a multitude of biomaterials constructed into 3D structures using microextrusion, including polymers, polymer–ceramic composites, living cells, proteins and biomolecules, growth factors, as well as hybrid structures (Sklare et al. 2015).

Fused deposition modelling (FDM), the most commonly recognised method of solid free-form fabrication, is a type of microextrusion and offers a means of fabricating tissue-engineered scaffolds. A key advantage of FDM is that it permits high-resolution geometric control and, in turn, the ability to design scaffold porosity. Although the technique does not permit the encapsulation of cells within the polymeric material, and hence is not considered bioprinting, it remains a valuable component in many multi-head bioprinters and is the precursor to many tissue engineering scaffolds. In this context, FDM is the extrusion of a molten biocompatible/biodegradable thermopolymer using pressure, where the material travels through a temperature-controlled nozzle and onto a solid substrate in order to fabricate a suitable tissue engineering construct. FDM is commonplace in tissue engineering as it provides significant benefit over previously ascribed methods for the 3D fabrication of bioresorbable polymers. To date, various methods of generating the necessary pressure to deposit the thermopolymer exist, including rollers and a material filament, direct pneumatic extrusion and Archimedes screw. For many bioprinting applications, the polymeric structure fabricated via FDM provides structural and mechanical rigidity alongside which the softer cell-laden bioinks are co-printed.

Cell encapsulation within a hydrogel is the most common platform for the microextrusion of a cell-laden polymer. Under pressure, the rheological properties of the hydrogel should permit its extrusion through a microscale nozzle. Aside from viscosity, physical and chemical properties, extrusion pressure, print speed and nozzle internal diameter are critical parameters that will affect print fidelity. The following section will underpin future sections, wherein the material and biological properties of unique cell-laden polymers are discussed, by introducing the principles behind the hardware exploited for pressure-driven extrusion.

8.2.1 Working Principles

In microextrusion, a bioink is extruded as single droplets or a continuous filament. There are two main configurations of microextrusion-based bioprinting: mechanical and pneumatic. Within the mechanical subdivision, there are screw-driven systems and piston-driven systems, both of which rely on mechanical actuation. Within the pneumatic branch, categorisation is based on the method of controlling the air pressure, namely, the valve system employed. There is the possibility of adding a 'floating' piston to the pneumatic barrel. The addition of the piston does not make the method of extrusion mechanical, as incorrectly labelled in several publications, as the pressure is still derived pneumatically. Consequently, we propose branding the addition of this ancillary piston to a pneumatically driven extrusion method as piston-assisted, maintaining the title 'piston dispensing' exclusively for a direct mechanically driven actuation (Fig. 8.1).

The pressure exerted in mechanical systems is generated by a stepper motor-driven screw or piston. The digital control over the motor position and actuation speed presides over material extrusion, and mechanically driven extrusion circumvents the delay associated with pneumatic systems (Murphy and Atala 2014). A direct comparison of the two mechanical systems has demonstrated that pistons typically generate more control over extrusion than screw systems, whilst screw systems are more suited to the extrusion of higher viscosity materials (Murphy and Atala 2014; Ozbolat and Hospodiuk 2016). However, pressure gradients present in mechanically driven systems are greater than in pneumatic systems, which could prove harmful to cells encapsulated within the bioink, negatively impacting viability (Murphy and Atala 2014). Subsequently, in an attempt to mitigate these

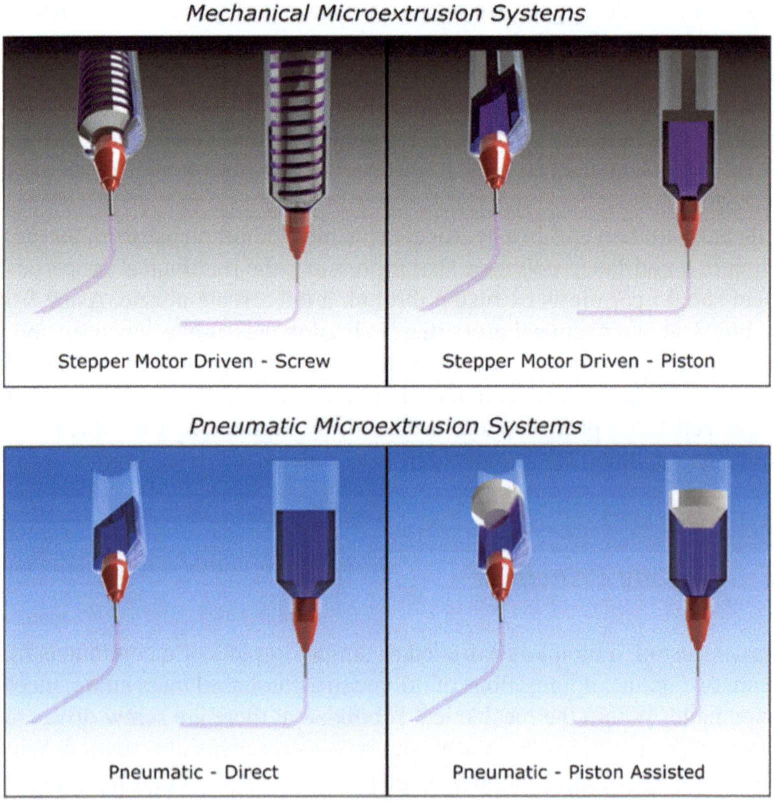

Fig. 8.1 Representation of methods for continuous microextrusion bioprinting. Mechanical systems are subdivided into screw and piston systems. In both cases, actuation is provided by a stepper motor and forces the bioink (purple) from the holding syringe and through the needle orifice. Pneumatic systems, both direct and piston assisted, use air pressure to control the extrusion of the material from the syringe

detriments, significant consideration is taken when designing the unique gearing which couples the stepper motor to the screw or piston within these systems (Malda et al. 2013; Ozbolat and Hospodiuk 2016). Despite screw-driven extrusion offering better spatial control over its mechanical counterpart (Ozbolat and Hospodiuk 2016), one potential pitfall is the exacerbation of these aforementioned pressure drops generated at the nozzle. As a result, screw systems are generally reserved for the extrusion of acellular thermoplastics (Ozbolat and Hospodiuk 2016), where pressure differentials, indicative of mechanical microextrusion, are less problematic and the mechanism design required is less bespoke.

Although this scheme generally holds true, as is common in bioprinting, hardware configurations can be case dependent. For example, Shim et al. (2012) utilised a direct piston controlled by a stepper motor for hydrogel extrusion and a direct pneumatic system for thermopolymer deposition when structuring of a multicomponent osteochondral construct. Moreover, mechanically driven pistons have proven useful in the extrusion of cell aggregates and spheroids (Jakab et al. 2007; Marga et al. 2012) an avenue of investigation within tissue engineering which is generating increased interest. Clearly, in case of specific examples, the shortcomings associated with mechanical extrusion, specifically the adverse pressure gradients can be justified. In doing so, some research groups have found this approach, and its associated benefits, advantageous in bioprinting.

In pneumatic systems, control over extrusion is via a valve. The valve mechanism could be pneumatic (Nair et al. 2009) or solenoid (Horváth et al. 2015), the former uses pressurised air to open a spring-loaded valve and in turn releases the pressurised material. The latter relies on the electromagnetic deflection of the opening mechanism; once the valve is opened, it allows pressurised air to flow through the system and force the material through the nozzle. The digital control over the opening and closing of the valve, in conjunction with the translation of the build platform and printhead, generates the spatial and temporal controls over material deposition. Generally, pneumatic microextrusion can facilitate a wider range of material viscosities at the cost of deposition control. Mechanical systems, involving smaller, more complex drive systems, inherently provide improved spatial control but with a limited force capacity, whereas the simpler drive mechanisms associated with pneumatic systems means the force is limited by the compressor used and valve capabilities (Murphy and Atala 2014).

The majority of commercial printers currently opt for a piston-assisted pneumatic system, most likely, due to its simplicity, versatility and amenability to disposables. As previously mentioned, in cases where resolution is paramount, the benefit of employing a complex mechanical piston would be significant. In such cases, one might preferentially employ a piston over a screw to benefit from the lower pressure gradients, whilst maintaining the accuracy and control of the mechanical system, avoiding the associated lag with using pressurised air. For example, a mechanically driven piston set-up has been used to print self-healing hydrogels to develop bifurcated perusable channel networks (Highley et al. 2015). The wide range of material viscosities printable using pneumatic microextrusion and the resolution, albeit low, offers the benefit of depositing large volumes of materials, high cell densities and cell aggregates with high viability, which are undeniable assets of the technology. Consequently, pneumatic approaches collectively remain more appealing to the majority of commercial and academic microextrusion bioprinters where flexibility and scalability take precedent.

8.2.2 Associated Hardware

The biologics encapsulating hydrogels (termed *bioinks*) selected for bioprinting are chosen based both on their biological properties and the physical/chemical properties which makes them amenable to 3D structuring. The mechanism of hydrogel gelation and the printing technique employed are intrinsically tied, markedly so in the case of microextrusion (Kirchmajer et al. 2015). As a result, in order to capitalise on these material properties, and improve the printability of the material, hardware modification is often necessary.

Typically, microextrusion exploits shear thinning hydrogels. This non-Newtonian behaviour is defined as the inverse relationship between shear force, resulting from the pressure applied during microextrusion, and the material viscosity (Guvendiren et al. 2012). The reduction in viscosity during extrusion allows the material to flow through the nozzle. Once ejected from the nozzle, there is a rapid removal of the shear force, which is predominantly generated as the material enters and travels through the orifice. This causes viscosity to rapidly increase back to equilibrium, and the material to hold its given form (Murphy and Atala 2014; Chimene et al. 2016). Significant work investigating the shear forces experienced by cells during printing processes and its impact on viability has been undertaken (Kong et al. 2003; Nair et al. 2009; Chung et al. 2013; Billiet et al. 2014). To this end, work has focused on the relationship between material rheology and printing pressures. Findings indicated that cell viability varies depending on both the magnitude and exposure time to shear stresses (Cheng et al. 2008), and that the translated stress from gel to cell can be minimised by altering bioink molecular weight (Kong et al. 2003). Moreover, cartridge and nozzle design have been identified as limitations to current microextrusion systems predominantly for the suboptimal shear stress applied to cells (Ozbolat et al. 2017). A simple change from a conventional straight nozzle to a conical nozzle can drastically reduce the exposure time to elevate shear stress, as well as reduce the force required for extrusion (Franco et al. 2010; Billiet et al. 2014), the implication of which could be the improved cell viability whilst using more viscous hydrogels. Interestingly, the formulation of the bioink can often result in more cell death than the printing procedure itself, for example, where the preparation of a viscous gel or pre-crosslinking mixing causes elevated stresses (Xu et al. 2005).

Alone, shear thinning gels many have a viscosity suited to producing structures with a high print fidelity. However, certain materials may require specific crosslinking mechanisms, or have physical or chemical sol–gel transitions. In these cases, modification of the traditional microextrusion system has facilitated their 3D structuring. For example, hydrogels such as agarose and gelatin will undergo a reversible gel–sol transition when heated. To structure such gels, a heated printhead and cooled build platform are typically employed (Maher et al. 2009; Duarte Campos et al. 2012; Bertassoni et al. 2014). Equally, gels that require ionotropic crosslinking such as alginate can be (1) pre-crosslinked prior to extrusion to increase viscosity (Tabriz et al. 2015), (2) crosslinked in a coaxial needle during printing (Zhang et al. 2013; Ozbolat et al. 2014) and (3) crosslinked via a vapour of an ionic solution within the build enclosure (Lee et al. 2013). Finally, chemical modification of bioinks is frequently employed, and methacrylation of polymeric chains permits the photo-induced free-radical covalent crosslinking of the construct post printing (Schuurman et al. 2013; Billiet et al. 2014).

A variant of bioprinting is bioplotting. Unlike bioprinting, during bioplotting, extrusion of the hydrogel takes place directly into a plotting medium. Consequently, although the process uses microextrusion apparatus, it is distinct from 3D bioprinting by virtue of the use of this plotting medium. This medium is typically a bath or pool of crosslinker with a density similar to that of the deposited material (Hinton et al. 2015). The temperature, plotting medium rheology and process parameters can all be varied to adjust the final outcome (Pfister et al. 2004; Ozbolat and Hospodiuk 2016). The construct stays in the bath until the entire process is completed yielding a noteworthy difference to bioprinting, as the structuring process is carried out in an aqueous medium. The buoyancy compensation associated with bioplotting allows for architectures to be fabricated without support structures and particularly suits the structuring of low-viscosity hydrogels (Pfister et al. 2004). Research areas such as vascular tissue engineering have identified the material flexibility and capacity to generate complex architectures and internal structures of bioplotting as a potential alternative to bioprinting (Stosich et al. 2009; Hinton et al. 2015).

8.2.3 Technique Evaluation

Microextrusion bioprinting has several advantages, justifying its inclusion within many commercial and research printers. In modern, commercially available systems, pneumatic extrusion is employed almost exclusively over mechanical (Ozbolat et al. 2017). Although there are notable material control and subsequent resolution benefits associated with the use of mechanical extrusion, the amenability of pneumatic systems to a wider range of gel viscosities generally favours its use. Moreover, the low cost of pneumatic microextrusion, due to the affordable and reliable implementation of its hardware, makes it an appealing bioprinting technology.

In comparison to some of the other available bioprinting modalities, microextrusion is amenable to printing large volumes of material at relatively high speed, making it attractive for scaling-up. Furthermore, the technique has demonstrated great versatility, being applied to a range of bioink materials, with the inclusion of several cellular configurations and biological enhancement (Malda et al. 2013; Ozbolat and Hospodiuk 2016). The high biological flexibility, ease of use and low levels of process-related cellular damage has ensured the inclusion of microextrusion in a multitude of bioprinting applications (Ozbolat and Hospodiuk 2016). Further motivation for its incorporation within multi-head bioprinters is its compatibility with accessory hardware, facilitating the printing of photo-, ionic- and thermal gels. This tunability makes it an ideal candidate for multi-head printers, which may include heated and cooled printheads and build substrates, in combination with photopolymerisation/crosslinking and chemical tools (Mandrycky et al. 2016). Although it offers a platform to print a wide range of hydrogels, microextrusion is limited to the deposition of cells encapsulated in a hydrogel ink. For certain applications, the need to encapsulate cells into a hydrogel carrier to facilitate printing can be considered a limitation (Ferris et al. 2013b).

Print resolution is often considered the overarching limitation of microextrusion, which limits its use for certain applications. It should be noted, however, that such a low resolution (minimum features in the order of 100–200 μm (Smith et al. 2004; Malda et al. 2013)) does facilitate the bioprinting of large tissue-engineered constructs in relatively short time periods (Visser et al. 2013). With finer resolution, the print time for such objects would be incompatible with the incorporation of cells as they would be devoid of nutrients and moisture for excessive periods of time. Another indirect benefit of a relatively coarse print resolution is that it facilitates the extrusion of relatively high cell densities with high cell viability. Furthermore, the bioprinting of multicellular spheroids is also possible using microextrusion (Jakab et al. 2007; Norotte et al. 2009; Murphy and Atala 2014; Li et al. 2016; Ozbolat and Hospodiuk 2016; Ozbolat et al. 2017).

As such, microextrusion proves an indispensable tool for many bioprinting research groups. In particular, the technique has found particular traction for engineering tissues of scale, where spatial organisation of large cell numbers or cell spheroids at high viability within clinically relevant sizes in a realistic time frame is required. Additionally, the flexibility of the hardware makes a myriad of materials compatible with microextrusion.

8.3 Droplet-Based Bioprinting

Droplet-based bioprinting (DBB) is recognised for providing, via well-defined positioning of cells and other biological materials, the generation of spatially heterocellular constructs (Gudapati et al. 2016). The technology, originating from reprographic ink-jet printing, was first introduced as a biological tool in 1988. In this pioneering study, Klebe (1988) spatially organised cells to 25 μm precision based on the deposition of a cell adhesion protein (fibronectin) via a modified conventional Hewlett Packard thermal ink-jet printer. Further modification of conventional ink-jet printheads in 2003 facilitated the deposition of viable large mammalian cells (Wilson and Boland 2003). Early reports of the application of this cell-friendly ink-jet droplet deposition in a bioprinting context were its application with an alginate hydrogel, wherein crosslinking solution was spatially deposited to build up a 3D structure suitable for cell encapsulation (Boland et al. 2007). To date, several modalities, including acoustic, electrohydrodynamic and ink-jet DBB, have been developed. The following section will describe the various working mechanisms behind droplet dispensing and the unique hardware and technology features that have been developed for this purpose.

8.3.1 Ink-Jet Bioprinting

Ink-jet bioprinting represents the earliest version of DBB, derived from the modification of commercially available desktop machines. The non-contact method is capable of reproducibly depositing precise volumes, as low as picolitres (Li et al. 2016), of biological material onto a solid substrate or hydrogel bio-paper (Xu et al. 2005; Li et al. 2016). Successful use of ink-jet as a means of patterning proteins and cells (Wilson and Boland 2003; Roth et al. 2004) has led to suggestion that the deposition of biomaterials, cells, growth factors and other important components with an appropriate temporal, spatial and volumetric control for organ printing is achievable (Boland et al. 2006). The common methods of generating the droplets in ink-jet bioprinting are described below.

8.3.1.1 Thermal Drop-on-Demand

The earliest incarnations of ink-jet bioprinting were thermal drop-on-demand systems. During thermal ink-jet printing, the bioink is locally heated when a voltage pulse is applied to a thermal actuator (a thin film resistor). This rapid heating generates a vapour bubble, the expedited expansion and collapse of which generates a pressure pulse within the fluid which is sufficient to overcome the liquid surface tension at the narrow nozzle opening, and the bioink is ejected (Gudapati et al. 2016).

In order to vaporise a micrometre-deep layer of liquid in contact with the thin film resistor, local temperatures can reach 250–350 °C. Despite concerns that such high temperatures could damage or lyse cells during printing, it has been demonstrated that high cell viability (>90%) can be achieved using thermal drop-on-demand (Xu et al. 2005). It was found that the microsecond timescale involved in droplet ejection was insufficient to allow the thermal energy generated to diffuse into the bulk medium. Consequently, the temperature change experienced by the cells within the fluid chamber is between 4 and 10 °C (Xu et al. 2005). The authors further demonstrated that decreasing the volume of the cells via osmotic shrinkage can provide thermal protection and aid in preventing nozzle blockage (Xu et al. 2005). Aside from thermal and shear concerns, others have identified a lack of regularity in droplet trajectory and size, as well as an intermittent printing caused by nozzle clogging (Jana and Lerman 2015). Due to their strong interdependence between the phenotype of mammalian and their environment stimulus, studies have presented not only viability data but also proof that the capacity for lineage commitment has not been lost. For example, the stability of human chondrocytes printed into an ex vivo chondral defect has been demonstrated by consistent gene expression and biochemical analysis (Cui et al. 2012b). Similar results have been reported for cardiac (Xu et al. 2009), vascular (Cui and Boland 2009) and neural (Xu et al. 2006) cells.

Thermal drop-on-demand systems have also been used to deliver genetic cargo to cells. It has been demonstrated that pores with an estimated size of 105 Å were generated during the printing process and healed within a 2-h time period. The creation of this transient membrane damage could be used for targeted gene delivery, and with a 30% transfection efficiency was achieved without significant optimisation. This simultaneous approach of transfection and deposition avoided the need for expensive and time-consuming procedures and equipment (Cui et al. 2010).

8.3.1.2 Piezoelectric Drop-on-Demand

Piezoelectric drop-on-demand systems function by the application of a voltage to a piezo element in the pressure channel of the printer head. This electrical voltage is converted into a mechanical deformation (which can be categorised into squeeze, bend, push and shear), which generates a pressure wave (Wijshoff 2010). This pressure waveforms a droplet at the nozzle and its expulsion. Piezoelectric ink-jet (PIJ) systems have been described as being more problematic to modify and maintain for bioprinting when compared to thermal systems with reduced biocompatibility (Cui et al. 2012a). Despite this, the piezoelectric approach is devoid of excessive heating; consequently, it has been suggested to be more suitable to tissue engineering applications than its thermal counterpart (Nakamura et al. 2005; Li et al. 2016). However, by nature of their working mechanism, the frequencies at which PIJ systems operate have been identified as overlapping with the well-documented sonification frequencies (15–25 kHz) known to damage cell membranes (Cui et al. 2012a). Despite this, studies have shown comparable cell survival rates (94%) in PIJ systems to thermal systems (Saunders et al. 2008). However, within this study the frequency employed was 10 kHz, and the investigation focused on droplet velocity and mass in relation to applied voltage, and its overall influence on cell behaviour (Saunders et al. 2008). Generally, the frequencies of pulses generated within PIJ for bioprinting applications follow suit and are kept below the detrimental range outlined above, ranging from 1 to 10 kHz (Saunders et al. 2004, 2008; Lorber et al. 2014).

Sedimentation of cells, resulting in the inconsistent cell number dispensing and nozzle clogging, is a commonly reported shortcoming experience with PIJ printheads (Yamaguchi et al. 2012; Lorber et al. 2014). Biocompatible modifications to the system, such as agitation, the addition of surfactants to the bioink (Parsa et al. 2010) or mitigation of cell sedimentation through neutral buoyancy (Chahal et al. 2012), have been shown to improve droplet consistency.

8.3.1.3 Microvalve Bioprinting

Nowadays, ink-jet technology, in terms of bioprinting, is custom designed to handle biological materials. Valve-based ink-jet printing, considered one of the gentlest cell printing methods, is an example of a custom designed addition to droplet dispensing (Faulkner-Jones et al. 2013). Microvalve ink-jet (MIJ) printing is the only ink-jet method requiring pneumatic pressure. The general working principle for the printing mechanism is as follows. The bioink is reserved in a pressurised barrel, in a multivalve system each container is independently pressurised (Lee et al. 2009). The nozzle orifice is gated by an electromechanical valve; this solenoid valve is opened when the valve coil generates an electromagnetic field. The field pulls the occluding plunger up from the nozzle opening and the bioink is ejected if the back pressure is sufficiently large to overcome the surface tension (Gudapati et al. 2016). Tuning the frequency of the valve-opening and magnitude of the back pressure, MIJ can operate both as a drop-on-demand and as a continuous method of material extrusion, as a miniaturised version of microextrusion.

Of all the methods of discretised cell droplet methods, MIJ generates the largest droplet size which impacts its single-cell control and spatial resolution (Tasoglu and Demirci 2013). Hence, where fine resolution is required, other drop-on-demand methods are often preferred (Ozbolat et al. 2017). Despite this, MIJ has several benefits which have ensured its inclusion in a wide range of bioprinting applications including cell and biomaterial printing (Moon et al. 2010; Faulkner-Jones et al. 2013, 2015; Ferris et al. 2013a, b; Xu et al. 2013). For example, as a pneumatically driven system, MIJ is capable of dispensing liquids with higher viscosities than other ink-jet systems (Lee et al. 2009). Equally, improved cell viability is associated with MIJ over PIJ as a lower pneumatic operating pressure required (Lee et al. 2009). Satellite droplets, small ancillary droplets that follow in the wake of the primary droplet, have diminished the printing precision of many conventional ink-jet printers. This feature carries over when desktop printers are converted for biological application. In contrast to TIJ and PIJ, operators of microvalve systems have not reported such repercussions (Ozbolat et al. 2017).

8.3.2 Acoustic Bioprinting

In acoustic bioprinting, ejection of a droplet occurs from a gentle acoustic field. The acoustic waves are sourced from a piezoelectric substrate with circular interdigitated gold rings. The acoustic focal point, situated at the air–bioink interface adjacent to the microfluidic channel exit(s), is where the converging acoustic waves interfere constructively. These exits, generally arranged singularly or in a two-dimensional array, provide the necessary surface tension to hold the bioink in situ; should the force exerted by the acoustic wave be sufficient to overcome said surface tension a droplet will be expelled (Demirci 2006). This open-pool, nozzle-free ejection system completely avoids the generation of shear stresses commonly imposed on cells extruded using nozzle systems (Demirci and Montesano 2007). Acoustic bioprinting emerged as a significant field, aiming to avoid the aforementioned mechanical and thermal stressors, poor droplet directionality and uniformity, possible nozzle occlusion from high cell densities and the inversely related encapsulation inhomogeneity resulting from reducing the cell density, associated with the aforementioned ink-jet methods (Demirci and Montesano 2007).

The nozzle-less system has provided a gentler means of depositing cells. Droplet size has been shown to be a function of both actuation frequency and wavelength (Demirci 2006). In bioprinting, the wavelength is tailored to provide the most harmless ejection of cells. To do so, the wavelength chosen should be comparable to, or much lower than the cell diameter. This constraint becomes increasingly imperative at higher frequencies (Demirci and Montesano 2007). There have only been a relatively small number of studies exploring the amenability of the acoustic ejection methods to bioprinting applications (Gudapati et al. 2016). These studies have demonstrated that droplets of reproducible size, containing viable (>89%) cells, can be deposited at rates of 1–10,000 droplets per second (Demirci and Montesano 2007).

Despite this success, the approach is not immune to drawbacks. For instance, it has been associated with undesirable disturbances, introduced during movement, resulting in the loss of droplet expulsion (Gudapati et al. 2016). This loss of ejection fundamentally limits the application of acoustic systems in bioprinting unless properly addressed due to the need to translate the printhead or substrate during the 3D fabrication process (Gudapati et al. 2016; Ozbolat et al. 2017). Other drop-on-demand techniques have not reported similar sensitivity and intermittence during printing. Equally, acoustic systems are generally confined to the deposition of low-viscosity materials, which may limit its universal employment (Gudapati et al. 2016), although bioinks of varying viscosities have been printed using acoustic wave deposition (Demirci 2006). The upper dynamic viscosity quoted within these publications as printable, 18 mPa s, is not comparable to commonly used hydrogel bioinks. For example, commonly used bioink (such as alginate) has a dynamic viscosity that ranges between 90 and 30,000 mPa s (He et al. 2016). This suggests that acoustic ejection is only suited to the dispensing of liquid cell suspensions. Up until now, acoustic bioprinting systems have had limited exposure. However, their hypothetical application is targeted at biosensors, high-throughput screening and microarrays (Demirci and Montesano 2007). The commonality between all these avenues of investigation is high resolution, requiring frequent picolitre dispensing.

8.3.3 Electrohydrodynamic Jetting

A pragmatic limitation to ink-jet printing is the achievable volumes of liquid that can be ejected, and as a result the minimum feature size. By decreasing the internal diameter of the nozzle, smaller droplets can be dispensed. However, this too becomes impractical for smaller nozzles because the pressure required to overcome the capillary forces becomes extremely high (Onses et al. 2015). Electrohydrodynamic

(EHD) jetting exploits a different mechanism of expulsion to aforementioned ink-jet methods. The method of droplet dispensing is analogous to electrospraying/spinning and relies on 'pulling' the material from the nozzle tip rather than 'pushing' it. By virtue of an electric field, the liquid is pulled from the orifice and obviates the need for excessively high pressures, allowing much lower nozzle diameters and subsequently a finer resolution.

A typical EHD jetting configuration uses back pressure to deliver the ink to the nozzle tip, which can be as small as ~100 nm in diameter. The back pressure is sufficient to cause a pendant meniscus at the nozzle opening, but should not surpass the liquids surface tension. The application of an electric field leads to the accrual of mobile ions at the ink surface, and in close proximity to one another, the Coulombic repulsion of the free ions causes the meniscus to deform into a Taylor cone. The ejection of droplets from the tip of the Taylor cone occurs when the electrical field is sufficiently high and the surface tension is surpassed by the electrostatic stresses. The underpinning physics of the technique permits its superior resolution in comparison with ink-jet methods, specifically lower nozzle diameters, smaller droplet size and droplet focusing (minimised lateral variation) by the distribution of electric field lines (Onses et al. 2015). For a complete review of EHD jetting, including the central factors in jet formation and application outside of bioprinting, the authors direct readers to Onses et al. (2015). In brief, the jetting phenomenon proceeds, generally, as follows. Low voltage and low back pressure result in material dripping, increasing the back pressure, or slightly increasing the voltage will cause jetting but the resolution will be poor and the droplets excessively large. This jetting is a result of the repeated formation of a Taylor cone and its relaxation and is assisted by gravity. At low pressures, increasing the voltage will form proper cone jetting, where the continuous stream of liquid is ejected from the pinnacle of the Taylor cone that is ever-present at the tip of the nozzle. Excessive voltage causes complex jetting, such as tilted or multiple jets. A pulsating jet mode is possible, where micro- or nano-sized individual droplets are ejected. This can be achieved either using the natural frequency of the jetting or a pulsed electrical field (Onses et al. 2015).

The technical features of EHD jetting make it the most competent jetting method for the deposition of small droplets with concentrated cell suspensions. The capacity to process concentrated suspensions from nozzles that are able to generate droplets a few micrometres in size without the risk of clogging, whilst maintaining cell viability and homogeneity, delineates this method from others (Jayasinghe et al. 2006). Intuitively, whilst relatively course structures are created by ink-jet printing, smaller features in applications such as DNA and protein microarrays, spatial positioning of cells and biomaterials to influence cell–substrate interactions for tissue engineering, drug assays and fundamental cell behaviour studies have all demonstrated the use of EDH jetting (Onses et al. 2015). The process, although conceptually simple, can be complex to set up and sensitive to ink viscosity, surface tension, electrical conductivity and evaporation rate. Additionally, environmental factors such as humidity and equipment features like nozzle and substrate surface properties can also have a significant influence and are often overlooked (Onses et al. 2015).

8.3.4 Technique Evaluation

There are several ascribed methods for droplet dispensing. Some of the aforementioned techniques offer many benefits, such as affordability, ease of use, high resolution and fast print speeds. However, generally DBB offers a limited viscosity range for dispensable liquids (in droplet form). This is particularly detrimental when attempting to build in 3D as the discrete liquid droplets must form a functional structure without losing spatial organisation. This process can be assisted by utilising materials with unique gelation properties (crosslinked via chemical, pH or UV mechanisms); despite this, it remains extremely challenging. As reviewed elsewhere, this not only increases print time, but chemical modification of biomaterials to generate the appropriate gelation kinetics can influence chemical and material properties as well as alter cell behaviour and viability (Murphy and Atala 2014).

The capacity to print multiple materials or cell types simultaneously is possible by combining multiple ejectors or printheads in an array format. Because of DBB systems are suited to this multifaceted approach, they are highly adept at generating highly complex heterocellular cellular constructs. Achieving a similar spatial control over multiple biologics using extrusion- or laser-based bioprinting would be challenging (Gudapati et al. 2016). Frequently, for scaled-up tissue engineering applications, where large porous constructs are necessary, DBB is offered in combination with another printhead. Typically, a microextrusion system is better suited to deposition of high-viscosity, high-density materials which are prone to clog a DBB printhead. Alone, DBB is employed for the creation of microarrays for high-throughput screening applications, where small, rapidly deposited volumes, in a highly reproducible fashion, are required. A wide range of bioink materials used in microextrusion printing, such as cell aggregates, micro-carriers and high-viscosity gels, are not compatible with DBB. As a non-contact printing technique, DBB methods from have been suggested for in situ bioprinting where biologics are applied directly on or into injured sites (Gudapati et al. 2016), one avenue of clinical translation that microextrusion systems will be unable to exploit. Ultimately, DBB methods are not yet capable of fabricating clinically relevant human organ replacements, although they are proving effective in drug discovery and development, disease modelling and multicellular–microenvironment studies (Gudapati et al. 2016).

8.4 Laser-Assisted Bioprinting

An alternative nozzle-free bioprinting approach to acoustic systems is laser-based direct-writing techniques. Like acoustic bioprinting, the technique is not susceptible to clogging. However, it boasts compatibility with a wider range of printable viscosities than its nozzle-free counterpart (Malda et al. 2013), and higher viscosities than ink-jet and in some cases extrusion-based bioprinters (Dababneh and Ozbolat 2014). Given that systems operate with the high-precision characteristics of a laser, they permit the precise, user-defined positioning of a variety of cell types and biomaterials. In this fashion, the technique is suggested to be capable or reproducing the smallest features of an organ or for the study of the highly sensitised cellular responses to specific printed microenvironments (Schiele et al. 2010; Dababneh and Ozbolat 2014). To date, the field has focused on maintaining cell viability and functionality whilst patterning in a spatially defined manner (Schiele et al. 2010). This has not been without its challenges, with high print resolutions often associated with complications and the need for bioinks with rapid gelation kinetics in order to produce 3D structures. Additionally, the fine resolution and low flow rate of amenable bioinks make the fabrication of clinically relevant constructs extremely challenging, hindering the technologies widespread uptake in tissue engineering applications (Malda et al. 2013). Herein, an outline of the most prolific laser-assisted bioprinting techniques is provided, with a description of their working principles and unique features. A complete review of all cell printing utilising laser-based systems can be found elsewhere (Schiele et al. 2010).

8.4.1 Volatilisation Methods

Four main laser-assisted bioprinting methods share a distinct homology. They are laser-induced forward transfer (LIFT), absorbing film-assisted laser-induced forward transfer (AFA-LIFT), biological laser processing (BioLP) and matrix-assisted pulsed laser evaporation direct writing (MAPLE DW). The overarching methodology that connects these modalities is the propulsion of cells to the receiving substrate via the rapid volatilisation of a support layer.

8.4.1.1 Laser-Induced Forward Transfer (LIFT)

The printer set-up involves two, coplanar glass slides and a laser. The lower of the former is referred to as the collector slide. Typically, the slide is coated with a medium, such as a thin layer of hydrogel or cell culture media, used to provide a hydrated environment and prevent cells drying out during prolonged processing times. Additionally, this supplementary substrate layer can offer cushioning but is not essential for cell viability, and in the case of hydrogel collection plates assist in retaining cell position. The donor slide is positioned between the collector slide and the high-powered pulsed laser. This slide, typically glass, is coated on the underside with a thin layer, 55–60 nm in thickness, of light-absorbing metal (such as gold) or another biocompatible laser-absorbing material. Below this material, the cells of interest are uniformly suspended and spread across the surface in either a culture medium or suitable hydrogel. This entire donor slide configuration is often referred to as a print ribbon. When the laser is focused through the donor slide, the local energy absorbed by the coating material causes its rapid evaporation. The local reaction generates a large gas pressure which propels the subjacent cell compound from the print ribbon, towards the collection slide and the chosen receiving substrate (Koch et al. 2010; Ovsianikov et al. 2010; Schiele et al. 2010).

Control over the laser-transferred droplet size is via the laser pulse energy, the gap distance between the print ribbon and collection substrate (nominally 300–500 μm apart), the spot size of the focused laser (~45 μm) and the thickness of the energy-absorbing compound (55–60 nm) as well as the cell carrier layer (~50 μm). With the given parameters, droplets between 80 and 140 μm are achievable at a rate of 1200 droplets per minute (Koch et al. 2010; Ovsianikov et al. 2010). Furthermore, human bone marrow-derived mesenchymal stem cells (hMSCs) have been printed using the aforementioned method without detriment to their viability, proliferation and phenotype (Koch et al. 2010).

8.4.1.2 Matrix-Assisted Pulsed Laser Evaporation Direct Writing (MAPLE DW)

A single matrix is used as both the sacrificial and transfer materials, making MAPLE DW unique in comparison to other forms of laser direct writing. The ribbon of coating material, within which the biologic to be printed is suspended, is typically a polymer of hydrogel such as Martigel™, cell culture media, a 5% glycerol solution or gelatin (Barron et al. 2004a, b; Dias et al. 2014). The transfer of sufficient laser energy at the interface between the quartz slide and the biolayer achieves droplet dispensing. The water within the aqueous biolayer absorbs the focused laser pulse and is rapidly heated causing the vaporisation of some of the material and the formation of a vapour pocket. The rapid expansion of the vapour pocket results in the propulsion of the subjacent biological suspension towards the collection plate. Like LIFT, varying spot size of the laser, laser energy, biomaterial properties and thickness will influence the area of biomaterial effected by the laser and consequently the volume of biomaterial ejected (Barron et al. 2004a, b).

Generally, the laser utilised in MAPLE DW is low powered and UV or near-UV, to help prevent excessive laser penetration and damage to the suspended cells or proteins (Schiele et al. 2010). However, concerns exist that large amounts (99%) of light within the UV wavelength (193 nm) are transmissible through liquids. Furthermore, the local vaporisation of the biolayer could transfer detrimental heat energy (Schiele et al. 2010).

8.4.1.3 Absorbing Film-Assisted Laser-Induced Forward Transfer (AFA-LIFT) and Biological Laser Processing (BioLP™)

The underpinning principles of LIFT apply to AFA-LIFT, whereby an energy-absorbent material is vaporised upon exposure to a focused pulse laser, which in turn ejects a droplet of cell-laden material from the surface. However, AFA-LIFT minimises the potential photonic damage to the biological material and diminishes the quantity of transferred impurity during the printing process. To do so, quartz is used instead of glass in the donor slide, which has a higher absorption coefficient, and the thickness of the print ribbon is thicker than the penetration depth of the laser. Additionally, the quartz plate is coated with silver and the layer thickness is approximately doubled (100 nm). The cell compound/solution is also thicker in AFA-LIFT, between 140 and 160 μm in thickness. Finally, the distance between the collector and donor plate is increased to 600 μm (Hopp et al. 2005).

Barron and colleagues (2004a, b, c, 2005) developed BioLP™, as a non-contact, orifice-free, forward transfer technique capable of rapidly depositing biological material, cells and functional proteins, in volumes of femtolitre (fL) to nanolitre (nL) scale with a spatial accuracy of below 5 μm. The technique, a version of AFA-LIFT, requires an analogous intermediate absorption layer placed between the laser and the biomaterial, which mitigates direct interaction between the laser energy and the biological material. The technique boasts improved transfer reproducibility compared to MAPLE DW, as application of the intermediate material normalises the laser interaction. Like AFA-LIFT, the laser target is an optically transparent quartz slide which is coated with a metal or metal oxide (such as Au, Ti, TiO_2) around 100 nm thick. Onto the sacrificial metal layer, between 10 and 100 μm of biomaterial is coated. The laser is focused on the quartz–absorption layer interface where the energy is absorbed. Unlike other methods (MAPLE DW), the laser–material interaction in BioLP™ results in the controlled ejection of a stream of biomaterial from the donor plate. This jetting phenomenon has been observed using high-speed imaging and indicates that the absorption metal acts as an energy conversion material (Barron et al. 2005). This conduction occurs via absorption of the radiant laser energy and its conduction into heat. The heat travels radially to the interface between the metal/metal oxide layer and the biolayer. Depending on the absorption material employed, the temperature gain can fluctuate from a few hundred Kelvin to over a thousand. In a model where the interfacial temperature gain was 1400 K, the thermal penetration into the aqueous biolayer was deemed negligible. The maximum thermal penetration into the aqueous material within this model was found to be 1.2 μm, although the model was an upper limit version and did not account for thermal cooling and vaporisation which would further reduce the thermal penetration (Barron et al. 2004a, b). An estimation of the gain in temperature above ambient within the biolayer is between 2% and 10%, although it is much more likely that far less of the material is influenced (Barron et al. 2004a, b). In addition, the damage incurred to the biological material via incidence radiation from the laser source is eliminated as the laser penetration is significantly lower than the total layer thickness of the absorption material (Barron et al. 2005). Calculation of the percentage of laser energy absorbed in the sacrificial interlayer of the print ribbon has revealed that greater than 99.9% of the incident laser energy is absorbed prior to interaction with the biomaterial layer (Barron et al. 2004a, b). BioLP™ has been used to successfully deposit proteins and mammalian cells in two and three dimensions (Barron et al. 2004a, c, 2005).

The introduction of thicker metallic laser-absorbing layers has mitigated the risk of rupture during vaporisation and hence, reduced the thermal and mechanical shock transferred to the biological material. However, the transfer of metallic residues or debris to the final construct remains a concern (Murphy and Atala 2014). Attempts to avoid such contamination have included the use of thick photodecomposing polymers, and even the invent of laser backward transfer (Kattamis et al. 2007; Duocastella et al. 2010). The former, named blister-actuated LIFT, negates the risk of debris as the propulsion of the biomaterial forms via the rapid deformation of the absorption polymer rather than its ablation (Guillemot et al. 2011).

8.4.2 Laser-Guided Methods

In laser-guided direct writing (LG DW), the laser beam is weakly focused into a bath of cells suspended in culture medium. The bean continuously captures cells as they are carried, through the media, via the natural convection within the fluid. The gradient force exerted by the beam pulls proximal cells into the centre of its path, the region of highest intensity, where an axial pushing force propels them towards the target substrate (Odde and Renn 2000). This method is slow and works over a very small distance, depositing ~2.5 cells/min, with a working distance of ~300 μm. However, the beam divergence and loss of optical forces can be improved by the use of a hollow optical fibre, which is capable of transporting cells up to 7 mm (Odde and Renn 2000). The rate of deposition could be increased by increasing the cell suspension or augmenting the laser power.

A challenging limitation is the techniques' reliance on the cells' refractive index. The axial pushing force exerted on a cell is proportionally linked to its refractive index. As a result, a 4% decrease in refractive index between cell types will cause a 25-fold decrease in the pushing force. The addition of optical handles, such as polystyrene particles, is one method of increasing the cells' relative refractive index and improving the system performance (Nahmias et al. 2005). However, in order to attach these beads to cells, they must be modified and incubated with cells prior to printing, which may be arduous. The beads are released over a period of time post-processing; however, they may elicit an undesirable effect on the cells or impede cell–microenvironment interactions. In summation, although the technique has demonstrated high resolution and spatial accuracy, and shown no initial detriment to cell viability and functionality, the deposition time limits the application of LG DW to investigation of cell–cell and cell–ECM interactions rather than scaled-up tissue engineering (Pirlo et al. 2006).

8.4.3 Technique Evaluation

Laser-assisted bioprinting can print high cell densities with a microscale resolution at high speed (Guillotin et al. 2010). The additional benefits of being a nozzle-free system circumvent cell or material clogging which can prove problematic in other bioprinting modalities. Despite this, ribbon preparation, required for the printing of every cell or biomaterial type, is time-consuming and could become impractical if co-deposition is required (Murphy and Atala 2014). The nature of ribbon cell coating means gravitational and random settling of cells in the precursor solutions largely dictates the accuracy and deposition reproducibility of cell-laden droplets. Furthermore, long fabrication times associated with the gelation kinetics, which frequently require a photocrosslinkable bioink, necessary to maintain 3D structural fidelity is a considerable limitation. Heat generated from the energy of the laser, although refuted by some, may cause damage to cells and prevent their ability to communicate and aggregate within the final construct (Dababneh and Ozbolat 2014). Additionally, the speed of deposition is very high (122 m/s), and this has been associated with cell death in regions of the droplet with the highest velocity (Hopp et al. 2005). Finally, the high cost of the printer equipment limits its application in many tissue engineering applications (Murphy and Atala 2014). Consequently, laser-assisted bioprinting has gained the most traction in managing tissue complexity, via the creation of in vitro microenvironments, and the manipulation of these artificial niches. Translation to the creation of macroscopic tissue-engineered constructs would require a laser-based technology to work synergistically with other manufacturing processes (Guillemot et al. 2011).

8.5 Lithography Bioprinting

Stereolithography is a solid free-form fabrication technique. The 3D structuring process utilises photopolymerisation to fabricate parts, via the solidification of a liquid resin, from a computer-aided design (CAD) file. Traditionally, the structuring of a light-sensitive polymer has been devoid of cells due to the use of bio-incompatible epoxies and acrylic precursor materials, in conjunction with cytotoxic photocuring agents (Li et al. 2016). The application of stereolithography in tissue engineering, until recently, was limited to the fabrication of cell-free structural scaffold using polymers, ceramics and composite materials (Melchels et al. 2010; Wang et al. 2015). A complete review of the application stereolithography in biomedical applications is beyond the remit of this chapter and is reviewed elsewhere (Melchels et al. 2010). Instead, along with the fundamentals of stereolithography, the recent technical developments that have given rise to the photopolymerised structuring of cell-laden hydrogels are reviewed.

8.5.1 *Working Principles*

The process of building a 3D object via stereolithography operates via one of the two methods, the first uses a computer-controlled laser beam to 'draw' the pattern for each layer onto the photocurable resin and the second uses a digital light projector to illuminate the entire pattern for each layer in one instance. In both versions, the light source is typically within the ultraviolet (UV) wavelength of the light spectrum and the pattern solidifies/gels to a specific depth.

8.5.2 Photopolymerisation of Cell-Laden Hydrogels

Of the two methods previously outlined, the digital light projection (DLP) of an entire layer using a digital micromirror device (DMD) is more commonplace in bioprinting. In short, a DMD is constituted of a rectangular array of millions of mirrors which can be independently rotated to an on or off position. Through this micro-actuation, the light projected onto the DMD can be selectively patterned into a pixelated image. The entire image for one layer is projected, via the on mirrors, onto the resin and the material cured in a spatially defined manner. As the pattern is made of pixels, the surface finish of a construct fabricated via DLP is typically rougher than one fabricated using a 2D raster scanning style.

The functionalisation of commonly used bioinks, permitting their covalent photo-crosslinking, has triggered their application in DLP bioprinting (Melchels et al. 2010). Two commonly used hydrogels are methacrylated gelatin (GelMA) (Gauvin et al. 2012; Grogan et al. 2013; Wang et al. 2015) and poly(ethylene glycol) diacrylate (PEGDA) (Lin et al. 2013; Huang et al. 2014). A high resolution is achievable using photolithographic systems, with the creation of features from tens to hundreds of microns frequently reported (Gauvin et al. 2012; Grogan et al. 2013; Huang et al. 2014; Mandrycky et al. 2016). Asides from the resolution, the technique permits the fabrication of complex geometries that would otherwise be beyond the capacity of other additive biofabrication techniques. With the development of novel bioinks and less cytotoxic photoinitiators, the capacity to encapsulate cells within a hydrogel via photopolymerisation is becoming increasingly feasible, with cell viability being reviewed at >85% (Wang et al. 2015). Although its application in tissue engineering is in its infancy, due to the low mechanical strength of the constructs fabricated using DLP bioprinting, it is best suited to the development of in vitro models and soft tissue applications.

For most bioprinting modalities, increased resolution would sacrifice print speed. Typically, as we have seen in previous sections, this inverse relationship would limit the potential application of high-resolution systems in scaled-up disciplines. However, the working mechanism of DLP means that regardless of the complexity of each layer, they can be printed at the same speed at consistently high resolution. Additionally, DLP is demarcated from other bioprinting modalities, as its potential for printing complex geometries is far superior. As the entire layer is being printed at the same time, DLP systems only require actuation in one direction. The build stage movement in only the *Z*-plane significantly simplifies the printing process and diminishes undesirable print artefacts caused by movement during dispending. Additionally, DLP systems are not expensive like the aforementioned laser-assisted bioprinters. Normally, DLP bioprinters are either designed and made in-house or open-source commercial systems. However, novel reapplication of existing technologies has provided proof that an inexpensive, simple DPL system for bioprinting is obtainable. In this example, a commercially available beam projector was employed as the light source to 3D structure a polyethylene glycol (PEG)–GelMA copolymer (Wang et al. 2015).

8.6 Summary

To date, a myriad of 3D bioprinting technologies have been developed. The flexibility of microextrusion bioprinting has ensured its widespread inclusion within commercial bioprinters and popularity in academic research (Ozbolat et al. 2017). Droplet-based and lithographic bioprinting, although less commercialised, have demonstrated strengths that may make them an attractive alternative. With continued investigation, the amalgamation of several bioprinting technologies into more complex multi-printhead systems will be realised. Moreover, the realisation of non-planar 3D bioprinting will improve the biofabrication of complex anatomical structures and could result in improved clinical translation of large bioprinted constructs. Aside from the biofabrication of engineered tissues and organs, 3D bioprinting technologies have been adopted by pharmaceutical and cancer research, where the high-resolution automated deposition of biologics is advantageous for high-throughput screening. Although each bioprinting technology suffers from limitations, they also offer unique strengths. As such, they have generated considerable hype, which has yet to result in a significant clinical contribution. Despite this, we recognise the significant progress made within the field of bioprinting and predict that the advancement of bioprinting technologies, in tandem with the development of novel bioinks, will ensure 3D bioprinting's inclusion in emerging state-of-the-art medical treatments.

References

Barron, J. A., et al. (2004a). Application of laser printing to mammalian cells. *Thin Solid Films, 453–454*, 383–387. https://doi.org/10.1016/J.TSF.2003.11.161.

Barron, J. A., et al. (2004b). Biological laser printing: A novel technique for creating heterogeneous 3-dimensional cell patterns. *Biomedical Microdevices, 6*(2), 139–147. https://doi.org/10.1023/B:BMMD.0000031751.67267.9f.

Barron, J. A., Spargo, B. J., & Ringeisen, B. R. (2004c). Biological laser printing of three dimensional cellular structures. *Applied Physics A: Materials Science and Processing, 79*, 1027–1030. https://doi.org/10.1007/s00339-004-2620-3.

Barron, J. A., et al. (2005). Printing of protein microarrays via a capillary-free fluid jetting mechanism. *Proteomics, 5*(16), 4138–4144. https://doi.org/10.1002/pmic.200401294.

Bertassoni, L. E., et al. (2014). Hydrogel bioprinted microchannel networks for vascularization of tissue engineering constructs. *Lab on a Chip, 14*(13), 2202–2211. https://doi.org/10.1039/C4LC00030G.

Billiet, T., et al. (2014). The 3D printing of gelatin methacrylamide cell-laden tissue-engineered constructs with high cell viability. *Biomaterials, 35*(1), 49–62. https://doi.org/10.1016/j.biomaterials.2013.09.078.

Boland, T., et al. (2006). Application of inkjet printing to tissue engineering. *Biotechnology Journal, 1*(9), 910–917. https://doi.org/10.1002/biot.200600081.

Boland, T., et al. (2007). Drop-on-demand printing of cells and materials for designer tissue constructs. *Materials Science and Engineering C, 27*(3), 372–376. https://doi.org/10.1016/j.msec.2006.05.047.

Chahal, D., Ahmadi, A., & Cheung, K. C. (2012). Improving piezoelectric cell printing accuracy and reliability through neutral buoyancy of suspensions. *Biotechnology and Bioengineering, 109*(11), 2932–2940. https://doi.org/10.1002/bit.24562.

Cheng, J., et al. (2008). Rheological properties of cell-hydrogel composites extruding through small-diameter tips. *Journal of Manufacturing Science and Engineering, 130*(2), 021014. https://doi.org/10.1115/1.2896215.

Chimene, D., et al. (2016). Advanced bioinks for 3D printing: A materials science perspective. *Annals of Biomedical Engineering, 44*(6), 2090–2102. https://doi.org/10.1007/s10439-016-1638-y.

Chung, J. H. Y., et al. (2013). Bio-ink properties and printability for extrusion printing living cells. *Biomaterials Science, 1*(7), 763–773. https://doi.org/10.1039/c3bm00012e.

Cui, X., & Boland, T. (2009). Human microvasculature fabrication using thermal inkjet printing technology. *Biomaterials, 30*(31), 6221–6227. https://doi.org/10.1016/j.biomaterials.2009.07.056.

Cui, X., et al. (2010). Cell damage evaluation of thermal inkjet printed Chinese hamster ovary cells. *Biotechnology and Bioengineering, 106*(6), 963–969. https://doi.org/10.1002/bit.22762.

Cui, X., Boland, T., et al. (2012a). Thermal inkjet printing in tissue engineering and regenerative medicine. *Recent Patents on Drug Delivery & Formulation, 6*(2), 149–155. https://doi.org/10.2174/187221112800672949.

Cui, X., Breitenkamp, K., et al. (2012b). Direct human cartilage repair using three-dimensional bioprinting technology. *Tissue Engineering Part A, 18*(11–12), 1304–1312. https://doi.org/10.1089/ten.tea.2011.0543.

Dababneh, A. B., & Ozbolat, I. T. (2014). Bioprinting technology: A current state-of-the-art review. *Journal of Manufacturing Science and Engineering, 136*(6), 061016. https://doi.org/10.1115/1.4028512.

Demirci, U. (2006). Acoustic picoliter droplets for emerging applications in semiconductor industry and biotechnology. *Journal of Microelectromechanical Systems, 15*(4), 957–966. https://doi.org/10.1109/JMEMS.2006.878879.

Demirci, U., & Montesano, G. (2007). Single cell epitaxy by acoustic picolitre droplets. *Lab on a Chip, 7*(9), 1139–1145. https://doi.org/10.1039/b704965j.

Dias, A. D., Kingsley, D. M., & Corr, D. T. (2014). Recent advances in bioprinting and applications for biosensing. *Biosensors, 4*, 111–136. https://doi.org/10.3390/bios4020111.

Duarte Campos, D. F., et al. (2012). Three-dimensional printing of stem cell-laden hydrogels submerged in a hydrophobic high-density fluid. *Biofabrication, 5*(1), 015003. https://doi.org/10.1088/1758-5082/5/1/015003.

Duocastella, M., et al. (2010). Novel laser printing technique for miniaturized biosensors preparation. *Sensors and Actuators B: Chemical, 145*(1), 596–600. https://doi.org/10.1016/j.snb.2009.11.055.

Faulkner-Jones, A., et al. (2013). Development of a valve-based cell printer for the formation of human embryonic stem cell spheroid aggregates. *Biofabrication, 5*(1), 015013. https://doi.org/10.1088/1758-5082/5/1/015013.

Faulkner-Jones, A., et al. (2015). Bioprinting of human pluripotent stem cells and their directed differentiation into hepatocyte-like cells for the generation of mini-livers in 3D. *Biofabrication, 7*(4), 044102. https://doi.org/10.1088/1758-5090/7/4/044102.

Ferris, C. J., Gilmore, K. J., et al. (2013a). Bio-ink for on-demand printing of living cells. *Biomaterials Science, 1*(2), 224–230. https://doi.org/10.1039/c2bm00114d.

Ferris, C. J., Gilmore, K. G., et al. (2013b). Biofabrication: An overview of the approaches used for printing of living cells. *Applied Microbiology and Biotechnology, 97*(10), 4243–4258. https://doi.org/10.1007/s00253-013-4853-6.

Franco, J., et al. (2010). Direct write assembly of calcium phosphate scaffolds using a water-based hydrogel. *Acta Biomaterialia, 6*(1), 218–228. https://doi.org/10.1016/j.actbio.2009.06.031.

Gauvin, R., et al. (2012). Microfabrication of complex porous tissue engineering scaffolds using 3D projection stereolithography. *Biomaterials, 33*(15), 3824–3834. https://doi.org/10.1016/j.biomaterials.2012.01.048.

Grogan, S. P., et al. (2013). Digital micromirror device projection printing system for meniscus tissue engineering. *Acta Biomaterialia, 9*(7), 7218–7226. https://doi.org/10.1016/j.actbio.2013.03.020.

Gudapati, H., Dey, M., & Ozbolat, I. (2016). A comprehensive review on droplet-based bioprinting: Past, present and future. *Biomaterials, 102*, 20–42. https://doi.org/10.1016/j.biomaterials.2016.06.012.

Guillemot, F., et al. (2011). Laser-assisted bioprinting to deal with tissue complexity in regenerative medicine. *MRS Bulletin, 36*(12), 1015–1019. https://doi.org/10.1557/mrs.2011.272.

Guillotin, B., et al. (2010). Laser assisted bioprinting of engineered tissue with high cell density and microscale organization. *Biomaterials, 31*(28), 7250–7256. https://doi.org/10.1016/j.biomaterials.2010.05.055.

Guvendiren, M., Lu, H. D., & Burdick, J. A. (2012). Shear-thinning hydrogels for biomedical applications. *Soft Matter, 8*(2), 260–272. https://doi.org/10.1039/C1SM06513K.

He, Y., et al. (2016). Research on the printability of hydrogels in 3D bioprinting. *Scientific Reports, 6*(1), 29977. https://doi.org/10.1038/srep29977.

Highley, C. B., Rodell, C. B., & Burdick, J. A. (2015). Direct 3D printing of shear-thinning hydrogels into self-healing hydrogels. *Advanced Materials, 27*(34), 5075–5079. https://doi.org/10.1002/adma.201501234.

Hinton, T. J., et al. (2015). Three-dimensional printing of complex biological structures by freeform reversible embedding of suspended hydrogels. *Science Advances, 1*(9), e1500758. https://doi.org/10.1126/sciadv.1500758.

Hopp, B., et al. (2005). Survival and proliferative ability of various living cell types after laser-induced forward transfer. *Tissue Engineering, 11*(11–12), 1817–1823. https://doi.org/10.1089/ten.2005.11.1817.

Horváth, L., et al. (2015). Engineering an in vitro air-blood barrier by 3D bioprinting. *Scientific Reports, 5*(1), 7974. https://doi.org/10.1038/srep07974.

Huang, T. Q., et al. (2014). 3D printing of biomimetic microstructures for cancer cell migration. *Biomedical Microdevices, 16*(1), 127–132. https://doi.org/10.1007/s10544-013-9812-6.

Jakab, K., et al. (2007). Tissue engineering by self-assembly of cells printed into topologically defined structures. *Tissue Engineering, 14*(3), 110306233438005. https://doi.org/10.1089/ten.2007.0173

Jana, S., & Lerman, A. (2015). Bioprinting a cardiac valve. *Biotechnology Advances, 33*(8), 1503–1521. https://doi.org/10.1016/j.biotechadv.2015.07.006.

Jayasinghe, S. N., Qureshi, A. N., & Eagles, P. A. M. (2006). Electrohydrodynamic jet processing: An advanced electric-field-driven jetting phenomenon for processing living cells. *Small, 2*(2), 216–219. https://doi.org/10.1002/smll.200500291.

Kattamis, N. T., et al. (2007). Thick film laser induced forward transfer for deposition of thermally and mechanically sensitive materials. *Applied Physics Letters, 91*(17), 171120. https://doi.org/10.1063/1.2799877.

Kirchmajer, D. M., Gorkin, R., III, & in het Panhuis, M. (2015). An overview of the suitability of hydrogel-forming polymers for extrusion-based 3D-printing. *Journal of Materials Chemistry B, 3*(20), 4105–4117. https://doi.org/10.1039/C5TB00393H.

Klebe, R. (1988). Cytoscribing: A method for micropositioning cells and the construction of two- and three-dimensional synthetic tissues. *Experimental Cell Research, 179*(2), 362–373. https://doi.org/10.1016/0014-4827(88)90275-3.

Koch, L., et al. (2010). Laser printing of skin cells and human stem cells. *Tissue Engineering. Part C, Methods, 16*(5), 847–854. https://doi.org/10.1089/ten.TEC.2009.0397.
Kong, H. J., Smith, M. K., & Mooney, D. J. (2003). Designing alginate hydrogels to maintain viability of immobilized cells. *Biomaterials, 24*(22), 4023–4029. https://doi.org/10.1016/S0142-9612(03)00295-3.
Lee, W., et al. (2009). Multi-layered culture of human skin fibroblasts and keratinocytes through three-dimensional freeform fabrication. *Biomaterials, 30*(8), 1587–1595. https://doi.org/10.1016/j.biomaterials.2008.12.009.
Lee, H., et al. (2013). Cell(MC3T3-E1)-printed poly(ϵ-caprolactone)/alginate hybrid scaffolds for tissue regeneration. *Macromolecular Rapid Communications, 34*(2), 142–149. https://doi.org/10.1002/marc.201200524.
Li, J., et al. (2016). Recent advances in bioprinting techniques: Approaches, applications and future prospects. *Journal of Translational Medicine, 14*(1), 1–15. https://doi.org/10.1186/s12967-016-1028-0.
Lin, H., et al. (2013). Application of visible light-based projection stereolithography for live cell-scaffold fabrication with designed architecture. *Biomaterials, 34*(2), 331–339. https://doi.org/10.1016/j.biomaterials.2012.09.048.
Lorber, B., et al. (2014). Adult rat retinal ganglion cells and glia can be printed by piezoelectric inkjet printing. *Biofabrication, 6*(1), 015001. https://doi.org/10.1088/1758-5082/6/1/015001.
Maher, P. S., et al. (2009). Construction of 3D biological matrices using rapid prototyping technology. *Rapid Prototyping Journal, 15*(3), 204–210. https://doi.org/10.1108/13552540910960307.
Malda, J., et al. (2013). 25th Anniversary article: Engineering hydrogels for biofabrication. *Advanced Materials, 25*(36), 5011–5028. https://doi.org/10.1002/adma.201302042.
Mandrycky, C., et al. (2016). 3D bioprinting for engineering complex tissues. *Biotechnology Advances, 34*(4), 422–434. https://doi.org/10.1016/j.biotechadv.2015.12.011.
Marga, F., et al. (2012). Toward engineering functional organ modules by additive manufacturing. *Biofabrication, 4*(2), 022001. https://doi.org/10.1088/1758-5082/4/2/022001.
Melchels, F. P., Feijen, J., & Grijpma, D. W. (2010). A review on stereolithography and its applications in biomedical engineering. *Biomaterials, 31*(24), 6121–6130. https://doi.org/10.1016/j.biomaterials.2010.04.050.
Moon, S. et al. (2010). Layer by layer three-dimensional tissue epitaxy by cell-laden hydrogel droplets. *Tissue Engineering Part C, Methods, 16*(1), 157–166. https://doi.org/10.1089/ten.tec.2009.0179.
Murphy, S. V., & Atala, A. (2014). 3D bioprinting of tissues and organs. *Nature Biotechnology, 32*(8), 773–785. https://doi.org/10.1038/nbt.2958.
Nahmias, Y., et al. (2005). Laser-guided direct writing for three-dimensional tissue engineering. *Biotechnology and Bioengineering, 92*(2), 129–136. https://doi.org/10.1002/bit.20585.
Nair, K., et al. (2009). Characterization of cell viability during bioprinting processes. *Biotechnology Journal, 4*(8), 1168–1177. https://doi.org/10.1002/biot.200900004.
Nakamura, M., et al. (2005). Biocompatible inkjet printing technique for designed seeding of individual living cells. *Tissue Engineering, 11*(11–12), 1658–1666. https://doi.org/10.1089/ten.2005.11.1658.
Norotte, C., et al. (2009). Scaffold-free vascular tissue engineering using bioprinting. *Biomaterials, 30*(30), 5910–5917. https://doi.org/10.1016/j.biomaterials.2009.06.034.
Odde, D. J., & Renn, M. J. (2000). Laser-guided direct writing of living cells. *Biotechnology and Bioengineering, 67*(3), 312–318. https://doi.org/10.1002/(SICI)1097-0290(20000205)67:3<312::AID-BIT7>3.0.CO;2-F.
Onses, M. S., et al. (2015). Mechanisms, capabilities, and applications of high-resolution electrohydrodynamic jet printing. *Small, 11*(34), 4237–4266. https://doi.org/10.1002/smll.201500593.
Ovsianikov, A., et al. (2010). Laser printing of cells into 3D scaffolds. *Biofabrication, 2*(1). https://doi.org/10.1088/1758-5082/2/1/014104.
Ozbolat, I. T., & Hospodiuk, M. (2016). Current advances and future perspectives in extrusion-based bioprinting. *Biomaterials, 76*, 321–343. https://doi.org/10.1016/j.biomaterials.2015.10.076.

Ozbolat, I. T., Chen, H., & Yu, Y. (2014). Development of "Multi-arm Bioprinter" for hybrid biofabrication of tissue engineering constructs. *Robotics and Computer-Integrated Manufacturing, 30*(3), 295–304. https://doi.org/10.1016/j.rcim.2013.10.005.

Ozbolat, I. T., Moncal, K. K., & Gudapati, H. (2017). Evaluation of bioprinter technologies. *Additive Manufacturing, 13*, 179–200. https://doi.org/10.1016/j.addma.2016.10.003.

Parsa, S., et al. (2010). Effects of surfactant and gentle agitation on inkjet dispensing of living cells. *Biofabrication, 2*(2), 025003. https://doi.org/10.1088/1758-5082/2/2/025003.

Pfister, A., et al. (2004). Biofunctional rapid prototyping for tissue-engineering applications: 3D bioplotting versus 3D printing. *Journal of Polymer Science Part A: Polymer Chemistry, 42*(3), 624–638. https://doi.org/10.1002/pola.10807.

Pirlo, R. K., et al. (2006). Cell deposition system based on laser guidance. *Biotechnology Journal, 1*(9), 1007–1013. https://doi.org/10.1002/biot.200600127.

Roth, E., et al. (2004). Inkjet printing for high-throughput cell patterning. *Biomaterials, 25*(17), 3707–3715. https://doi.org/10.1016/j.biomaterials.2003.10.052.

Saunders, R., Gough, J., & Derby, B. (2004). Ink Jet printing of mammalian primary cells for tissue engineering applications. *MRS Proceedings, 845*, AA2.8. https://doi.org/10.1557/PROC-845-AA2.8.

Saunders, R. E., Gough, J. E., & Derby, B. (2008). Delivery of human fibroblast cells by piezoelectric drop-on-demand inkjet printing. *Biomaterials, 29*(2), 193–203. https://doi.org/10.1016/j.biomaterials.2007.09.032.

Schiele, N. R., et al. (2010). Laser-based direct-write techniques for cell printing. *Biofabrication, 2*(3), 032001. https://doi.org/10.1088/1758-5082/2/3/032001.

Schuurman, W., et al. (2013). Gelatin-methacrylamide hydrogels as potential biomaterials for fabrication of tissue-engineered cartilage constructs. *Macromolecular Bioscience, 13*(5), 551–561. https://doi.org/10.1002/mabi.201200471.

Shim, J.-H., et al. (2012). Bioprinting of a mechanically enhanced three-dimensional dual cell-laden construct for osteochondral tissue engineering using a multi-head tissue/organ building system. *Journal of Micromechanics and Microengineering, 22*(8), 085014. https://doi.org/10.1088/0960-1317/22/8/085014.

Sklare, S. C., et al. (2015). The power of CAD/CAM laser bioprinting at the single-cell level: Evolution of printing. In *3D bioprinting and nanotechnology in tissue engineering and regenerative medicine* (pp. 79–103). Amsterdam: Elsevier. https://doi.org/10.1016/B978-0-12-800547-7.00004-7.

Smith, C. M., et al. (2004). Three-dimensional BioAssembly tool for generating viable tissue-engineered constructs. *Tissue Engineering, 10*(9–10), 1566–1576. https://doi.org/10.1089/ten.2004.10.1566.

Stosich, M. S., et al. (2009). Bioengineering strategies to generate vascularized soft tissue grafts with sustained shape. *Methods, 47*(2), 116–121. https://doi.org/10.1016/j.ymeth.2008.10.013.

Tabriz, A. G., et al. (2015). Three dimensional bioprinting of complex cell laden alginate hydrogel structures. *Biofabrication, 7*(4), 045012. https://doi.org/10.1088/1758-5090/7/4/045012.

Tasoglu, S., & Demirci, U. (2013). Bioprinting for stem cell research. *Trends in Biotechnology, 31*, 10–19. https://doi.org/10.1016/j.tibtech.2012.10.005.

Visser, J., et al. (2013). Biofabrication of multi-material anatomically shaped tissue constructs. *Biofabrication, 5*(3), 035007. https://doi.org/10.1088/1758-5082/5/3/035007.

Wang, Z., et al. (2015). A simple and high-resolution stereolithography-based 3D bioprinting system using visible light crosslinkable bioinks. *Biofabrication, 7*(4), 045009. https://doi.org/10.1088/1758-5090/7/4/045009.

Wijshoff, H. (2010). The dynamics of the piezo inkjet printhead operation. *Physics Reports, 499*, 77–177. https://doi.org/10.1016/j.physrep.2010.03.003.

Wilson, W. C., & Boland, T. (2003). Cell and organ printing 1: Protein and cell printers. *Anatomical Record - Part A Discoveries in Molecular, Cellular, and Evolutionary Biology, 272*(2), 491–496. https://doi.org/10.1002/ar.a.10057.

Xu, T., et al. (2005). Inkjet printing of viable mammalian cells. *Biomaterials, 26*(1), 93–99. https://doi.org/10.1016/j.biomaterials.2004.04.011.

Xu, T., et al. (2006). Viability and electrophysiology of neural cell structures generated by the inkjet printing method. *Biomaterials, 27*(19), 3580–3588. https://doi.org/10.1016/j.biomaterials.2006.01.048.

Xu, T., et al. (2009). Fabrication and characterization of bio-engineered cardiac pseudo tissues. *Biofabrication, 1*(3), 035001. https://doi.org/10.1088/1758-5082/1/3/035001.

Xu, T., et al. (2013). Hybrid printing of mechanically and biologically improved constructs for cartilage tissue engineering applications. *Biofabrication, 5*(1), 015001. https://doi.org/10.1088/1758-5082/5/1/015001.

Yamaguchi, S., et al. (2012). Cell patterning through inkjet printing of one cell per droplet. *Biofabrication, 4*(4), 045005. https://doi.org/10.1088/1758-5082/4/4/045005.

Zhang, Y., et al. (2013). Characterization of printable cellular micro-fluidic channels for tissue engineering. *Biofabrication, 5*(2), 025004. https://doi.org/10.1088/1758-5082/5/2/025004.

Chapter 9
Bioinks and Their Applications in Tissue Engineering

Jessica Nulty, Rossana Schipani, Ross Burdis, and Daniel J. Kelly

Abstract Three-dimensional (3D) bioprinting is a promising technology which utilises additive manufacturing technologies to produce functional tissues and organs. A bioink is the term used to describe any biomaterial, containing cells or other biologics, which is compatible with 3D bioprinting processes. The ideal bioink is one which is biocompatible, biodegradable, can achieve high shape fidelity during and after the bioprinting process and can provide a microenvironment which can accommodate the formation of a functional biological tissue. This chapter will discuss the composition of multiple naturally derived bioinks and synthetic bioinks which are commonly used in 3D bioprinting. It will also discuss how bioink stiffness and composition can influence and direct cell fate. Finally, this chapter will demonstrate how bioinks have advanced the field of tissue engineering by highlighting some key strategies involving the 3D bioprinting of tissues such as bone, vascular tissue and liver.

Keywords 3D bioprinting · Biofabrication · Bioink · Hydrogel · Decellularised matrix components · Cell fate · Tissue and organ biofabrication

J. Nulty · R. Schipani · R. Burdis
Trinity Centre for Bioengineering, Trinity Biomedical Sciences Institute, Trinity College Dublin, Dublin, Ireland

Department of Mechanical and Manufacturing Engineering, School of Engineering, Trinity College Dublin, Dublin, Ireland

D. J. Kelly (✉)
Trinity Centre for Bioengineering, Trinity Biomedical Sciences Institute, Trinity College Dublin, Dublin, Ireland

Department of Mechanical and Manufacturing Engineering, School of Engineering, Trinity College Dublin, Dublin, Ireland

Department of Anatomy, Royal College of Surgeons in Ireland, Dublin, Ireland

Advanced Materials and Bioengineering Research Centre (AMBER), Royal College of Surgeons in Ireland and Trinity College Dublin, Dublin, Ireland
e-mail: kellyd9@tcd.ie

D. M. Devine (ed.), *Polymer-Based Additive Manufacturing*,
https://doi.org/10.1007/978-3-030-24532-0_9

9.1 Introduction

The term bioink is a term that, although universally used in many papers and review articles, is not universally defined. For some a bioink is simply a class of biomaterials that are utilised for the fabrication of 3D bioprinted constructs (Williams et al. 2018), for others a bioink is "a formulation of cells that is suitable to be processed by an automated biofabrication technology" (Groll et al. 2018). For the purpose of this chapter, we define a bioink as a biomaterial containing a biologic (a cell, nucleic acid or biomolecule) which can be dispensed through a printhead whilst maintaining the viability or functionality of the contained biologic. Bioinks can vary widely, possessing different viscosities, crosslinking mechanisms and bioactivity. The selection of a bioink is dependent on the methodology of bioprinting as well as the desired end application. Different target tissues will require bioinks with specific physical properties and cell-bioink interactions. For this reason, the "ideal" bioink biomaterial is one which possesses the appropriate rheological and crosslinking properties to achieve high shape fidelity and cell viability post-printing, is biocompatible, provides adequate mechanical support in the short-term, a microenvironment capable of facilitating appropriate extracellular matrix deposition and ultimately biodegrades at a rate that facilitates vascularisation upon implantation and replacement of the bioink with a functional biological tissue.

The rheological properties required for the ideal bioink greatly depends on the methodology of bioprinting. For instance, inkjet printing requires bioinks to be much less viscous (~10 mPa s) compared to bioinks used in extrusion-based bioprinting which can accommodate a wide range of viscosities (typically between 30 and 6×10^7 mPa s) (Ji and Guvendiren 2017). It should be noted that the polymer concentration, molecular weight, cell density and cell type all greatly affect bioink viscosity.

The degradation rate of a chosen bioink is also vital to the success of many tissue engineering strategies. Ideally, the rate of degradation of the bioink should approximate the rate of new tissue growth. Furthermore, as most tissues are vascularised, bioink degradation may be required to facilitate vascularisation of the printed construct in vivo. Again, the polymer concentration, molecular weight, cell density and cell type all greatly affect bioink degradability.

Bioink materials can be broadly divided into two categories: natural (e.g. collagen) and synthetic (e.g. polyethylene glycol) materials. Naturally derived bioinks often possess inherent cell binding sites which can facilitate cell attachment, proliferation and differentiation; however, they usually display weak or poorly controlled mechanical properties. Synthetically derived bioinks, although often lacking cell binding sites in their unmodified state, can be functionalised and their mechanical properties can be easily tuned for specific applications.

9.2 Naturally Derived Bioinks

Natural bioink biomaterials commonly used for bioprinting include alginate, collagen, gelatin, fibrin or the extracellular matrix derived from a specific tissue. This section will focus on these five classes of common bioink biomaterial, although other natural polymers have been used as bioinks.

9.2.1 *Alginate-Based Bioinks*

Alginate is a naturally occurring biopolymer which is isolated from the cell walls of brown algae (Phaeophyceae). It is a negatively charged linear polysaccharide consisting of (1,4)-linked α-L-guluronic acid (G) and β-D-mannuronic acid (M) residues. The M and G residues are arranged in homogeneous blocks which are interspersed with heterogenous MG blocks. Alginate forms a gel through ionic crosslinking between multivalent cations and the carboxylate groups of the G blocks on distinct alginate chains typically represented by the egg-box model. It has been shown that the solubility and size of the ion involved can greatly affect both the gelation time and gel strength of alginate respectively (Gwon et al. 2015). The pore size of alginate gels varies between 5 and 200 μm, and is dependent on the concentration of G-blocks contained on the alginate backbone, higher G-block content leads to bigger pore size.

Alginate has been widely used in the field of bioprinting due to its biocompatibility, non-immunogenicity and its relatively fast gelation time which enable fast encapsulation of cells. Alginate also has the capacity to draw in water and other molecules through capillary action which in turn can diffuse out over time (Axpe et al. 2016).

One of the major disadvantages of alginate is that it has a relatively slow and unpredictable degradation rate. Alginate lysases catalyse the degradation of alginate; however, this enzyme is not produced in mammals, therefore, in vivo, alginate degrades through random dissolution (Boonthookul et al. 2005; Loibl et al. 2014). There are several approaches to increase the degradation rate of alginate in vivo including oxidation and gamma irradiation. Oxidation occurs when alginate reacts with sodium periodate cleaving the carbon-carbon bond of the cis-diol group in the uronate residue resulting in the formation of two aldehyde functional groups. These aldehyde functional groups react spontaneously with the closest hydroxyl groups of the adjacent unoxidised uronic residues in the polymer chain and form hemiacetals rings (Gomez et al. 2007). The higher the degree of oxidation, the higher the susceptibility it has to hydrolysis. Therefore, by altering the degree of oxidation, the degradation rate of alginate, through hydrolytic scission, can be controlled.

An alternative method of increasing the degradation rate of alginate is by decreasing the molecular weight of the alginate polymer backbone. The most common method to reduce the molecular weight of alginate is through gamma irradiation

(Alsberg et al. 2003; Lee et al. 2003; LUAN et al. 2009). Gamma irradiation decreases the molecular weight of the alginate by breaking the polymer backbone of alginate within MG blocks through free radical-induced scission of glycosidic bond which increases the degradation rate (Kong et al. 2002).

Although these methods greatly increase the degradation rate of alginate bioinks, they also greatly decrease their mechanical properties. For this reason, many have turned to bioink blends to leverage the printability of alginate with the biocompatibility properties of other bioinks such as gelatin (Di Giuseppe et al. 2018; Jiang et al. 2017; Huang et al. 2016) or collagen (Yang et al. 2018).

9.2.2 Collagen-Based Bioinks

The extracellular matrix (ECM) is the non-cellular constituent of all tissues, primarily comprising of water and an assortment of proteins and polysaccharides. It is synthesised during development by the tissue's residing cells who continue to maintain and repair the matrix throughout their lifespan. The ECM, in turn, influences cell function and regulates cell behaviour (Bissell and Aggeler 1987). Each tissue's ECM has a unique combination of matrix macromolecules and topography which determines mechanical properties and bioactivity of that tissue. The most abundant proteins in the ECM are a family of structural proteins known as collagens. Collagens vary substantially in size, function and tissue distribution, but all contain a characteristic triple helix comprising of three parallel polypeptide chains coiled around one another in a rope-like structure (Shoulders and Raines 2009). Twenty-eight different isotypes of collagen, typically numbered with Roman numerals, have been identified and are divided into four major subfamilies:

1. Fibril-forming collagens; comprising of collagen I, II, III, V, XI. XXIV and XXVII.
2. Fibril-associated collagens with interrupted triple helices known as FACITs; comprising of collagen IX, XII, XIV, XVI, XIX, XX, XXI and XXII.
3. Network-forming collagens; comprising of collagen IV, VI, VIII and X.
4. Membrane-associated collagen with interrupted triple helices known as MACITs; comprising of collagen XIII, XVII, XXIII and XXV
5. Multiplexins; comprising of XVand XVIII

The majority of collagens used as a biomaterial for tissue engineering applications fall into the category of fibril-forming collagens. These are an attractive material for many tissue engineering applications since collagen fibrils and the networks they form regulate and define most tissues in the body (Cen et al. 2008). In terms of bioinks, collagen is rarely used alone and is most often incorporated into bioink blends to help recapitulate a tissue-specific ECM and improve cell compatibility (Wu et al. 2016a; Yang et al. 2018). Methods for printing collagen include lowering the pH to around 4 to ensure that collagen remains uncrosslinked during the printing process (Lee et al. 2010b) and using high collagen concentrations (15 and 17.5 mg/ml)

(Rhee et al. 2016). However, these harsh pH conditions are not compatible with most cell-laden printing and the slow gelation rate of collagen can lead to inhomogeneity in cell distribution (Smith et al. 2004). For this reason, collagen is often used in tandem with other bioinks to improve their cell compatibility and better mimic the native ECM (Wu et al. 2016b).

9.2.3 Gelatin-Based Bioinks

Gelatin is a water-soluble protein derived from collagen. It is produced by the denaturation of the triple-helix structure of collagen into single strand molecules through hydrolysis. It retains the natural cell binding Arg-Gly-Asp (RGD) motifs from collagen and is an attractive biomaterial due to its high cell compatibility and biodegradability. Gelatin is thermosensitive. At 37 °C, it dissolves as a colloidal sol and at lower temperatures forms a gel in a concentration-dependent manner. Some bioprinting strategies have taken advantage of this sol-gel transition using gelatin to temporarily stabilise another biomaterial during the printing process (Kang et al. 2016; Zhang et al. 2013). The most common use of gelatin as a bioink, however, comes in its methacrylated form, gelatin methacryloyl (GelMA). GelMA forms covalently crosslinked hydrogels through photo-initiated radical polymerisation.

9.2.4 Fibrin-Based Bioinks

Fibrin is a fibrous, non-globular protein involved in wound healing and blood clotting (Mosesson 2005). It is formed by the proteolytic cleavage of fibrinogen by the serine protease thrombin and further stabilised through the action of Factor XIIIa.

Fibrin has multiple applications and is used clinically as a sealant (Jackson 2001) and as a drug delivery system (Spicer and Mikos 2010). It is used extensively for tissue engineering applications due to its biocompatibility, biodegradability and its capacity to bind cells, growth factors and other ECM proteins. Furthermore, because fibrinogen is a large glycoprotein produced by the liver and circulates in the blood system, it can be relatively easily isolated from blood plasma. This enables autologous fibrin being able to be polymerised from the patient's own blood.

Fibrin has been shown to promote cell growth and tissue regeneration and has had success in multiple tissue engineering approaches including vascular tissue engineering (Lesman et al. 2011), musculoskeletal engineering (Perka et al. 2000; Eyrich et al. 2007) and cardiac tissue engineering (Jockenhoevel et al. 2001). In terms of its use as a bioink, fibrin is usually printed as two separate components, fibrinogen solution and thrombin solution. For its use in extrusion-based bioprinting, fibrinogen solution is usually stabilised in some form, until thrombin can polymerise it into its solid fibrin form. One successful method for this is the aforementioned gelatin carrier method (Kang et al. 2016), where the inclusion of gelatin into the fibrinogen bioink maintains the architecture until thrombin polymerisation stabilises the fibrin gel. Another successful method of bioprinting fibrin involves a bioprinting technique termed freeform reversible embedding of suspended hydrogels (FRESH) (Hinton et al. 2015). This method involves fibrin being printed within a second thermoreversible hydrogel support bath containing thrombin. Once polymerisation has occurred, fibrin constructs can be easily extracted from the support bath.

Fibrin has also been incorporated into several inkjet printing approaches. However, often these approaches do not use direct fibrin/fibrinogen printing but use thrombin solutions as the bioink and print directly onto a fibrinogen substrate (Cui and Boland 2009; XU et al. 2006).

9.2.5 *ECM-Based Bioinks*

The complex composition of ECM contains intrinsic cues to promote cell proliferation, differentiation and ultimately tissue homeostasis and repair. The success of many individual ECM components as bioinks (e.g. collagen) has motivated the formulation of decellularised ECM bioinks to better mimic native cell environments in 3D bioprinted constructs. One of the major advantages of using decellularised ECM as a bioinks is that ECM can be derived directly from specific tissues; therefore, a bioink can be tailored to the organ type which is being tissue engineered and therefore contains the appropriate signals to regulate cell functions. Multiple sources have been utilised to produce decellularised ECM products including porcine, bovine and caprine and even human sources (Saldin et al. 2017; Johnson et al. 2014) and multiple decellularisation protocols have been undertaken including the use of Triton X and sodium dodecyl sulphate (SDS). The method of decellularisation can affect the type of proteins and growth factors remaining in the tissue, so the methods are usually tailored depending on the application. To date, the ECM derived from articular cartilage, meniscus, liver and cardiac tissue has been used to generate printable and instructive bioinks (Critchley and Kelly 2017; Romanazzo et al. 2018; Pati et al. 2014; Lee et al. 2017). Depending on the tissue source of the ECM-derived bioink, some contain the intrinsic mechanical strength to be directly printed (Pati et al. 2014) whilst others require a support gel such as alginate (Gao et al. 2017) or polyethylene glygol (PEG) (Skardal et al. 2015). Overall, decellularised ECM is a promising source for emerging bioink formulations.

9.3 Synthetic Bioinks

Although natural bioinks have high cell compatibility, biodegradability and can promote cell proliferation and differentiation, they often have less than desirable mechanical properties and often unpredictable degradation rates. Synthetic polymers, however, can be easily tuned to control their mechanical properties and rate of degradation. Furthermore, such polymers can be functionalised to include peptides such as the arginine-glycine-aspartic (RGD) peptide to promote specific cell-matrix interactions.

9.3.1 Polyethylene Glycol (PEG)

One of the most commonly used synthetic bioinks used in bioprinting approaches is poly(ethylene glycol) (PEG). PEG is a bioinert, hydrophilic polyether compound which can be easily tailored to improve mechanical properties and easily modified with cell binding peptides and other bioactive molecules. PEG is suitable for many tissue engineering applications due to its biocompatibility, hydrophilicity which enables the diffusion of nutrients and gases to and from encapsulated cells and its tuneable mechanical properties. Pure PEG has a relatively low viscosity and is therefore challenging to bioprint alone. However, PEG can be easily functionalised with diacrylate and dimethacrylate to create the photo-crosslinkable PEG derivatives PEGDA and PEGDMA.

9.3.2 Poly(N-isopropylacrylamide) (PNIPAAM)

Poly(*N*-isopropylacrylamide) (PNIPAAM) is another synthetic polymer which has been used as a bioink. PNIPAAM is thermoresponsive and gels at physiological temperatures through a combination of hydrophobic interaction and coil-globular gelation mechanisms. However, similar to other synthetic bioinks, PNIPAAM has relatively low biocompatibility and biodegradability. For this reason, PNIPAAM is most often used in combination with other biopolymer and bioinks such as hyaluronic acid (Kesti et al. 2015) and alginate (Tan et al. 2012).

9.3.3 Pluronic

Pluronic block copolymers of poly(ethylene oxide-b-propylene oxide-b-ethylene oxide) is a promising emerging 'bioink'. Although not technically a bioink because of its limited cell supporting capacity, it warrants a mention due to the impact it is having on the field through its use as a supporting and sacrificial ink. This fugitive ink is biologically inert and undergoes thermally reversible gelation. Between 4 and 22 °C the fugitive ink is stiff and exhibits a solid-like response and a strong shear thinning response. Below 4 °C the ink liquefies and can be removed from the construct (Kolesky et al. 2014). This can be utilised as a support structure to enable the bioprinting of complex architectures such as a human ear (Kang et al. 2016), as a sacrificial ink to create channels (Daly et al. 2018) or vessels (Kolesky et al. 2014) within bioprinted constructs and to improve printability and create micron-sized pores within another bioink (Armstrong et al. 2016).

9.4 Cell Fate in Printed Constructs

The utility of bioprinting will ultimately depend on our ability to direct the fate of bioprinted cells in order to produce a functional tissue. This section will review how the fate of cells within bioprinted constructs depends on factors such as the stiffness and composition of the bioink.

9.4.1 Bioink Stiffness

The identification of a suitable bioink represents a major challenge when bioprinting cell-encapsulating constructs. The bioink biomaterial needs to possess the appropriate mechanical and structural properties in order to prevent cell damage during the fabrication process, as well as providing the appropriate local environment to direct cell proliferation and differentiation towards the development of specific tissues. Hydrogels have been shown to support MSCs viability, differentiation and tissue-specific matrix formation especially in the presence of exogenous biochemical or biophysical cues. Although the poor mechanical properties of hydrogels limit their use in load-bearing environments, they still have an inherent impact on regulating the phenotype of the generated tissue. It has been shown this is determined in part by the stiffness of the hydrogels themselves. In general, cells are more adherent on rigid substrates than soft ones, and this stiffness-dependent cell adhesion regulates migration, morphology, gene expression and ultimately cell function (Yeung et al. 2005; Engler et al. 2006; Guilak et al. 2009). Engler et al. (2006) demonstrated that MSCs differentiation could be modulated by the stiffness of the local substrate. When culturing hMSCs on type I collagen-coated polyacrylamide gels with variable compliance, it was observed that the stiffest substrates (25–50 kPa) enhanced the expression of osteogenic markers and cells exhibited the typical cuboidal osteoblast morphology. The softest substrates (0.1–1 kPa) upregulated neurological markers and cells showed a dendritic morphology, while gels mimicking the stiffness of smooth muscle tissue (8–17 kPa) favoured myogenesis and adopted a spread cell morphology. Similar stiffness-dependent changes in MSC differentiation have been reported in 3D hydrogels (Pek et al. 2010).

Several strategies have been explored to improve the mechanical properties of hydrogel scaffolds for tissue engineering applications. For example, the stiffness of hydrogels such as alginate has been shown to depend strongly on the efficiency of ionic crosslinking (Li et al. 2018). Four alginate-gelatin bioink formulations were prepared with increasing phosphate buffered saline (PBS) concentrations. Despite all the groups demonstrating good printing resolution, excessively high PBS ionic strength limited macromolecule interactions, thereby preventing the crosslinking of the alginate. This limited the availability of sites for embryonic stem cells (ESCs) to spread within the 3D printed constructs, which instead formed isolated clusters. On the other hand, lower cell proliferation and aggregation was observed in the stiffest constructs (low ionic strength).

Other measures to enhance the mechanical performance of hydrogels include increasing crosslinker density, introduction of covalent crosslinks, modulation of material molecular weight (MW) and preculture to promote matrix deposition before the implantation of a tissue-engineered graft (Anseth et al. 1996). The mechanical strength of hydrogels can usually be tuned by using lower or higher MW hydrogels or by increasing or decreasing the number of crosslinking sites. Freeman and Kelly (2017) demonstrated that by using the same alginate bioink and calcium sulphate ($CaSO_4$) as a crosslinker, it was possible to spatially control the stiffness from the core to the periphery of MSC-laden bioprinted constructs by varying the material to crosslinker ratio (Fig. 9.1a, b). Dual Oil Red O (ORO) and Alkaline Phosphatase (ALP) staining showed that MSCs in the stiffer region (periphery) differentiated preferentially towards the osteogenic lineage, whereas in the softcore cells underwent adipogenesis (Fig. 9.1c, d). Although the introduction

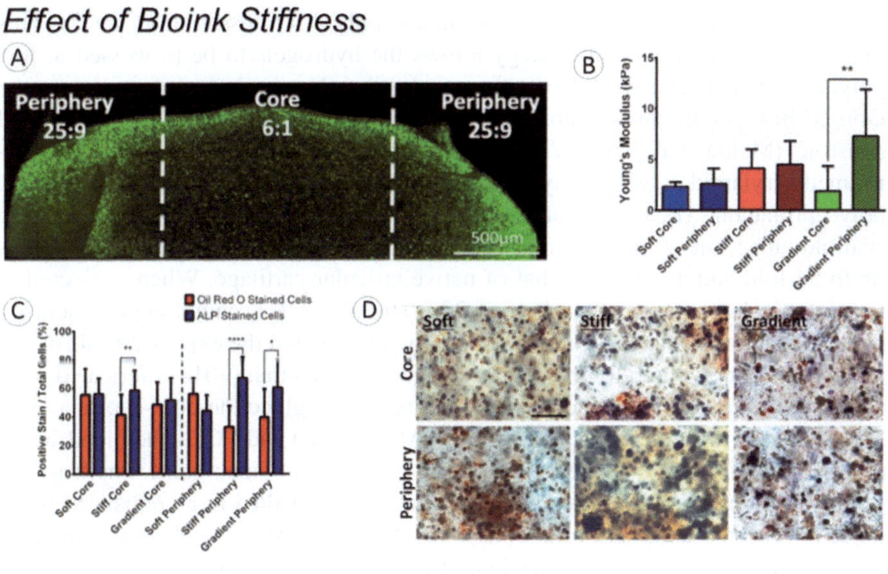

Fig. 9.1 Effect of bioink stiffness on cell fate in printed constructs. (**a**) Representation of a MSC-laden bioprinted construct when varying the stiffness from the core to the periphery of the construct (scale bar 500 μm). (**b**) Young's Modulus of the experimental groups when varying the material to crosslinker ratio. $**p < 0.001$ vs. periphery of printed construct. (**c**) Quantitative and (**d**) qualitative analysis of Oil Red O (red) and ALP (blue) dual staining of the core and periphery of the bioprinted constructs after 7 days in culture. $*p < 0.05$, $**p < 0.001$ and $****p < 0.0001$ vs. positive ALP stained cells (Freeman and Kelly 2017)

of high polymer concentration or chemical crosslinking can enhance hydrogels mechanical stability, this does not necessarily represent an optimum solution. High polymer concentration can limit nutrients diffusion, and crosslinking agents are often toxic to cells. Moreover, despite the above-mentioned modifications, the mechanical properties of the gels remain inadequate for many tissue engineering applications.

As cells in native tissues are responsive to different types of mechanical stimuli, it is hypothesised that the performance of hydrogel scaffolds is superior when they possess tissue specific-like mechanical properties. More recently, the combination of biomimetic and bioinspired strategies to create advanced composites is being explored.

9.4.1.1 Fibre-Reinforced Hydrogels

Many native tissues architecture is characterised by the presence of a fibrous framework which not only provides structural integrity, but also anisotropy and depth-dependent mechanical properties (Aspden 1994; Visser et al. 2015). Hydrogels have been reinforced with thermoplastic polymer fibres obtained via melt electrospinning writing (MEW) (Bas et al. 2015, 2017; Visser et al. 2015), solution electrospun (Eslami et al. 2014), woven (Moutos et al. 2007; Schuurman et al. 2011; Liao et al. 2013; Boere et al. 2014) and non-woven scaffolds (Marijnissen et al. 2002). This multiple-tool biofabrication strategy allows the hydrogels to be processed at low polymer concentrations, while the thermoplastic polymer network features can be tailored in order to obtain stable structures and secure the strength of the overall construct (Malda et al. 2013). Visser et al. (2015) mechanically reinforced soft gelatin methacrylamide (gelMA) hydrogels with organised, highly porous, 3D printed polycaprolactone (PCL) microfibre networks. Compared to the hydrogel or fibre scaffold alone, the composite constructs showed a synergistic increase in stiffness up to 54-fold and approached that of native articular cartilage. When subjected to physiological compressive loading of 20% strain at 1 Hz, chondrocytes incorporated in the composite scaffolds significantly upregulated the expression of ACAN and COL1A1 compared to non-reinforced gels. Boere et al. (2014) fabricated reinforced scaffolds using a blend of poly(hydroxymethylglycolide-co-ε-caprolactone) functionalised with methacrylate groups (pMHMGCL) and PCL. The blend-based constructs were then covalently linked to a chondrocyte-laden gelMA hydrogel to improve the integration between the materials. This resulted in an enhanced resistance to repeated axial and rotational loads. Furthermore, the embedded chondrocytes showed significant increases in cartilage-like matrix production both in vitro and in vivo. Moutos et al. (2007) infused 3D woven polyglycolic acid structures with either agarose or fibrin gels which showed anisotropic, viscoelastic and tension-compression non-linearity properties similar to native articular cartilage. The hydrogel components seemed to be responsible for the viscoelastic behaviour, whereas the presence of the fibre network resulted in tension-compression non-linear properties which are believed to play a key role in the load-bearing capacity under compression.

3D printed PCL-based constructs have been used not only in an attempt to mimic the mechanical properties of native tissues, but also to provide mechanically stable implants that would allow a better integration with the surrounding environment in vivo. Osteochondral defects in a rabbit model were treated by 3D printing PCL networks supporting hyaluronic acid and collagen-based bioinks containing hMSCs in the cartilage and bone region of the construct respectively (Shim et al. 2016). After 8 weeks, regeneration of thick neocartilage tissue was observed which integrated well with the adjacent tissue. Lee et al. (2010a) 3D printed a PCL-hydroxyapatite scaffold mimicking the geometry of a rabbit shoulder for synovial joint regeneration. The fabricated constructs were infused with TGF-β3 to promote host cells infiltration. After 4 months of implantation, bone growth and some cartilage repair were observed from histological analysis and the animals showed no impaired movement. Another emerging application of 3D bioprinted mechanically reinforced constructs is the fabrication of developmentally inspired templates for tissue or organ regeneration (Jukes et al. 2008; Scotti et al. 2010; Daly et al. 2016). In the case of bone tissue regeneration, vertebrae-shaped hypertrophic cartilage templates were engineered in vitro by co-printing MSC laden RGD-y irradiated alginate bioink and a network of PCL microfibres (Daly et al. 2016). When implanted in vivo, the bioprinted construct matured into a vascularised bone organ.

9.4.1.2 Interpenetrating Networks Hydrogels

Hydrogels used in tissue engineering and 3D bioprinting commonly consist of only a single polymer, but in recent years there has been a shift to the use of more complex systems characterised by combining multiple polymer networks. The latter often exhibit superior mechanical properties compared to the single polymer networks and are known as Interpenetrating Network (IPN) hydrogels. IPNs are composed by separate polymer networks, physically entangled with each other where different chemistries are used to encourage the individual networks to crosslink only with themselves (Chimene et al. 2016). The two networks cannot be separated unless chemical bonds are broken. Partial inter-network binding can occur depending on the type of crosslinking reaction and polymer used. For example, Jeon et al. (2017) engineered highly elastic and tough IPNs based on alginate and gelatin. Mechanically stimulation of encapsulated hMSCs in the IPN hydrogel resulted in enhanced proliferation and osteogenic differentiation.

9.4.1.3 Double Networks Hydrogels

Double Networks (DNs) hydrogels are a subset of IPNs which consist of two polymer networks. In general, the primary network consists of an elastic and more flexible component, whereas the second network is characterised by a brittle, high stiffness polymer at low concentration (Chimene et al. 2016; Vega et al. 2017). This characteristic makes DNs appealing for tissue engineering applications since cartilage and other skeletal tissues ECM, for example, naturally use double networks to achieve robust and functional mechanical properties. DN hydrogels are synthesised via a two-step polymerisation. First, the primary hydrogel network is obtained by covalently crosslinking different polymer chains. The second step comprises of secondary polymer monomers dispersion throughout the primary network to be finally crosslinked (Chimene et al. 2016). Although the above-mentioned synthesis method has been widely used, it is not necessarily suitable for 3D printing applications as the hydrogel formation rate is too slow. Alternatively, ionic-covalent entanglement (ICE) gels have recently been explored for 3D printing. The gels are both chemically and physically crosslinked and the DNs formation occurs at a suitable rate for 3D printing. Hong et al. (2015) fabricated elastomeric ICE hydrogels from poly (ethylene glycol) (PEG) and alginate-nanoclay composite that could be stretched to a strain of 400% with no fracture. Encapsulated hMSCs maintained high cell viability over 7 days of culture.

9.4.2 Bioink Composition

As already mentioned in the previous section, use of single-component hydrogels has yielded some success for various tissue engineering applications (Fedorovich et al. 2007; Skardal et al. 2010; Blaeser et al. 2013; Inzana et al. 2014; Ouyang et al. 2017), although the use of harsh crosslinking agents or the lack of appropriate cells-material interactions can limit new matrix formation (Fedorovich et al. 2007, 2008). Traditional hydrogels are not able to mimic the complexity of the natural ECM; therefore, they cannot provide the adequate microenvironmental conditions to recreate the unique cellular niche of living tissues. Recently, the use of a new generation of bioinks has enabled the engineering of more complex structures and microenvironments (Chimene et al. 2016). These advanced bioinks include IPNs (discussed in the previous section), ECM-derived and/or functionalised materials, nanocomposites and multibioink hydrogels. A number of advantages of these novel materials are shear-thinning characteristics, fabrication of structures with high shape fidelity, and the ability to modulate mechanical properties as well as cellular functions (Chimene et al. 2016).

9.4.2.1 ECM-Based and/or Functionalised Bioinks

ECM-based constructs have been widely investigated for tissue engineering applications (Brown and Badylak 2015). The tissue undergoes a decellularisation process and the ECM is typically micronised or solubilised before use. The perceived advantages of ECM-based constructs are that they generally not rejected by the body and are able to maintain some chemical as well as biological cues that can direct cell proliferation, migration and differentiation (Almeida et al. 2015). The ECM can be printed on its own or can be incorporated into an existent bioink such as gelMA or alginate in the form of particles (hyaluronic acid, cartilage pieces, collagen, etc.) (Pati et al. 2014; Kim et al. 2017). Pati et al. (2014) demonstrated the advantage of utilising the second approach, showing that printable bioinks can be developed by decellularising and solubilising heart, adipose and cartilage ECM (Fig. 9.2a, b) The ECM-derived bioinks provided tissue-specific factors that directed the differentiation of embedded MSCs. In another study, Lee et al. developed collagen/ECM/alginate-based bioinks from the matrix synthesised by preosteoblast cells cultured on gelled collagen. The addition of the alginate component increased the bioink printability. This bioink provided the embedded preosteoblasts with a more suitable microenvironment compared to pure alginate, which led to improved osteoblastic differentiation and higher levels of mineralisation. For cardiac tissue engineering applications, Jang et al. (2017) developed a 3D printed pre-vascularised patch by spatially controlling the deposition of MSCs and cardiac progenitor cells (CPCs) through a heart tissue-derived decellularised ECM (hdECM) bioink. When compared to the control bioink (collagen hydrogel), the maturation of the CPCs in the hdECM gel (Bioink I) was significantly improved. Moreover, MSCs supplemented with VEGF in the hdECM gel (Bioink II) favoured vascularisation after only 5 days of culture. The therapeutic effect of the 3D printed patches was tested also in vivo in mice and rat models, where improved cardiac function and a decrease of negative left ventricle remodelling was observed.

Bas et al. (2017) functionalised four-armed maleimide terminated PEG with Heparin (sPEG/Hep) to create a bioink that emulates the negatively charged proteoglycan matrix of articular cartilage. The soft sPEG/Hep bioink was reinforced with melt electrospun fibrous networks to mimic the biomechanical behaviour of articular cartilage. Above all, the high negative charge density and strong water retention

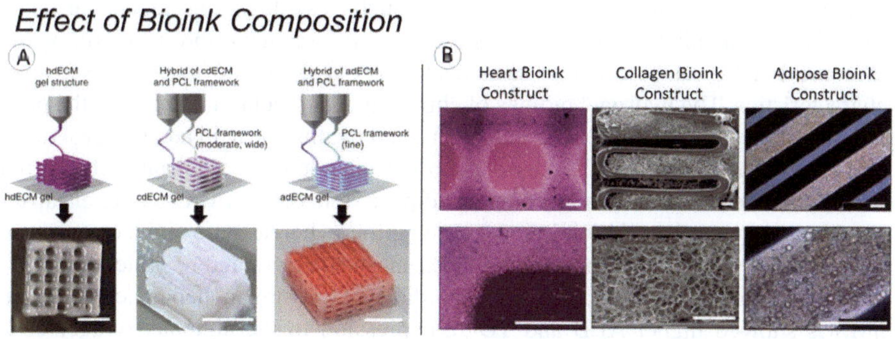

Fig. 9.2 Effect of bioink composition on cell fate in printed constructs. (**a**) Printing of decellularised heart ECM bioink (hdECM) alone, cartilage and adipose ECM (cdECM and adECM) bioinks combined with a PCL frame. (**b**) Representative microscopic images of the hdECM, cdECM and adECM constructs (scale bar 400 μm). (Adapted from Pati et al. 2014)

capacity of the bioink led to a charge-driven osmosis similar to native articular cartilage, that resulted into the enhanced biological performance as demonstrated by high chondrocyte viability and differentiation under physiological mechanical stimulation.

9.4.2.2 Nanocomposite Bioinks

Nanocomposite bioinks are characterised by the addition of nanoparticles able to alter various physical and chemical characteristics such as stiffness, degradation and shear-thinning properties (Markstedt et al. 2015; Xavier et al. 2015; Jaiswal et al. 2016). Depending on the type of nanoparticles added to the hydrogel network, unique properties can be incorporated including electrical conductivity, bioactivity, controlled drug release, etc. (Akhilesh K. Gaharwar et al. 2015). Gao et al. (2014) evaluated the bioactivity of hydroxyapatite (HA) and bioactive glass (BG) nanoparticles, which were incorporated in 3D printed poly(ethylene glycol) dimethacrylate (PEGDMA) scaffolds with embedded hMSCs. Compared to PEG-BG, the PEG-HA group promoted higher levels of collagen production and ALP activity.

9.4.2.3 Multibioink Strategies

Multibioink approaches involve the use of printing systems that allow the deposition of a number of bioinks simultaneously such as conventional reinforcing polymers (e.g. PCL, PLA, PLGA), hydrogels to provide biological function, and sacrificial materials used as fugitive templates to achieve complex shapes (Liu et al. 2017). This strategy enables the extrusion of continuous materials with tunable properties that could be useful for printing zonal structures like in articular cartilage. Kang et al. (2016) used a multi-head printer system to fabricate functional vascularised tissues at relatively large scales such as skeletal muscle, ear cartilage, calvarial and mandible bone. PCL was used as reinforcement fibrous matrix, while Pluronic F-127 was selected as sacrificial material thanks to its thermosensitive characteristics. The hydrogel bioinks of choice included gelatin to improve the print fidelity, fibrinogen to provide cell adhesions, and hyaluronic acid and glycerol to act as plasticisers. Yeo et al. (2016) used a core-sheath needle to 3D print multilayered cells-embedded mesh structures. Human adipose stem cells (hASCs) were encapsulated in a collagen-based bioink in the core region, whereas pure alginate was used in the sheath region to support the collagen mesh structure and to protect the cells during the fabrication process. After 20 days of culture in hepatogenic media, hMSCs showed higher ALB and TDO2 expression in the alginate/collagen/cell-laden structure compared to control (alginate + cells).

9.4.3 *Delivery of Signalling Factors*

The behaviour of engineered tissue constructs can be also manipulated by controlling the release of growth factors or other biomolecules over time. The benefits of developing biomolecule delivery strategies based on 3D printed scaffolds are the ability to place the molecules at a target position, and to provide the sufficient dosage in a time-dependent manner by tailoring the structural features and degradation time of the construct (Park et al. 2016).

9.4.3.1 Homogeneous Immobilisation of Molecules

Biomolecules can be homogeneously immobilised either by coating the surface of the constructs post-printing or by incorporating them in the printed materials before fabrication. Surface coating of a construct occurs generally by chemical conjugation or physical adsorption. The former method allows to bind factors covalently to the carrier material via functional groups, whereas the latter utilises the intermolecular forces between molecules and solid adsorbents to immobilise the molecules (Chung and Park 2007; Park et al. 2016). Several materials such as gelatin, heparin, fibronectin and dopamine have been used as a surface coating to create biological bindings for the molecules. For example, Kang et al. (2011) modified the surface of 3D printed PCL/PLGA constructs by first dipping them in fibrin/hyaluronic acid hydrogel, followed by immobilisation of BMP-2 through absorption. The continuous release of BMP-2 from hASCs seeded-fibrin/hyaluronic acid-coated scaffolds resulted in significantly higher bone formation and mineralisation when implanted in a mouse model compared to the other groups. Du et al. (2015) 3D printed BMSC-laden gelMA scaffolds with collagen microfibers binding BMP-2 using a collagen binding domain (CBD) that allowed for controlled growth factor. These CBD-BMP-2-collagen microfibers were found to promote BMSCs differentiation into osteocytes. In general, synthetic polymers have been used as carriers because of their slow degradation and associated sustained release of loaded biomolecules. On the other hand, they lack some biological properties which are found in naturally derived polymers. Because natural polymers are characterised by fast degradation, high hydrophilicity and porosity they induce rapid release of molecules (Silva and Mooney 2010; Freeman and Kelly 2017; Leijten et al. 2017). By combining synthetic and natural polymers, structural and spatiotemporal properties can be tailored accordingly. For example, a 3D printed PCL/PLGA/β-TCP supporting matrix was infused with collagen hydrogel as a carrier of recombinant human BMP-2 to improve the healing process in bone defects (Zhao et al. 2015).

Despite the advantage of immobilising molecules by surface coating for localised factor release, limitations include initial burst release of the molecules caused by the exposure of the coated material to wet conditions (Park et al. 2016). It is important then to select materials with appropriate degradation rates that are able to entrap growth factors and release them in a controlled-temporal manner. Use of encapsulation methods to control the release process has showed some success, but they can fail sometimes in retaining the bioactivity of the molecules after the fabrication process. Several printing techniques use radiation (e.g. LAB, SLA) or heating (e.g. SLS, FDM, MEW) that could degrade or denature some factors. To solve these issues, auxiliary carriers such as microspheres or nanoparticles can be added in the scaffold to protect the biomolecules during the printing process. Poldevaart et al. investigated the controlled release of VEGF using gelatin microparticles (GMPs) within 3D printed Matrigel/alginate scaffolds (Poldervaart et al. 2014). The vasculogenic capacity of the printed constructs, containing VEGF and human endothelial progenitor cells (EPCs), was investigated in vivo in nude mice. One week after implantation, vessel formation was superior in constructs that released VEGF in a sustained manner compared to the fast release group.

9.4.3.2 Heterogeneous Immobilisation of Molecules

Due to the heterogenous nature of native tissues, homogeneous immobilisation of biomolecules likely does not represent the optimal strategy to promote the development of a specific tissue phenotype. 3D printing offers the flexibility to develop constructs with incorporated biological cues that can be deposited in complex patterns in order to promote tissue regeneration in a spatiotemporal manner. Specifically, scaffolds characterised by heterogeneous architectures and molecule patterning can be engineered to influence cell fate by controlling the dose release rate spatiotemporally (Park et al. 2016). 3D printing was used to fabricate a temporomandibular joint (TMJ) disc scaffold for a rabbit model. The implant was made of PCL filaments blended with PLGA microspheres loaded with either CTGF and TGF-β3 to mimic the multiphase fibrocartilage of the native tissue. After 4 weeks of implantation, the CTGF/TGF-β3-microspheres embedded scaffolds resulted in the regeneration of superior tissue compared to the degenerated TMJ disc in the control group where the scaffold contained empty microspheres. For meniscus tissue engineering applications, Lee et al. (2014a) developed a 3D printed PCL meniscus-shaped implant loaded with PLGA microspheres to control the release of both CTGF and TGF-β3. The spatial controlled release of CTGF and TGF-β3 induced cells to form region-specific collagen type I and II matrices, similar to native tissues. When grafted in place of a resected meniscus in an ovine model, the growth factor-releasing construct was found to facilitate the regeneration of the tissue with zone-specific matrix type: type I collagen in the outer region and type II collagen in the inner region, and to restore mechanical functionality 3 months after implantation. Park et al. (2015) developed large-volume structures for treating bone defects by 3D printing human dental pulp stem cells (DPSCs) with BMP-2 in the peripheral region of the construct and VEGF in the central region which was more hypoxic. BMP-2 was encapsulated in collagen type I hydrogel to obtain a sustained release of the growth factor to induce bone regeneration, whereas VEGF was rapidly released through an alginate-gelatin hydrogel to induce vascularisation. After 4 weeks implantation in the back of a mouse, new microvessels formation was observed in the central area of the scaffold while spontaneous angiogenesis occurred in the periphery of the implanted graft. In another study, inkjet printing technology was used to create combinatorial arrays of multiple growth factors by immobilising concentration gradients of BMP-2 and FGF-2 on fibrin-coated surfaces (Miller et al. 2009). This printing strategy was used to evaluate how different growth factor patterns influence the fate of mouse muscle-derived stem cells. It was observed that as the surface concentration of FGF-2 increased, the expression of ALP decreased on the BMP-2 patterns.

9.5 Strategies Towards the Bioprinting of Functional Tissues and Organs

Advances in 3D printing have enabled the production of structures with a defined microarchitecture, multiple cell types and tissue-like biological characteristics. The advantages of using 3D printing for developing organ/tissue models include (Pati et al. 2016).

- Automated and reproducible fabrication of cell-laden scaffolds
- Creation of tissue-like structural, mechanical and biological environments
- Generation of a cell-specific microenvironment
- Fabrication of vascular network
- Integration of microfluidics
- Evaluation/screening of drugs/chemicals on tissue-specific models

The following section will outline some of the key advances that may ultimately enable the printing of functional tissues and organs. In particular, we will describe developments to vascularise printed tissues, which is integral to the engineering of most tissues and organs.

9.5.1 Bioprinting Musculoskeletal Tissues

9.5.1.1 Bone

Musculoskeletal bioprinting is one of the most established applications of bioprinting. To date, it has primarily focused on bone and cartilage regeneration (Li et al. 2016). The capacity to fabricate anatomically accurate, patient-specific implants is particularly appealing in musculoskeletal applications. Despite efforts to improve the mechanical properties of bioinks, they are generally considered too mechanically weak for load-bearing applications. As such, multi-material bioprinting is increasingly employed to reinforce the 'soft' bioink (Daly et al. 2017).

In one co-printing approach, human amniotic fluid-derived stem cells (hAFSCs) were bioprinted within a composite hydrogel alongside a mixture of PCL and tricalcium phosphate (TCP). After 10 days of osteogenic culture, the constructs were

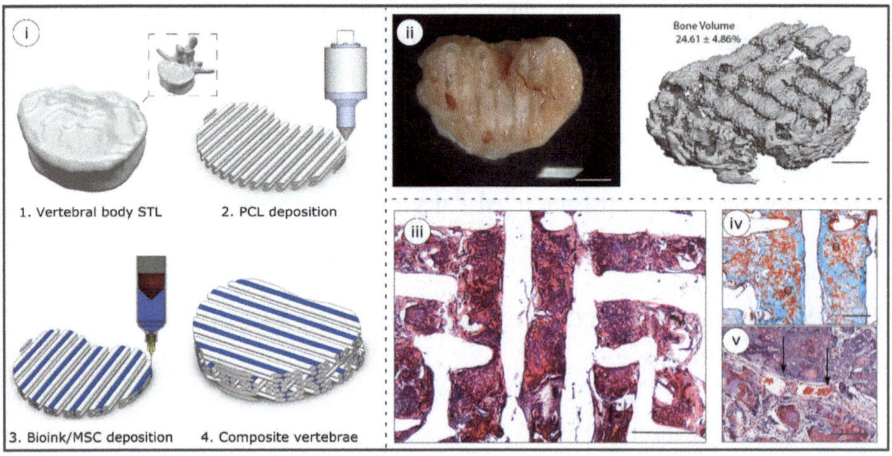

Fig. 9.3 Strategies towards bioprinting functional bone organs. (i) A 3D co-printing strategy for bone tissue engineering. Depiction of the sequential deposition of a PCL support framework and a cell-laden hydrogel to form a reinforced composite structure in the shape of a vertebrae. (ii) Development of a vascularised bone tissue after 12 weeks in vivo, which mimicked the geometry of a vertebral body, demonstrated via macroscopic and μCT Imaging (Scale Bar, 2 mm). (iii) H&E staining of construct seen in (ii) at quarter section (Scale Bar 1 mm). (iv) goldners trichrome staining of construct (red; non-mineralised bone, green; mineralised bone) (scale bar 400 μm). (v) H&E staining demonstrating perfused vasculature within the construct (Black arrows indicate blood vessel) (scale Bar 200 μm). (Daly et al. 2016)

implanted into a calvarial bone defect, where they showed newly formed, vascularised bone throughout the defect after 5 months (Kang et al. 2016). Similarly, co-printing of PCL and a MSC laden alginate bioink, using piston-assisted pneumatic microextrusion, has been used to biofabricate a construct that mimicked a human vertebral body. The construct developed into a whole bone organ in vivo via endochondral ossification of a hypertrophic cartilage rudiment (Fig. 9.3) (Daly et al. 2016).

9.5.1.2 Cartilage

3D bioprinting's capacity to spatially organise biologics has been proposed as a means of recreating the regional heterogeneity of native cartilage tissues. Mechanical piston microextrusion has been used to bioprint cell-laden, thermally responsive collagen hydrogels into anatomically accurate, spatially heterogeneous constructs. The precise 3D deposition permitted the biofabrication of distinct regions with unique compositions and mechanical properties, whilst maintaining high cell viability (~90%) throughout (Rhee et al. 2016). Thermal ink-jet bioprinting has been demonstrated as a means of in situ bioprinting for direct articular cartilage repair (Cui et al. 2012). Poly(ethylene glycol) dimethacrylate (PEGDMA), deposited into an ex vivo human chondral defect, was simultaneously crosslinked using UV light. The result was precise immobilisation of the seeded chondrocytes and improved cell viability. Articular cartilage exhibits a highly organised, depth-dependent cellular distribution (Daly et al. 2017), the capacity to localise cells demonstrated using ink-jet bioprinting could prove advantageous for recapitulating the tissue's zonal structure. Ren et al. substantiated the putative benefit associated with bioprinting a stratified cell distribution, demonstrating that a gradient of ECM components can be established using a biomimetic cell distribution throughout the construct's depth (Ren et al. 2016).

9.5.1.3 Composite Tissues

In addition to bioprinting single tissues, bioprinting has been employed in recent attempts to biofabricate multiple tissue types within organ-scale constructs. The osteochondral unit is a highly organised organ, exhibiting a complex interface between bone and overlying articular cartilage. One attempt to bioprint an osteochondral unit utilised a structural thermo-responsive polymer poly(L-lactic acid-co-glycolic acid), and a cell-laden alginate enriched with either a cartilage-derived ECM, or hydroxyapatite for the chondral and osseous regions respectively. The two enhanced bioinks were deposited using a screw-driven microextrusion system into discrete layers of the osteochondral implant. The addition of the ECM components of bone and cartilage caused the upregulation of chondrogenic and osteogenic markers of human stem cells in the cartilage and bone layers respectively, indicating that bioprinting had not negatively impacted the efficacy of the biological cargo or the plasticity of the encapsulated stem cells (Yang et al. 2015). Osteogenic (recombinant human bone morphogenic protein—BMP-2) and chondrogenic (transforming growth factor—TGF-β) growth factors have also been included within hydrogel bioinks to direct stem cell fate within a multi-layered osteochondral construct. The inclusion of TGF-β and BMP-2 in chondral and osseous bioinks respectively appeared to have a potent impact both in vitro and in vivo, resulting in heterogeneous tissue development and the effective repair of an osteochondral defect model (Shim et al. 2016). Aside from osteochondral tissue regeneration, another example of the recreation of a complex tissue interface using 3D bioprinting is the muscle-tendon functional unit. One method of bioprinting this composite tissue involved the co-printing of PCL support structures, pluronic F-127 as a sacrificial material, and a cell-laden composite hydrogel (Kang et al. 2016). The encapsulated mouse C2C12 myoblasts demonstrated unidirectional organisation along the longitudinal axis of the printed fibres. The bioprinted skeletal muscle was implanted ectopically in vivo and connected to the host nervous system. By 2 weeks the construct exhibited organised muscle fibres, innervating capability, and infiltration of the host nerve and vasculature, demonstrating that the bioprinted muscle construct was suitably robust to maintain its structural characteristics and induce integration (Kang et al. 2016).

Advancements in bioprinter technologies are set to further the capacity to control the spatial arrangement and presentation of multiple physical and chemical cues within a single construct. One recent, noteworthy technological development is a bioprinting apparatus capable of depositing up to seven bioinks through a single orifice (Liu et al. 2017). This continuous, multi-material pneumatic extrusion system could prove incredibly powerful for the biofabrication of composite tissues, giving the user control over deposition of homogenous bioinks, or mixtures in any composition (Bittner et al. 2018). In addition to technological advances, researchers have investigated bioprinting cell spheroids and 'tissue strands' intended as organoid building blocks for reaching the ambitious goal of bioprinting organs (Marga et al. 2012; Yu et al. 2016).

9.5.2 Bioprinting Vascular Tissue

The formation of blood vessels involves a complex set of cell-cell interactions. Cells must proliferate, elongate, undergo lumenogenesis and tubulogenesis; and finally, stabilisation. This process is relatively slow, the average growth rate of newly developing microvessels is only ~5 μm/h (Utzinger et al. 2015), and it also produces networks which are heterogeneous and unpredictable. A more efficient approach which many researchers are now undertaking is to directly pre-fabricate microchannels into tissue engineered constructs. This approach enables a faster and more controlled establishment of a vascular network as well as the possibility to design locations for either natural or surgical vascular anastomosis.

9.5.2.1 The Incorporation of Seeded and Unseeded Perfusable Channels

The ideal vascular network needs to ensure that all the cells are within 200 μm of a blood source. To achieve this, a very fine distribution of vessels is needed. In vivo, this has been achieved through the organisation of the vascular network into a vascular tree; large vessels branching into smaller vessels which in turn diverge into capillaries which are distributed throughout the tissue. Many approaches have been investigated to create these desired patterns of perfusable channels within tissue engineered constructs to mimic in vivo vasculature.

One method is to use a sacrificial material to fabricate a vascular network template which can be embedded within a hydrogel and later removed, thereby creating hollow channels to facilitate vascularisation. Bellan et al. created microchannels within a biocompatible gelatin hydrogel construct, using a sacrificial melt-spun shellac microfiber network. They exploited the pH-sensitive solubility of shellac by changing the pH of the composite construct by submerging in a warm ammonia bath. This enabled the dissolution of the sacrificial network without the use of organic solvents or extreme temperatures (Bellan et al. 2012). Miller et al. generated a cytocompatible means of creating densely populated tissue constructs containing perfusable vascular channels (Miller et al. 2012). Sacrificial carbohydrate glass was structured using a 3D printer to create a filamentous carbohydrate glass lattice. This lattice was then protectively coated using poly (D-lactide-co-glycolide) (PDLGA) and cast within a cell-laden hydrogel. The carbohydrate glass was then dissolved away in cell culture medium to create embedded channels (Fig. 9.4a). They demonstrated this method's versatility by using a wide range of both natural and synthetic hydrogels: agarose, alginate, PEG, fibrin and Matrigel, which exhibit a variety of crosslinking approaches including chain entanglements, ionic interactions, photo-polymerisation, enzymatic activity and protein precipitation respectively. Although this is a major step forward in the field, it requires a hybrid printed/moulded technique which may limit its application to relatively simple tissue geometries.

Kolesky et al. (2014) endeavoured to overcome these shortcomings by employing a 3D bioprinting method to produce complex vascularised tissue containing multiple cell types. They developed a custom-built 3D bioprinter and incorporated a fugitive ink (containing pluronic F127) into the printing process (Wu et al. 2011). GelMA was utilised as the cell carrier and bulk matrix. Using these two bioinks, vascular networks could be embedded within GelMA, photo-polymerised and then

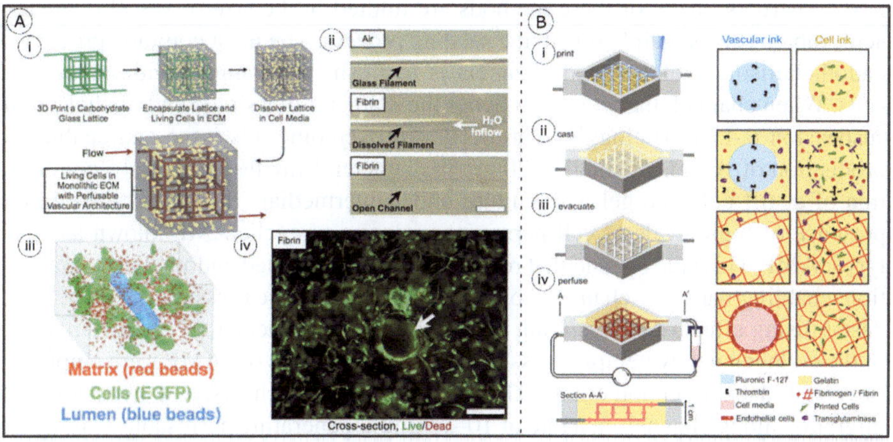

Fig. 9.4 (**a**) (i) Schematic overview of an open, interconnected, self-supporting carbohydrate-glass lattice is printed to serve as the sacrificial element for the casting of 3D vascular architectures (ii), A single carbohydrate-glass fibre (200 μm in diameter, top) encapsulated in a fibrin gel after which the carbohydrates can be dissolved resulting in a channel (middle). Removal of the filament yields an open perfusable channel (bottom, scale bar, 500 μm) (iii) Schematic of embedded cells surrounding channels. (iv) Representative cross-section image of unlabelled HUVEC (1×10^6 ml^{-1}) and 10T1/2 (1×10^6 ml^{-1}) co-cultures encapsulated in the interstitial space of a fibrin gel (10 mg ml^{-1}) with perfusable networks after 2 days in culture stained with a fluorescent live/dead assay (green, Calcein AM; red, Ethidium Homodimer). Adapted from (Miller et al. 2012). (**b**) Schematic illustration of the 3D vascularised tissue fabrication. (i) Fugitive (vascular) ink is printed within a 3D perfusion chip. (ii) ECM material is then cast over the printed inks. After casting, thrombin induces fibrinogen polymerisation into fibrin. TG diffuses from the molten casting matrix and slowly crosslinks the gelatin and fibrin. (iii) Upon cooling, the fugitive ink liquefies and is evacuated, leaving behind a pervasive vascular network, which is (iv) endothelialised and perfused via an external pump

cooled to allow the extraction of the pluronic by gentle vacuum leaving behind perfusable interconnecting vessels. The embedded vasculature was shown to be easily endothelialised by flushing the vessels with a HUVEC cell suspension. In a follow-up study, Kolesky et al. also utilised the fugitive ink Pluronic F-127 to fabricate cell-laden tissues exceeding 1 cm in thickness containing embedded vasculature using a gelatin-fibrin matrix which is crosslinked by a dual enzymatic, thrombin and transglutaminase (TG) strategy (Kolesky et al. 2016) (Fig. 9.4b).

Several other bioprinting compatible methods have been shown in literature to create hollow vessel-like channels within tissue engineered constructs. Lee et al. (2016) used gelatin as a sacrificial template, printed between collagen. The construct could then be simply heated to 37 °C which allowed the selective liquidation of the gelatin which could be drained out of the construct leaving behind channels. Hammer et al. used a coaxial flow of alginate and calcium chloride solution to fabricate microfibers with sizes ranging 150–200 mm. Cell-laden microfibers were encapsulated within 3D hydrogels and then exposed the EDTA to dissolve the alginate microfibers in order to leave behind microchannels containing cells along the luminal surfaces (Hammer et al. 2014).

Despite the successes of direct seeding as an approach to produce prevascularised constructs, many of these methods are limited in the size of vessel diameters achievable. This is mainly due to the seeding process. The most common process of endothelialising prefabricated channels is to flush the channels with endothelial cells. When channel diameter approaches the diameter of a single endothelial cell in suspension (~15 μm) clogging can occur. To overcome this issue and enable the production of vessels as small as 20 μm in diameter, Linville et al. used a combination of a stiff collagen gel with culture under intermediate reverse pressure in the presence of high levels of cell-permeant analogue of cyclic AMP known as dibutyryl cAMP. This study found that reverse pressure promoted endothelial cell migration, enabling narrower channels to be seeded. This reverse pressure however, had a destabilising effect of the resulting endothelium. To counteract this affect stiffening of the collagen gel by genipin crosslinking and high levels of cAMP were employed to re-stabilise the endothelium. This combined approach resulted in endothelial migration along channels at a rate of 10–15 μm/h, generating perfusable 1-mm-long vessels in 3 days or less (Linville et al. 2016).

Recently, a new surgical technique has been shown to directly connect host femoral arteries of Wistar rats to patterned microvessel networks (Sooppan et al. 2016). This demonstrates that 3D-printed vascularised constructs can be rapidly perfused immediately upon implantation in vivo and may help overcome the challenge of insufficient oxygen and nutrient delivery to engineered vascularised tissues implanted in vivo.

9.5.2.2 Direct Printing of Cell Patterning

Another more direct approach which is emerging in the field in order to achieve controlled vascular networks is to directly pattern cells within printed constructs. There are multiple methods employed to achieve this, including the 2D patterning of cells using cell sheet technology followed by the stacking of these of cells sheets to create 3D vascular networks (Tsuda et al. 2007). However, the process of 3D bioprinting is arguably the technique which has achieved the highest level of control of the production of 3D vascular systems. There are many bioprinting methods used in the field including stereolitography (Dhariwala et al. 2004; Albrecht et al. 2006), cell sheet lamination (Shimizu 2002; Haraguchi et al. 2012), inkjet printing (Boland et al. 2006, 2003; Xu et al. 2005, 2006), laser assisted (Barron et al. 2004; Ringeisen et al. 2004; Schiele et al. 2010; Yan et al. 2013) and extrusion based-printing (Khalil and Sun 2007; Yan et al. 2003; Duan et al. 2013; Ghorbanian et al. 2014; Beyer et al. 2013).

Xu et al. reported the production of 3D zigzag cellular tubes that were fabricated using a Z-shape platform-assisted inkjet printing system (Xu et al. 2012). This method, which the authors have coined scaffold-free bioprinting, mixed 3T3 fibroblast cells in 1% sodium alginate solution which was deposited drop-wise into a crosslinker pool, these sodium alginate droplets in turn fused to form tubular structures. This method however required very long printing times.

A novel 3D method has recently been reported which uses a coaxial nozzle to print hollow alginate filaments. These filaments enable the simultaneous fabrication of a high tensile strength (0.116 MPa) cell-laden hydrogel with incorporated microchannels. The coaxial nozzle enables sodium alginate solution to be dispensed through the outer tube of the coaxial nozzle, while the calcium chloride solution is dispensed through the inner of the coaxial nozzle. Where the two solutions intermix, the alginate solution can crosslink resulting in a hollow alginate filament being formed. The outer alginate, which is not in contact with the crosslinker, stays ungelled, thereby allowing two filaments printed adjacently to one another to fuse, thereby enabling the 3D assembly of these hollow fibres (Gao et al. 2015).

9.5.3 Emerging Applications

9.5.3.1 Liver

The liver has a key role in synthesising vital proteins and in the metabolism of xenobiotic (Bhatia et al. 2014). In general, it has a strong regeneration capacity, but this process takes time. Moreover, there is shortage of healthy donor when liver transplantation is needed (Bhatia et al. 2014; Yoon No et al. 2015; Li et al. 2016). The fabrication of an in vitro liver model represents a difficult task as this organ is characterised by a complex microenvironment and cell combinations (Bhatia et al. 2014). Ma et al. (2016) developed 3D assemblies mimicking the hexagonal liver lobule units which are known to be an important factor in regulating the liver functions (Yoon No et al. 2015). Human-induced pluripotent stem cells-derived hepatic progenitor cells (hiPSCs-HPCs) were embedded in a gelMA bioink that mimicked the stiffness of healthy liver tissues, whereas human umbilical vein endothelial cells (HUVECs) and human adipose-derived stem cells (hADSCs) were mixed with glycidal methacrylate-hyaluronic acid (GMHA)-gelMA hydrogels in order to promote vascularisation. The obtained hepatic model was able to mimic the morphological organisation, and promote metabolic product secretion and liver-specific gene expression. Chang et al. (2010) fabricated microfluidic devices containing 3D bioprinted liver cells within alginate hydrogels in defined micropatterns mimicking the cells microenvironment. This system can be used as a liver tissue analogue to assess drugs toxicity as well as to simulate in vitro experiments.

9.5.3.2 Disease Models

Clinically relevant disease models are essential to find advanced and effective treatments. In particular, cancer is one of the most common diseases causing death whose progression is affected by the interaction between a complex microenvironment, multiple type of cells and ECM molecules (Jang et al. 2016). 3D printing offers the opportunity of developing complex in vitro systems able to simulate pathophysiological microenvironments to enable to study diseases. As an example, Zhao et al. (2014) developed an in vitro cervical tumour model by 3D printing Hela cells-laden gelatin/alginate/fibrinogen hydrogels to mimic the ECM characteristics and cervical cancer microenvironment. Compared to a 2D culture model, Hela cells in the printed 3D environment showed higher proliferation, enhanced MMP protein expression and higher chemoresistance. This suggested that the 3D printed models represented a promising platform for studying 3D tumour biology.

9.6 Summary

In summation, we have defined a bioink as any biomaterial containing a biologic (a cell, nucleic acid or biomolecule) which can be dispensed through a printhead whilst maintaining the viability or functionality of the contained biologic. Although there are new bioink formulations continually emerging, the backbones of most bioinks stem from biomaterials which have previously been used in traditional tissue engineering approaches and can be classified into two broad categories: naturally derived bioinks and synthetic bioinks. There are many factors to take into consideration when choosing a bioink for any particular application. The success of 3D bioprinted tissue replacements is highly determined by the fate of cells within the bioprinted constructs. Factors such as bioink stiffness, composition and incorporation of biologics can play a key role in directing cell differentiation and proliferation towards a specific lineage. To date, significant work has been undertaken towards tuning functional bioinks specific to recapitulating the complex local environment of native tissues. The combination of such different factors offers a promising approach towards the fabrication of biomimetic implants. However, there are a number of challenges still to overcome before 3D bioprinting can enable the biofabrication of physiologically relevant tissues and organs.

References

Albrecht, D. R., et al. (2006). Probing the role of multicellular organization in three-dimensional microenvironments. *Nature Methods, 3*(5), 369–375. https://doi.org/10.1038/nmeth873.

Almeida, H. V., et al. (2015). Coupling freshly isolated CD44$^+$ infrapatellar fat pad-derived stromal cells with a TGF-β3 eluting cartilage ECM-derived scaffold as a single-stage strategy for promoting chondrogenesis. *Advanced Healthcare Materials, 4*(7), 1043–1053. https://doi.org/10.1002/adhm.201400687.

Alsberg, E., et al. (2003). Regulating bone formation *via* controlled scaffold degradation. *Journal of Dental Research, 82*(11), 903–908. https://doi.org/10.1177/154405910308201111.

Anseth, K. S., Bowman, C. N., & Brannon-Peppas, L. (1996). Mechanical properties of hydrogels and their experimental determination. *Biomaterials, 17*(17), 1647–1657. https://doi.org/10.1016/0142-9612(96)87644-7.

Armstrong, J. P. K., et al. (2016). 3D bioprinting using a templated porous bioink. *Advanced Healthcare Materials, 5*(14), 1724–1730. https://doi.org/10.1002/adhm.201600022.

Aspden, R. M. (1994). Fibre reinforcing by collagen in cartilage and soft connective tissues. *Proceedings of the Royal Society B: Biological Sciences, 258*(1352), 195–200. https://doi.org/10.1098/rspb.1994.0162.

Axpe, E., et al. (2016). Applications of alginate-based bioinks in 3D bioprinting. *International Journal of Molecular Sciences, 17*(12), 1976. https://doi.org/10.3390/ijms17121976.

Barron, J. A., et al. (2004). Biological laser printing: A novel technique for creating heterogeneous 3-dimensional cell patterns. *Biomedical Microdevices, 6*(2), 139–147. https://doi.org/10.1023/B:BMMD.0000031751.67267.9f.

Bas, O., et al. (2015). Enhancing structural integrity of hydrogels by using highly organised melt electrospun fibre constructs. *European Polymer Journal, 72*, 451–463. https://doi.org/10.1016/j.eurpolymj.2015.07.034.

Bas, O., et al. (2017). Biofabricated soft network composites for cartilage tissue engineering. *Biofabrication, 9*(2), 025014. https://doi.org/10.1088/1758-5090/aa6b15.

Bellan, L. M., et al. (2012). A 3D interconnected microchannel network formed in gelatin by sacrificial shellac microfibers. *Advanced Materials, 24*(38), 5187–5191. https://doi.org/10.1002/adma.201200810.

Beyer, S. T., et al. (2013). 3D alginate constructs for tissue engineering printed using a coaxial flow focusing microfluidic device. In *2013 Transducers & Eurosensors XXVII: The 17th International Conference on Solid-State Sensors, Actuators and Microsystems (TRANSDUCERS & EUROSENSORS XXVII)* (pp. 1206–1209). Washington: IEEE. https://doi.org/10.1109/Transducers.2013.6626990.

Bhatia, S. N., et al. (2014). Cell and tissue engineering for liver disease. *Science Translational Medicine, 6*(245), 245sr2. https://doi.org/10.1126/scitranslmed.3005975.

Bissell, M. J., & Aggeler, J. (1987). Dynamic reciprocity: how do extracellular matrix and hormones direct gene expression? *Progress in Clinical and Biological Research, 249*, 251–262.

Bittner, S. M., et al. (2018, March). Three-dimensional printing of multilayered tissue engineering scaffolds. *Materials Today, 21*, 861. https://doi.org/10.1016/j.mattod.2018.02.006.

Blaeser, A., et al. (2013). Biofabrication under fluorocarbon: A novel freeform fabrication technique to generate high aspect ratio tissue-engineered constructs. *BioResearch Open Access, 2*(5), 374–384. https://doi.org/10.1089/biores.2013.0031.

Boere, K. W. M., et al. (2014). Covalent attachment of a three-dimensionally printed thermoplast to a gelatin hydrogel for mechanically enhanced cartilage constructs. *Acta Biomaterialia, 10*(6), 2602–2611. https://doi.org/10.1016/j.actbio.2014.02.041.

Boland, T., et al. (2003). Cell and organ printing 2: Fusion of cell aggregates in three-dimensional gels. *The Anatomical Record, 272A*(2), 497–502. https://doi.org/10.1002/ar.a.10059.

Boland, T., et al. (2006). Application of inkjet printing to tissue engineering. *Biotechnology Journal, 1*(9), 910–917. https://doi.org/10.1002/biot.200600081.

Boontheekul, T., Kong, H.-J., & Mooney, D. J. (2005). Controlling alginate gel degradation utilizing partial oxidation and bimodal molecular weight distribution. *Biomaterials, 26*(15), 2455–2465. https://doi.org/10.1016/J.BIOMATERIALS.2004.06.044.

Brown, B. N., & Badylak, S. F. (2015). Extracellular matrix as an inductive scaffold for functional tissue reconstruction. *Translating Regenerative Medicine, 163*(4), 11–29. https://doi.org/10.1016/B978-0-12-800548-4.00002-4.

Cen, L., et al. (2008). Collagen tissue engineering: Development of novel biomaterials and applications. *Pediatric Research, 63*(5), 492–496. https://doi.org/10.1203/PDR.0b013e31816c5bc3.

Chang, R., et al. (2010). Biofabrication of a three-dimensional liver micro-organ as an in vitro drug metabolism model. *Biofabrication, 2*(4), 045004. https://doi.org/10.1088/1758-5082/2/4/045004.

Chimene, D., et al. (2016). Advanced bioinks for 3D printing: A materials science perspective. *Annals of Biomedical Engineering, 44*(6), 2090–2102. https://doi.org/10.1007/s10439-016-1638-y.

Chung, H. J., & Park, T. G. (2007). Surface engineered and drug releasing pre-fabricated scaffolds for tissue engineering. *Advanced Drug Delivery Reviews, 59*(4–5), 249–262. https://doi.org/10.1016/j.addr.2007.03.015.

Critchley, S. E., & Kelly, D. J. (2017). Bioinks for bioprinting functional meniscus and articular cartilage. *Journal of 3D Printing in Medicine, 1*(4), 269–290. https://doi.org/10.2217/3dp-2017-0012.

Cui, X., & Boland, T. (2009). Human microvasculature fabrication using thermal inkjet printing technology. *Biomaterials, 30*(31), 6221–6227. https://doi.org/10.1016/J.BIOMATERIALS.2009.07.056.

Cui, X., et al. (2012). Direct human cartilage repair using three-dimensional bioprinting technology. *Tissue Engineering Part A, 18*(11–12), 1304–1312. https://doi.org/10.1089/ten.tea.2011.0543.

Daly, A. C., et al. (2016). 3D bioprinting of developmentally inspired templates for whole bone organ engineering. *Advanced Healthcare Materials, 5*(18), 2353–2362. https://doi.org/10.1002/adhm.201600182.

Daly, A. C., et al. (2017). 3D bioprinting for cartilage and osteochondral tissue engineering. *Advanced Healthcare Materials, 6*, 1700298. https://doi.org/10.1002/adhm.201700298.

Daly, A. C., et al. (2018). 3D printed microchannel networks to direct vascularisation during endochondral bone repair. *Biomaterials, 162*, 34–46. https://doi.org/10.1016/J.BIOMATERIALS.2018.01.057.

Dhariwala, B., Hunt, E., & Boland, T. (2004). Rapid prototyping of tissue-engineering constructs, using photopolymerizable hydrogels and stereolithography. *Tissue Engineering, 10*(9–10), 1316–1322. https://doi.org/10.1089/ten.2004.10.1316.

Di Giuseppe, M., et al. (2018). Mechanical behaviour of alginate-gelatin hydrogels for 3D bioprinting. *Journal of the Mechanical Behavior of Biomedical Materials, 79*, 150–157. https://doi.org/10.1016/J.JMBBM.2017.12.018.

Du, M., et al. (2015). 3D bioprinting of BMSC-laden methacrylamide gelatin scaffolds with CBD-BMP2-collagen microfibers. *Biofabrication, 7*(4), 044104. https://doi.org/10.1088/1758-5090/7/4/044104.

Duan, B., et al. (2013). 3D bioprinting of heterogeneous aortic valve conduits with alginate/gelatin hydrogels. *Journal of Biomedical Materials Research. Part A, 101*(5), 1255–1264. https://doi.org/10.1002/jbm.a.34420.

Engler, A. J., et al. (2006). Matrix elasticity directs stem cell lineage specification. *Cell, 126*(4), 677–689. https://doi.org/10.1016/j.cell.2006.06.044.

Eslami, M., et al. (2014). Fiber-reinforced hydrogel scaffolds for heart valve tissue engineering. *Journal of Biomaterials Applications, 29*(3), 399–410. https://doi.org/10.1177/0885328214530589.

Eyrich, D., et al. (2007). Long-term stable fibrin gels for cartilage engineering. *Biomaterials, 28*(1), 55–65. https://doi.org/10.1016/j.biomaterials.2006.08.027.

Fedorovich, N. E., et al. (2007). Hydrogels as extracellular matrices for skeletal tissue engineering: State-of-the-art and novel application in organ printing. *Tissue Engineering, 13*(8), 1905–1925. https://doi.org/10.1089/ten.2006.0175.

Fedorovich, N. E., et al. (2008). Three-dimensional fiber deposition of cell-laden, viable, patterned constructs for bone tissue printing. *Tissue Engineering Part A, 14*(1), 127–133. https://doi.org/10.1089/ten.a.2007.0158.

Freeman, F. E., & Kelly, D. J. (2017). Tuning alginate bioink stiffness and composition for controlled growth factor delivery and to spatially direct MSC Fate within bioprinted tissues. *Scientific Reports, 7*(1), 1–12. https://doi.org/10.1038/s41598-017-17286-1.

Gaharwar, A. K., Peppas, N. A., & Khademhosseini, A. (2015). Nanocomposite hydrogels for biomedical applications. *Biotechnology and Bioengineering, 111*(3), 441–453. https://doi.org/10.1002/bit.25160.Nanocomposite.

Gao, G., et al. (2014). Bioactive nanoparticles stimulate bone tissue formation in bioprinted three-dimensional scaffold and human mesenchymal stem cells. *Biotechnology Journal, 9*(10), 1304–1311. https://doi.org/10.1002/biot.201400305.

Gao, Q., et al. (2015). Coaxial nozzle-assisted 3D bioprinting with built-in microchannels for nutrients delivery. *Biomaterials, 61*, 203–215. https://doi.org/10.1016/j.biomaterials.2015.05.031.

Gao, G., et al. (2017). Tissue engineered bio-blood-vessels constructed using a tissue-specific bioink and 3D coaxial cell printing technique: A novel therapy for ischemic disease. *Advanced Functional Materials, 27*(33), 1700798. https://doi.org/10.1002/adfm.201700798.

Ghorbanian, S., et al. (2014). Microfluidic direct writer with integrated declogging mechanism for fabricating cell-laden hydrogel constructs. *Biomedical Microdevices, 16*(3), 387–395. https://doi.org/10.1007/s10544-014-9842-8.

Gomez, C. G., Rinaudo, M., & Villar, M. A. (2007). Oxidation of sodium alginate and characterization of the oxidized derivatives. *Carbohydrate Polymers, 67*(3), 296–304. https://doi.org/10.1016/J.CARBPOL.2006.05.025.

Groll, J., et al. (2018). A definition of bioinks and their distinction from biomaterial inks. *Biofabrication, 11*(1), 013001. https://doi.org/10.1088/1758-5090/aaec52.

Guilak, F., et al. (2009). Control of stem cell fate by physical interactions with the extracellular matrix. *Cell Stem Cell, 5*(1), 17–26. https://doi.org/10.1016/j.stem.2009.06.016.

Gwon, S. H., et al. (2015). Gelation dynamics of ionically crosslinked alginate gel with various cations. *Macromolecular Research, 23*(12), 1112–1116. https://doi.org/10.1007/s13233-015-3151-9.

Hammer, J., et al. (2014). A facile method to fabricate hydrogels with microchannel-like porosity for tissue engineering. *Tissue Engineering. Part C, Methods, 20*(2), 169–176. https://doi.org/10.1089/ten.TEC.2013.0176.

Haraguchi, Y., et al. (2012). Scaffold-free tissue engineering using cell sheet technology. *RSC Advances, 2*(6), 2184. https://doi.org/10.1039/c2ra00704e.

Hinton, T. J., et al. (2015). Three-dimensional printing of complex biological structures by freeform reversible embedding of suspended hydrogels. *Science Advances, 1*(9), e1500758. https://doi.org/10.1126/sciadv.1500758.

Hong, S., et al. (2015). 3D printing: 3D printing of highly stretchable and tough hydrogels into complex, cellularized structures. *Advanced Materials, 27*(27), 4034. https://doi.org/10.1002/adma.201570182.

Huang, S., et al. (2016). 3D bioprinted extracellular matrix mimics facilitate directed differentiation of epithelial progenitors for sweat gland regeneration. *Acta Biomaterialia, 32*, 170–177. https://doi.org/10.1016/J.ACTBIO.2015.12.039.

Inzana, J. A., et al. (2014). 3D printing of composite calcium phosphate and collagen scaffolds for bone regeneration. *Biomaterials, 35*(13), 4026–4034. https://doi.org/10.1016/j.biomaterials.2014.01.064.

Jackson, M. R. (2001). Fibrin sealants in surgical practice: An overview. *American Journal of Surgery, 182*(2 Suppl), 1S–7S.

Jaiswal, M. K., et al. (2016). Mechanically stiff nanocomposite hydrogels at ultralow nanoparticle content. *ACS Nano, 10*(1), 246–256. https://doi.org/10.1021/acsnano.5b03918.

Jang, J., Yi, H.-G., & Cho, D.-W. (2016). 3D printed tissue models: Present and future. *ACS Biomaterials Science & Engineering, 2*(10), 1722–1731. https://doi.org/10.1021/acsbiomaterials.6b00129.

Jang, J., et al. (2017). 3D printed complex tissue construct using stem cell-laden decellularized extracellular matrix bioinks for cardiac repair. *Biomaterials, 112*, 264–274. https://doi.org/10.1016/j.biomaterials.2016.10.026.

Jeon, O., et al. (2017). Highly elastic and tough interpenetrating polymer network-structured hybrid hydrogels for cyclic mechanical loading-enhanced tissue engineering. *Chemistry of Materials, 29*(19), 8425–8432. https://doi.org/10.1021/acs.chemmater.7b02995.

Ji, S., & Guvendiren, M. (2017). Recent advances in bioink design for 3D bioprinting of tissues and organs. *Frontiers in Bioengineering and Biotechnology, 5*, 23. https://doi.org/10.3389/fbioe.2017.00023.

Jiang, T., et al. (2017). Directing the self-assembly of tumour spheroids by bioprinting cellular heterogeneous models within alginate/gelatin hydrogels. *Scientific Reports, 7*(1), 4575. https://doi.org/10.1038/s41598-017-04691-9.

Jockenhoevel, S., et al. (2001). Fibrin gel – advantages of a new scaffold in cardiovascular tissue engineering. *European Journal of Cardio-Thoracic Surgery, 19*(4), 424–430. https://doi.org/10.1016/S1010-7940(01)00624-8.

Johnson, T. D., et al. (2014). Human versus porcine tissue sourcing for an injectable myocardial matrix hydrogel. *Biomaterials Science, 2*(5), 735–744. https://doi.org/10.1039/C3BM60283D.

Jukes, J. M., et al. (2008). Endochondral bone tissue engineering using embryonic stem cells. *Proceedings of the National Academy of Sciences of the United States of America, 105*(19), 6840–6845. https://doi.org/10.1073/pnas.0711662105.

Kang, S. W., et al. (2011). Surface modification with fibrin/hyaluronic acid hydrogel on solid-free form-based scaffolds followed by BMP-2 loading to enhance bone regeneration. *Bone, 48*(2), 298–306. https://doi.org/10.1016/j.bone.2010.09.029.

Kang, H.-W., et al. (2016). A 3D bioprinting system to produce human-scale tissue constructs with structural integrity. *Nature Biotechnology, 34*(3), 312–319. https://doi.org/10.1038/nbt.3413.

Kesti, M., et al. (2015). A versatile bioink for three-dimensional printing of cellular scaffolds based on thermally and photo-triggered tandem gelation. *Acta Biomaterialia, 11*, 162–172. https://doi.org/10.1016/j.actbio.2014.09.033.

Khalil, S., & Sun, W. (2007). Biopolymer deposition for freeform fabrication of hydrogel tissue constructs. *Materials Science and Engineering: C, 27*(3), 469–478. https://doi.org/10.1016/j.msec.2006.05.023.

Kim, B. S., et al. (2017). Decellularized extracellular matrix: A step towards the next generation source for bioink manufacturing. *Biofabrication, 9*(3), 034104. https://doi.org/10.1088/1758-5090/aa7e98.

Kolesky, D. B., et al. (2014). 3D bioprinting of vascularized, heterogeneous cell-laden tissue constructs. *Advanced Materials, 26*(19), 3124–3130. https://doi.org/10.1002/adma.201305506.

Kolesky, D. B., et al. (2016). Three-dimensional bioprinting of thick vascularized tissues. *Proceedings of the National Academy of Sciences of the United States of America, 113*(12), 3179–3184. https://doi.org/10.1073/pnas.1521342113.

Kong, H.-J., Lee, K. Y., & Mooney, D. J. (2002). Decoupling the dependence of rheological/mechanical properties of hydrogels from solids concentration. *Polymer, 43*(23), 6239–6246. https://doi.org/10.1016/S0032-3861(02)00559-1.

Lee, D. W., et al. (2003). Effect of γ-irradiation on degradation of alginate. *Journal of Agricultural and Food Chemistry, 51*(16), 4819–4823. https://doi.org/10.1021/jf021053y.

Lee, C. H., et al. (2010a). Regeneration of the articular surface of the rabbit synovial joint by cell homing: A proof of concept study. *The Lancet, 376*(9739), 440–448. https://doi.org/10.1016/S0140-6736(10)60668-X.

Lee, Y.-B., et al. (2010b). Bio-printing of collagen and VEGF-releasing fibrin gel scaffolds for neural stem cell culture. *Experimental Neurology, 223*(2), 645–652. https://doi.org/10.1016/J.EXPNEUROL.2010.02.014.

Lee, C. H., et al. (2014a). Protein-releasing polymeric scaffolds induce fibrochondrocytic differentiation of endogenous cells for knee meniscus regeneration in sheep. *Science Translational Medicine, 6*(266), 266ra171. https://doi.org/10.1126/scitranslmed.3009696.

Lee, J. B., et al. (2016). Development of 3D microvascular networks within gelatin hydrogels using thermoresponsive sacrificial microfibers. *Advanced Healthcare Materials, 5*, 781. https://doi.org/10.1002/adhm.201500792.

Lee, H., et al. (2017). Development of liver decellularized extracellular matrix bioink for three-dimensional cell printing-based liver tissue engineering. *Biomacromolecules, 18*(4), 1229–1237. https://doi.org/10.1021/acs.biomac.6b01908.

Leijten, J., et al. (2017). Spatially and temporally controlled hydrogels for tissue engineering. *Materials Science & Engineering R: Reports, 119*(September), 1–35. https://doi.org/10.1016/j.mser.2017.07.001.

Lesman, A., et al. (2011). Engineering vessel-like networks within multicellular fibrin-based constructs. *Biomaterials, 32*(31), 7856–7869. https://doi.org/10.1016/j.biomaterials.2011.07.003.

Li, J., et al. (2016). Recent advances in bioprinting techniques: Approaches, applications and future prospects. *Journal of Translational Medicine, 14*(1), 1–15. https://doi.org/10.1186/s12967-016-1028-0.

Li, Z., et al. (2018). Tuning alginate-gelatin bioink properties by varying solvent and their impact on stem cell behavior. *Scientific Reports, 8*(1), 8020. https://doi.org/10.1038/s41598-018-26407-3.

Liao, I.-C., et al. (2013). Composite three-dimensional woven scaffolds with interpenetrating network hydrogels to create functional synthetic articular cartilage. *Advanced Functional Materials, 23*(47), 5833–5839. https://doi.org/10.1002/adfm.201300483.

Linville, R. M., et al. (2016). Physical and chemical signals that promote vascularization of capillary-scale channels. *Cellular and Molecular Bioengineering, 9*(1), 73–84. https://doi.org/10.1007/s12195-016-0429-8.

Liu, W., et al. (2017). Rapid continuous multimaterial extrusion bioprinting. *Advanced Materials, 29*(3), 1–8. https://doi.org/10.1002/adma.201604630.

Loibl, M., et al. (2014). Direct cell-cell contact between mesenchymal stem cells and endothelial progenitor cells induces a pericyte-like phenotype in vitro. *BioMed Research International, 2014*, 395781. https://doi.org/10.1155/2014/395781.

LUAN, L. Q., et al. (2009). A study of degradation mechanism of alginate by gamma-irradiation. *Radioisotopes, 58*(1), 1–11. https://doi.org/10.3769/radioisotopes.58.1.

Ma, X., et al. (2016). Deterministically patterned biomimetic human iPSC-derived hepatic model via rapid 3D bioprinting. *Proceedings of the National Academy of Sciences of the United States of America, 113*(8), 2206–2211. https://doi.org/10.1073/pnas.1524510113.

Malda, J., et al. (2013). 25th anniversary article: Engineering hydrogels for biofabrication. *Advanced Materials, 25*(36), 5011–5028. https://doi.org/10.1002/adma.201302042.

Marga, F., et al. (2012). Toward engineering functional organ modules by additive manufacturing. *Biofabrication, 4*(2), 022001. https://doi.org/10.1088/1758-5082/4/2/022001.

Marijnissen, W. J. C. M., et al. (2002). Alginate as a chondrocyte-delivery substance in combination with a non-woven scaffold for cartilage tissue engineering. *Biomaterials, 23*(6), 1511–1517. https://doi.org/10.1016/S0142-9612(01)00281-2.

Markstedt, K., et al. (2015). 3D bioprinting human chondrocytes with nanocellulose-alginate bioink for cartilage tissue engineering applications. *Biomacromolecules, 16*(5), 1489–1496. https://doi.org/10.1021/acs.biomac.5b00188.

Miller, E., et al. (2009). Inkjet printing of growth factor concentration gradients and combinatorial arrays immobilized on biologically-relevant substrates. *Combinatorial Chemistry & High Throughput Screening, 12*(6), 604–618. https://doi.org/10.2174/138620709788681907.

Miller, J. S., et al. (2012). Rapid casting of patterned vascular networks for perfusable engineered three-dimensional tissues. *Nature Materials, 11*(9), 768–774. https://doi.org/10.1038/nmat3357.

Mosesson, M. W. (2005). Fibrinogen and fibrin structure and functions. *Journal of Thrombosis and Haemostasis, 3*(8), 1894–1904. https://doi.org/10.1111/j.1538-7836.2005.01365.x.

Moutos, F. T., Freed, L. E., & Guilak, F. (2007). A biomimetic three-dimensional woven composite scaffold for functional tissue engineering of cartilage. *Nature Materials, 6*(2), 162–167. https://doi.org/10.1038/nmat1822.

Ouyang, L., et al. (2017). A generalizable strategy for the 3D bioprinting of hydrogels from nonviscous photo-crosslinkable inks. *Advanced Materials, 29*(8), 1604983. https://doi.org/10.1002/adma.201604983.

Park, J. Y., et al. (2015). 3D printing technology to control BMP-2 and VEGF delivery spatially and temporally to promote large-volume bone regeneration. *Journal of Materials Chemistry B, 3*(27), 5415–5425. https://doi.org/10.1039/C5TB00637F.

Park, J. Y., et al. (2016). 3D printed structures for delivery of biomolecules and cells: tissue repair and regeneration. *Journal of Materials Chemistry B, 4*(47), 7521–7539. https://doi.org/10.1039/C6TB01662F.

Pati, F., Jang, J., Ha, D.-H., et al. (2014). Printing three-dimensional tissue analogues with decellularized extracellular matrix bioink. *Nature Communications, 5*, 3935. https://doi.org/10.1038/ncomms4935.

Pati, F., Gantelius, J., & Svahn, H. A. (2016). 3D bioprinting of tissue/organ models. *Angewandte Chemie, International Edition, 55*(15), 4650–4665. https://doi.org/10.1002/anie.201505062.

Pek, Y. S., Wan, A. C. A., & Ying, J. Y. (2010). The effect of matrix stiffness on mesenchymal stem cell differentiation in a 3D thixotropic gel. *Biomaterials, 31*(3), 385–391. https://doi.org/10.1016/j.biomaterials.2009.09.057.

Perka, C., et al. (2000). Segmental bone repair by tissue-engineered periosteal cell transplants with bioresorbable fleece and fibrin scaffolds in rabbits. *Biomaterials, 21*(11), 1145–1153. https://doi.org/10.1016/S0142-9612(99)00280-X.

Poldervaart, M. T., et al. (2014). Prolonged presence of VEGF promotes vascularization in 3D bioprinted scaffolds with defined architecture. *Journal of Controlled Release, 184*(1), 58–66. https://doi.org/10.1016/j.jconrel.2014.04.007.

Ren, X., et al. (2016). Engineering zonal cartilage through bioprinting collagen type II hydrogel constructs with biomimetic chondrocyte density gradient. *BMC Musculoskeletal Disorders, 17*(1), 301. https://doi.org/10.1186/s12891-016-1130-8.

Rhee, S., et al. (2016). 3D bioprinting of spatially heterogeneous collagen constructs for cartilage tissue engineering. *ACS Biomaterials Science & Engineering, 2*, 1800. https://doi.org/10.1021/acsbiomaterials.6b00288.

Ringeisen, B. R., et al. (2004). Laser printing of pluripotent embryonal carcinoma cells. *Tissue Engineering, 10*(3–4), 483–491. https://doi.org/10.1089/107632704323061843.

Romanazzo, S., et al. (2018). Meniscus ECM-functionalised hydrogels containing infrapatellar fat pad-derived stem cells for bioprinting of regionally defined meniscal tissue. *Journal of Tissue Engineering and Regenerative Medicine, 12*(3), e1826–e1835. https://doi.org/10.1002/term.2602.

Saldin, L. T., et al. (2017). Extracellular matrix hydrogels from decellularized tissues: Structure and function. *Acta Biomaterialia, 49*, 1–15. https://doi.org/10.1016/J.ACTBIO.2016.11.068.

Schiele, N. R., et al. (2010). Laser-based direct-write techniques for cell printing. *Biofabrication, 2*(3), 032001. https://doi.org/10.1088/1758-5082/2/3/032001.

Schuurman, W., et al. (2011). Bioprinting of hybrid tissue constructs with tailorable mechanical properties. *Biofabrication, 3*(2), 021001. https://doi.org/10.1088/1758-5082/3/2/021001.

Scotti, C., et al. (2010). Recapitulation of endochondral bone formation using human adult mesenchymal stem cells as a paradigm for developmental engineering. *Proceedings of the National Academy of Sciences of the United States of America, 107*(16), 7251–7256. https://doi.org/10.1073/pnas.1000302107.

Shim, J.-H., et al. (2016). Three-dimensional bioprinting of multilayered constructs containing human mesenchymal stromal cells for osteochondral tissue regeneration in the rabbit knee joint. *Biofabrication, 8*(1), 014102. https://doi.org/10.1088/1758-5090/8/1/014102.

Shimizu, T. (2002). Fabrication of pulsatile cardiac tissue grafts using a novel 3-dimensional cell sheet manipulation technique and temperature-responsive cell culture surfaces. *Circulation Research, 90*(3), e40–e48. https://doi.org/10.1161/hh0302.105722.

Shoulders, M. D., & Raines, R. T. (2009). Collagen structure and stability. *Annual Review of Biochemistry, 78*, 929–958. https://doi.org/10.1146/annurev.biochem.77.032207.120833.

Silva, E. A., & Mooney, D. J. (2010). Effects of VEGF temporal and spatial presentation on angiogenesis. *Biomaterials, 31*(6), 1235–1241. https://doi.org/10.1016/j.biomaterials.2009.10.052.

Skardal, A., Zhang, J., & Prestwich, G. D. (2010). Bioprinting vessel-like constructs using hyaluronan hydrogels crosslinked with tetrahedral polyethylene glycol tetracrylates. *Biomaterials, 31*(24), 6173–6181. https://doi.org/10.1016/j.biomaterials.2010.04.045.

Skardal, A., et al. (2015). A hydrogel bioink toolkit for mimicking native tissue biochemical and mechanical properties in bioprinted tissue constructs. *Acta Biomaterialia, 25*, 24–34. https://doi.org/10.1016/J.ACTBIO.2015.07.030.

Smith, C. M., et al. (2004). Three-dimensional BioAssembly tool for generating viable tissue-engineered constructs. *Tissue Engineering, 10*(9–10), 1566–1576. https://doi.org/10.1089/ten.2004.10.1566.

Sooppan, R., et al. (2016). In vivo anastomosis and perfusion of a three-dimensionally-printed construct containing microchannel networks. *Tissue Engineering. Part C, Methods, 22*(1), 1–7. https://doi.org/10.1089/ten.TEC.2015.0239.

Spicer, P. P., & Mikos, A. G. (2010). Fibrin glue as a drug delivery system. *Journal of Controlled Release, 148*(1), 49–55. https://doi.org/10.1016/j.jconrel.2010.06.025.

Tan, R., et al. (2012). Thermo-sensitive alginate-based injectable hydrogel for tissue engineering. *Carbohydrate Polymers, 87*(2), 1515–1521. https://doi.org/10.1016/J.CARBPOL.2011.09.048.

Tsuda, Y., et al. (2007). Cellular control of tissue architectures using a three-dimensional tissue fabrication technique. *Biomaterials, 28*(33), 4939–4946. https://doi.org/10.1016/j.biomaterials.2007.08.002.

Utzinger, U., et al. (2015). Large-scale time series microscopy of neovessel growth during angiogenesis. *Angiogenesis, 18*(3), 219–232. https://doi.org/10.1007/s10456-015-9461-x.

Vega, S. L., Kwon, M. Y., & Burdick, J. A. (2017). Recent advances in hydrogels for cartilage tissue engineering. *European Cells & Materials, 33*, 59–75. https://doi.org/10.22203/eCM.v033a05.

Visser, J., et al. (2015). Reinforcement of hydrogels using three-dimensionally printed microfibres. *Nature Communications, 6*, 6933. https://doi.org/10.1038/ncomms7933.

Williams, D., et al. (2018). A perspective on the physical, mechanical and biological specifications of bioinks and the development of functional tissues in 3D bioprinting. *Bioprinting, 9*, 19–36. https://doi.org/10.1016/J.BPRINT.2018.02.003.

Wu, W., Deconinck, A., & Lewis, J. A. (2011). Omnidirectional printing of 3D microvascular networks. *Advanced Materials, 23*(24), 178–183. https://doi.org/10.1002/adma.201004625.

Wu, Z., Su, X., Xu, Y., Kong, B., Sun, W., Hunt, N. C., et al. (2016a). Bioprinting three-dimensional cell-laden tissue constructs with controllable degradation. *Scientific Reports, 6*, 24474. https://doi.org/10.1038/srep24474.

Wu, Z., Su, X., Xu, Y., Kong, B., Sun, W., & Mi, S. (2016b). Bioprinting three-dimensional cell-laden tissue constructs with controllable degradation. *Scientific Reports, 6*(1), 24474. https://doi.org/10.1038/srep24474.

Xavier, J. R., et al. (2015). Bioactive nanoengineered hydrogels for bone tissue engineering: A growth-factor-free approach. *ACS Nano, 9*(3), 3109–3118. https://doi.org/10.1021/nn507488s.

Xu, T., et al. (2005). Inkjet printing of viable mammalian cells. *Biomaterials, 26*(1), 93–99. https://doi.org/10.1016/j.biomaterials.2004.04.011.

XU, T., et al. (2006). Viability and electrophysiology of neural cell structures generated by the inkjet printing method. *Biomaterials, 27*(19), 3580–3588. https://doi.org/10.1016/j.biomaterials.2006.01.048.

Xu, C., et al. (2012). Scaffold-free inkjet printing of three-dimensional zigzag cellular tubes. *Biotechnology and Bioengineering, 109*(12), 3152–3160. https://doi.org/10.1002/bit.24591.

Yan, Y., et al. (2003). Layered manufacturing of tissue engineering scaffolds via multi-nozzle deposition. *Materials Letters, 57*, 2623. https://doi.org/10.1016/S0167-577X(02)01339-3.

Yan, J., et al. (2013). Laser-assisted printing of alginate long tubes and annular constructs. *Biofabrication, 5*(1), 015002. https://doi.org/10.1088/1758-5082/5/1/015002.

Yang, S. S., et al. (2015). Fabrication of an osteochondral graft with using a solid freeform fabrication system. *Tissue Engineering and Regenerative Medicine, 12*(4), 239–248. https://doi.org/10.1007/s13770-015-0001-y.

Yang, X., et al. (2018). Collagen-alginate as bioink for three-dimensional (3D) cell printing based cartilage tissue engineering. *Materials Science and Engineering: C, 83*, 195–201. https://doi.org/10.1016/J.MSEC.2017.09.002.

Yeo, M. G., et al. (2016). An innovative collagen-based cell-printing method for obtaining human adipose stem cell-laden structures consisting of core-sheath structures for tissue engineering. *Biomacromolecules, 17*(4), 1365–1375. https://doi.org/10.1021/acs.biomac.5b01764.

Yeung, T., et al. (2005). Effects of substrate stiffness on cell morphology, cytoskeletal structure, and adhesion. *Cell Motility and the Cytoskeleton, 60*(1), 24–34. https://doi.org/10.1002/cm.20041.

Yoon No, D., et al. (2015). 3D liver models on a microplatform: well-defined culture, engineering of liver tissue and liver-on-a-chip. *Lab on a Chip, 15*(19), 3822–3837. https://doi.org/10.1039/c5lc00611b.

Yu, Y., et al. (2016). Three-dimensional bioprinting using self-assembling scalable scaffold-free “tissue strands” as a new bioink. *Scientific Reports, 6*, 28714. https://doi.org/10.1038/srep28714.

Zhang, T., et al. (2013). Mechanical characterization of bioprinted in vitro soft tissue models. *Biofabrication, 5*(4), 045010. https://doi.org/10.1088/1758-5082/5/4/045010.

Zhao, Y., et al. (2014). Three-dimensional printing of Hela cells for cervical tumor model in vitro. *Biofabrication, 6*(3), 035001. https://doi.org/10.1088/1758-5082/6/3/035001.

Zhao, P., et al. (2015). In vitro and in vivo drug release behavior and osteogenic potential of a composite scaffold based on poly(ε-caprolactone)-block-poly(lactic-co-glycolic acid) and β-tricalcium phosphate. *Journal of Materials Chemistry B, 3*(34), 6885–6896. https://doi.org/10.1039/C5TB00946D.

Chapter 10
Post-processing Considerations for Biomedical 3D Printing of Polymers

Katie Hardiman

Abstract In the last 10 years, additive manufacturing has progressed from being not just a prototyping technology but a manufacturing technology within the medical manufacturing industry; it has turned many medical manufacturing sectors completely on their heads such as the dental and hearing aid device making industries. 3D printing can create models of high organic intricacy in a relatively short period of time. It has greatly added to the medical industry advancements in visualising and creating complex human or animal anatomies and can be used as an aid for surgical planning and preparation, physician and patient pre-operative/treatment education, medical procedure training and upskilling, medical devices, medical device development, and custom medical device manufacturing. Additive manufacturing (AM) is rapidly growing with advances in new material development, higher resolution, and faster speeds, making it a cost-effective alternative manufacturing solution for many medical applications. The objective of this chapter is to provide some guidelines and considerations of the many steps involved in the manufacturing of biomedical polymers and the main post-processing considerations for individual types of industrial polymer technologies.

Keywords Post-processing · Additive manufacturing · Design for manufacturing

10.1 Introduction

Additive manufacturing (AM) consists of a complex sequence of operations that are necessary to make a part meet its design specification. Amongst the many crucial steps are tasks that occur after a part is built using an AM machine and before it is ready for final inspection, testing, and certification. These tasks as a group are called post-processing. Post-processing differs depending upon the material and technology used. Nevertheless, there are commonalities among all the different platforms. These include removing surplus material from the built part's external and internal

K. Hardiman (✉)
Irish Manufacturing Research, Mullingar, Ireland

D. M. Devine (ed.), *Polymer-Based Additive Manufacturing*,
https://doi.org/10.1007/978-3-030-24532-0_10

surfaces, freeing the part from the build plate or build cake, removing or dissolving supports, ultraviolet (UV) curing and thermal curing of material, machining of the parts to final dimensional tolerances depending on product or part requirements, and processing to attain the desired surface finish (America Makes and ANSI Additive Manufacturing Standardization Collaborative (AMSC) 2017).

Post-processing of polymeric AM components is usually performed to complete chemical reactions, crosslink the microstructure of photopolymer materials and reduce residual stresses, enhance surface finish, reduce surface porosity, and/or meet geometric tolerance requirements that are specified in the part manufacturing specification, compared to the initial built part (AMSC 2017).

Generalised post-processing tasks for polymers include the following:

- Support Material Removal
- Surface Texture Improvements
- Accuracy Improvements
- Aesthetic Improvements
- Property Enhancements using Non-Thermal Techniques
- Property Enhancements using Thermal Techniques

With respect to process control, post-processing must be performed identically from build to build to reach consistent part performance and finish for a given additive part. Furthermore, post-processing methods used during product development and qualification of the AM procedure parameters must be representative of the final component post-processing parameters and the methods to ensure that the performance of the final product matches that of the qualification specification (AMSC 2017).

Due to its effects on the stability of material and part performance, post-processing should be a key feature of calibration and qualification for additively manufactured products. Additive manufacturing currently relies heavily on the presence of human operators for many steps in the manufacturing process; one of these main steps is post-processing. A typical day for an additive operator consists of: removing parts and build material (supports) from parts, cleaning up and resetting the machine for the next build or batch. Although the automation of these tasks is almost certain in the future, currently many of these tasks are manual. Some of these tasks include delivering and loading of raw materials, removal of waste material, final post-processing, testing and quality check of the product produced (The National Research Council of the National Academies 2014).

It is advantageous to have an experienced additive operator since he or she is familiar with the process and can easily identify issues in parts and differences from batch to batch. Slight defects in parts depending on processes can predict issues with the machine that may need preventative maintenance or replacement of parts, e.g., recoater blade, laser power, laser window, etc. Operators are very much still required for inspection and post-processing although automated systems for various post-processing steps are under development. Limiting human error by using autonomous systems for process control, adds complexity by reducing the flexibility of additive manufacturing as a process. The limited freedom of additive manufacturing to be turned around and used for another type of product then restricts the entire manufacturing line. A trade-off decision should be made on process control and line flexibility. Currently, additive manufacturing may require a lot more monitoring during production compared to other more established forms of production systems (The National Research Council of the National Academies 2014).

Gage repeatability and reproducibility should be considered during all stages of product development to measure variables in post-processing from operator to operator and allow suitable tolerances to meet product requirements. Operator handling of the product can greatly influence whether a product is in or out of specification particularly with abrasive post-processing or by removing the part simply from the build platform or chamber. Some post-processing steps can be very labour intensive and require skilled and experienced operators.

Designing an efficient process with an effective process control approach is dependent on the process knowledge and understanding obtained from operators with good knowledge of additive systems and post-processing systems; process design can push operators to their limits and capabilities. Design of experiment (DOE) studies can help develop process knowledge by revealing relations between the variable inputs (e.g., component characteristics or process parameters) and the resulting outputs (e.g., in-process material, intermediates, the final product and handling) (Westphalen et al. 2003). Food and Drug Administration (FDA) does not generally expect manufacturers to develop and test the process until it fails. However, worst-case scenarios of potential product builds are expected to be tested. This can also be said for each worst case of every post-processing step required to get a product in specification (FDA 2017).

Due to the numerous approaches of building AM polymer parts and the diverse effects each may have on the final materials, ensuring a consistent and reliable method of post-processing calibration will provide a method of correlating these parts across AM machines. The need for post-processing standard operating procedures must be created in the development stages of a new product to be produced using AM. These can be helped along with results from DOE experiments, results from different AM machines (selective laser sintering or vat polymers) and material characteristics. Procedure instructions, which are provided with the material on delivery, should be followed for consistent post-processing to allow for qualification and production builds. These methods are process and material specific, and should seek to express minimum best practices for part qualification and production builds.

Currently, there are no available standards for additive polymers or additive polymer machines to follow, but standards for polymers are in development. The main ones in development currently are for polymer extrusion methods; there is still standardisation required for powder and vat polymer technologies although some metal powder guidelines could be followed for powder-based polymers: ISO/TC 61/SC 9 WG: Additive manufacturing for plastics, ISO/TC 261-ASTM F 42 group: standard specification for extrusion-based additive manufacturing of plastic materials (International Organization for Standardization n.d.).

10.2 Medical AM Industry

There are several motives to use AM for medical devices. Among the most prevalent to date are porous surfaces, lattice features, and patient-specific devices that can improve product function and the ability to perform tasks more efficiently and create features or novel designs that cannot be produced in any other way. Another main reason for the use of AM is the manufacturing product volume per annum. AM has the ability to create a product that can treat uncommon ailments that have a lower demand compared to other devices and can still be manufactured in a cost-effective way that makes it feasible to be produced and placed on the market.

International standards recognised by the FDA are vital tools to ensure that the best information is supplied for the assessment of medical devices. Standards for existing methods of design and manufacturing may not contain all the capabilities, parameters, and considerations required for AM. FDA guidance documents provide recommendations for device production and testing as well as regulatory submission obligations. Manufacturers can use recognised industry standards, established methods, or justified scientific conclusions with validated test methods to show the safety, effectiveness, or substantial equivalence of their medical devices (AMSC 2017).

According to FDA guidance, even if the supplier certifies the raw material, the device manufacturer is responsible for the qualification of the final device. Furthermore, according to ISO 13485:2016, the device manufacturer is responsible for establishing and maintaining procedures for the use and removal of manufacturing materials to ensure that the device's quality is not adversely affected (AMSC 2017). This is relevant to AM for a number of reasons. Many of the raw materials used are toxic in their uncured state, particularly photopolymers, and post-printing operations such as support structure removal, conventional machining, polishing operations, sterilisation, etc., can leave residues on the product. These residues may be unsafe to the patient, or they may adversely affect the device's performance that is critical to its quality. The process by which the device or product is made may also have to be FDA approved due to different finishing processes and materials' change in structure during the printing process. Getting an AM process approved by the FDA could take a considerable amount of time.

As there are currently no available standards for post-processing of AM medical polymers, the following standards should be pursued and proven for qualification:

- ISO 13485:2016 describes the need to ensure that the final finished device—including raw materials, pigments, contact materials, etc.—meets the design requirements and does not cause harm to the patient.
- 10993-1:2009—Biological evaluation of medical devices—Part 1: Evaluation and testing within a risk management process.

10.3 Support Removal

Complex curved surfaces, re-entrant features, lattice structures, small and delicate features easily designed and built by a printer can be extremely difficult to remove from a build plate or powder cake. The first step in obtaining a successful part is removing it from its building environment and passing it through a number of hurdles for part qualification. The most common type of post-processing step in AM is support removal. The support material can be broadly classified into two categories:

(a) Loose support material which surrounds the part during the build process such as unsintered powder or support material
(b) Solid supports structures which are designed and built to support, restrain or attach the part being built to a build platform that is attached to the product and usually made from the same material.

A support material is defined as a material that is added for stability during the additive manufacturing process and can either be a separate support material or made out of the final product material. When there are moving parts in an item, or small details or blind holes/cavities that are crucial for the functioning or quality of the part, the residues of support material or other contamination can render the object, or the whole item non-functional, leading to undesired economic losses or part failures. The cleaning of printed parts can be complex and time-consuming, taking up to 30% of the total production time (Verhaagen et al. 2016).

10.3.1 Loose Support

In processes where the part being built is fully enclosed in the build material, the part must be removed from the surrounding material before proceeding further. Processes which provide natural supports are primarily powder-based processes such as selective laser sintering (SLS) or polymer binder printing. After the part is built, it is typically necessary to allow the part to go through a cool-down stage to prevent warpage and significant shrinkage, particularly for SLS. The part should remain embedded inside the powder to minimise part distortion due to uncontrollable cooling. The cool-down time is dependent on the build material, the size of the part, its ability to retain heat and the overall time it takes to print the part. Once cool-down is complete, there are several methods used to remove the part from the surrounding loose powder. Typically, the entire build (made up of loose powder and fused parts) is removed from the machine as a block or powder cake and transported to a 'breakout' station where the parts are removed manually from the surrounding powder material. Nylon bead blasting is used to remove loosely adhered powder, usually for nylon 11 and nylon 12 and pressure can be varied depending on the powder cake (Gibson et al. 2010). Unsintered powder can be sieved and recycled back into the system; this is common for SLS technologies.

Care is needed to remove the powder cake; powder that is fused to a part must go into waste. It can be difficult to decide at what point the powder around the part is of sufficient quality to be recycled and what is for waste. A slight colour difference can be seen in powder closer to the sintered part, which is an indication that the material should be disposed of. As a rule of thumb, any material within 5 mm of the surface of a part should be put into waste. This distance can increase when large wall thicknesses, complex geometries with overhanging features, internal cavities, channels or fine features are used. Design of additive manufacturing should be considered to reduce powder removal time and to de-risk cleaning (Gibson et al. 2010).

10.3.2 Solid Supports

Processes which do not naturally support parts may need additional solid supports (Fig. 10.1) to be added to the part. Examples of such technologies in polymer applications are fused deposition modelling (FDM), SLA, digital light processing (DLP), material or polymer jetting and continuous liquid interface production (CLIP). Solid supports have two categories:

- Build material—final part material: FDM, SLA, DLP, CLIP
- Secondary material—wax or water-soluble material: FDM, material and polymer jetting

Solid supports are required for overhangs and adheres the part to the build plate; they also resist potential distortion. Secondary support materials simplify the removal of synthetic supports as these materials are either weaker, soluble in a liquid solution or melt at a lower temperature than the build material. The orientation of the part in the Z-axis significantly affects support generation and removal. If a thin part is laid flat along the build plate with overhangs that require supports, for

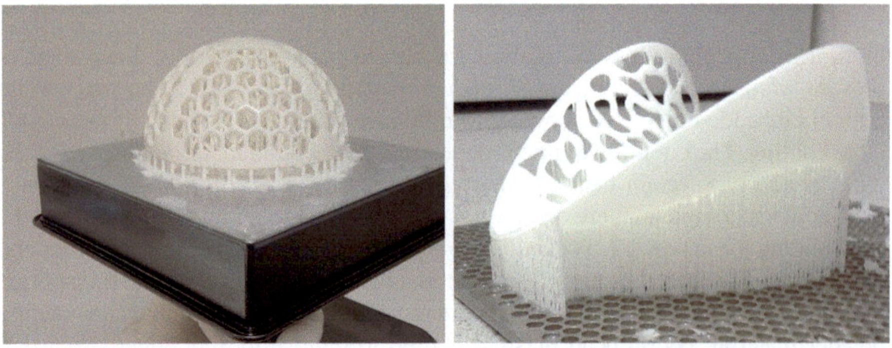

Fig. 10.1 (left) Solid supports on a Formlabs form 2 part and (right) solid supports on a 3D systems viper si2

instance, the amount of support material consumed can at times significantly exceed the amount of build material for the actual part. The orientation of supports also affects the surface finish of the part, as support removal typically leaves a surface defect or marks (small bumps or divots) where the supports were attached. Additionally, the use of supports in regions of small features may lead to these features being broken when the supports are removed. Clever part re-orientation and creating self-supporting features can reduce the amount of supports required in a part, speed up the removal process, and also reduce production costs (Gibson et al. 2010).

10.3.3 Solvent Washing

The removal of supports can be process dependent and can be left on for the washing stage and removed after or vice versa. These steps are very much material dependent. The material used for vat polymers will provide guidelines for solvent washing. Most materials will have to be washed in a solvent after printing to remove excess material; the solvent is usually isopropyl alcohol (IPA), acetone and more recently, dimethylacetamide (DMA) and vertrel XM. The parts are placed in a solvent bath and depending on the material requirements, the bath may be placed on a shaker or left to soak for a designated time listed on the material data sheet or material processing guidelines. Parts and part features should be robust enough to survive washing with automated washers or orbital shakers. Delicate parts can be soaked without movement but it is significantly difficult to remove excess material from these parts.

10.3.4 Ultrasonic Bath

For small parts, usually with internal shaft diameters, an ultrasonic bath can be used. Recommended cleaning solution for this is IPA. Specialist explosion-proof baths must be used to solvent wash with IPA or other solvents as the solvent gas could ignite. The ultrasonic bath should be kept in a well-ventilated area as fumes produced could be hazardous and appropriate personal protection equipment (PPE) should be worn.

Parts must be robust to withstand the ultrasonic bath as it could crack or split the part in the green uncured state where the part is most vulnerable; in post-processing, this is a particular risk for many flexible or durable mimicking photopolymers. It is known that the prolonged effect of bubbles collapsing close to surfaces can lead to erosion or demolition of the surface, depending on different factors, such as the type of cavitation: inertial or stable. For this reason, it is important to ensure that the duration of the cleaning process, and the applied process parameters, such as the pressure, amplitude and frequency, allow for a certain degree of control over the desired cleaning while avoiding damaging behaviour for different materials and different part geometries (Verhaagen et al. 2016).

Care is required during solvent washing as over-exposure of parts to solvents can lead to a reduction in tensile strength and an increase in ductility. Another crucial process observed in chemically loaded parts is chemical corrosion; this can lead to irreversible degradation of the material. Some polymers absorb and retain chemicals; this is checked by measuring the weight of the parts before and after exposure (Safka et al. 2016). Many companies have now created automated solutions for repeatable and consistent agitation of parts to be used in conjunction with their printers to decrease part failures in line and reduce risk to over-exposure.

In addition, an assessment of the toxicity resulting from uncured reagents in liquid resins used during these processes would also be warranted to ensure product and environmental safety during and after production (AMSC 2017).

The solvent being used to wash and agitate parts must also be checked; in some cases, the solvent can be re-used for several cleans before it needs to be changed for fresh solvent. Some materials require the solvent to be changed for every wash; this is essential for part cleanliness. If the solvent becomes too soiled, it becomes extremely difficult to remove excess build residue from parts and can lead to build residue being added to parts; this can result in further contaminated parts, which could then lead to solvent over-exposure of the polymer and cause corrosion and degradation.

10.3.5 Centrifugal Force Cleaning

An effective way to remove excess resins and residual solvents from the cleaning stages for vat polymer resins is to use centrifugal force to displace the contaminants remaining on or in the part. By utilising centrifugal force in a horizontal spinning device and depending on the viscosity and the density of the resin/solvent, centrifugal force and velocity should displace the resin from the part to the wall of the spinning device.

10.4 Post-curing Methods for Polymers

Some AM processes produce cured polymers that require a secondary post-cure operation to further advance crosslinking and improve the materials' functional properties in photopolymers.

These can include:

- increased stiffness
- better chemical resistance
- higher temperature stability
- reduced toxicity (due to reduction of unreacted constituents)
- increased strength
- increase hardness

10.4.1 Ultraviolet (UV) Curing of Photopolymers

AM technologies such as SLA, DLP, material or polymer jetting and CLIP or generic VAT polymer technologies, cures the deposited plastics using focused ultraviolet radiation by light projection or laser. During the build, parts do not achieve optimum polymerization and can dramatically affect the final part performance (i.e., degradation or warpage), especially if these materials are open to sunlight or other radiation sources during use or storage.

After solvent washing, parts will need to be exposed to a UV radiation source usually in a UV oven; this enables the part to meet its peak material properties and makes it more chemically stable. Curing guidelines on recommended cycle times, ovens and a suitable wavelength of UV exposure, usually between 365 and 405 nm, are provided with the material. Some photopolymers may require heat and UV combination to achieve optimum material properties.

UV post-curing of SLA printed resin parts can radically increase the mechanical strength due to the complete curing of the part according to Zguris (n.d.). A study by Dizon et al. found that using a 405 nm wavelength light source produces higher mechanical properties compared with using a lower wavelength of light source. Additionally, post-curing at higher temperatures could also lead to shorter curing time, resulting in higher mechanical properties. Tensile strength was shown to improve with increase in UV post-curing period (Dizon et al. 2018).

There are difficulties in achieving optimum material properties when additive photopolymer parts have varying wall thicknesses; parts may reach a fully cured state in some thinner areas and may not be fully cured in thicker sections. Currently, UV curing is labour intensive and requires an operator to insert the parts into a UV oven or a UV flood lamp, depending on the material manufacturer guidelines and machine guidelines. Parts are usually rotated in the UV oven for a given period; information regarding the maximum time will be supplied in the data sheet of the material. Half way through this cycle time, parts need to be manually taken out and flipped so that the underside of the part can now be UV cured. Cycle times for different photopolymers vary and can range from 60 s to up to 60 min depending on the material. It is important to follow curing guidelines carefully as over-curing can discolour and age the parts; this can affect the parts' mechanical performance and lead to unwanted product failures.

Another major reason why UV curing parts is essential during the additive manufacturing of photopolymers is the cytotoxicity of the polymers when they are not completely cured. In a study by Oskui et al., cytotoxicity testing was carried out on zebrafish. Embryos exposed to SLA-printed parts that were UV-cured fared much better than embryos exposed to parts that were not UV-cured. UV treatment appears to significantly reduce the toxicity of SLA-printed parts to zebrafish; however, it does not completely eliminate the toxicity of these parts (Oskui et al. 2016). Operators are required to take care when handling uncured polymer parts during removal, washing and UV curing for any photopolymers.

10.4.2 Thermal Curing and Heat Treatment of Polymer AM Parts

Some thermoset polymers may also need to be thermally cured to enhance the final polymer crosslinking stage and to increase the parts functional and mechanical performance. In this stage, the part needs to be heat treated in an oven. Temperatures and timeframes will be provided on the material data sheets provided by material manufacturers. It is recommended to submerge parts in salt or similar media so that the polymer part will not warp or distort during this process. For thermosetting, the heat cycle can be constant or variable over a period of time. Material manufacturers usually provide presets and oven recommendations for the given material. Thermal curing some of the photosensitive thermoset polymers increases the material stability and the part becomes less susceptible to UV degradation.

Heat treatment can also be used for parts made using semi-crystalline thermoplastic materials, such as AM polyamides. In this instance, the aim is to reduce residual stresses generated in the part by the AM building process, to minimize warpage, and improve dimensional stability. This is accomplished by allowing the time/temperature profile necessary for the maturation of the crystalline domains in the printed part. The physical and mechanical properties of the finished part can be considerably affected by the degree of crystallization of polymers, which can be controlled by the change of cooling rate during and after the AM process. Heat treatment is an effective method to modify the properties of AM built polymer parts (AMSC 2017).

Heat treatment could also be used as a way to alter a product after building; parts can be formed in moulds during a heat cycle. This can be an effective way to build a part cheaper and quicker and form it after it had been removed from the build chamber and sand blasted.

When heat treating polymer parts, where the product is placed within the oven must be carefully recorded and monitored. Testing and temperature mapping of ovens should be carried out as a precaution; oven may not retain the set temperature or stay constant at the pre-set heat temperature and fluctuate dramatically during a heat cycle.

10.5 Testing of Material Properties in AM Parts

According to the FDA, the type of performance testing that should be conducted on a device made using AM is generally the same as if the device was manufactured using a traditional manufacturing method. Depending on the device type, these may include material property testing such as modulus, yield strength, ultimate strength, creep/viscoelasticity, fatigue, or abrasive wear. Testing should be carried out on the final fully finished devices subjected to all post-processing techniques, cleaning and sterilization steps. As with any recommended testing, the final finished device should be used, or a justification should be provided explaining why a test coupon representative of the final finished device is used for testing.

In addition, the worst-case combinations of dimensions and features (e.g., holes, supports, porous regions) should be considered when determining the worst-case devices for performance testing. An explanation of how the worst-case devices were selected must also be entered in a lab book with recorded testing to back up the submission.

The orientation and build location can affect the final properties of the part and should be considered when conducting device mechanical testing. Orientation can vary based on the manufacturing technology used; a baseline study of the machine and the material combination used may be helpful in determining the degree to which the build orientation will affect mechanical properties. Coupons may be used for material property assessments if there is adequate justification provided for why the coupon is representative of the final device. This justification should consider critical design elements, post-printing processes, cleaning and sterilization as they relate to each type of testing (FDA 2017).

Orientation and build location should be decided at the early stages of design before the product goes to manufacture; carrying out repeated DOEs on the design and figuring out how it affects the part can speed up testing at a later stage. Results from orientation and build location experiments will help determine the sweet spot within any additive machine and will highlight parts that may be sub-standard. This information can then directly feed the justification for the worst-case selected parts. It is difficult to establish what can be tested with coupons and what must be tested on the part. The quality of the part can be strongly affected by the other parts in the build volume or positioning of parts in the build plate area, meaning that careful coupon planning must be considered. Clear guidelines are not yet available for these aspects of coupon use in AM for the medical field, particularly polymers.

While it is desirable to measure the total degree of cure and the crosslink density of the finished part, almost all methods (physical, chemical, mechanical and dielectric) depend on a destructive sampling; these should be carried out during the DOE stages of each of the aforementioned post-processing steps.

These methods include:

- glass transition temperature (Tg)
- differential scanning calorimetry (DSC)
- thermogravimetric analysis (TGA)
- thermomechanical analysis (TMA)
- dynamic mechanical analysis (DMA)

For a non-destructive testing of the cured state of the manufactured part, optical density measurements or surface Fourier transform infrared (FTIR), as used in certain cases of crosslinked polymers, may be applied. FTIR can also be valuable in finding any contaminants within the materials that should not be present. This can be particularly valuable when looking at recyclability of SLS material. Optical measurements would also help to characterize voids and void density and entrapments. A full implementation of this technique, however, would depend on the overall thickness/diameter of the parts (AMSC 2017).

10.6 Surface Finishing of Polymers

Parts built using AM processes may have undesirable or critical surface textures that may require post-build or post-process finishing. These include built surface asperities such as partially fused powder or support marks from vat polymer applications, a degree of striation or stair-stepping typical of the layer process on an inclining surface (AMSC 2017).

In addition, a disparity between core and surface (e.g., contour, upskin, downskin) beam scanning patterns, or non-optimized surface parameters, could potentially produce very small voids or areas filled with unmelted powder or unreacted resin. This can result in subsurface porosity and/or lack of sintering/curing (AMSC 2017). Both surface asperities and subsurface porosity significantly reduce fatigue and fracture properties (Evans et al. 2016). Surface asperities, surface breaking porosity or poorly fused particle boundaries may entrap post-processing solvents and therefore, complicate part cleaning and lead to material corrosion. Complex internal passages may inhibit the finishing and coating of internal surfaces while surface texture may entrain chemical, abrasive, or polishing media. The final surface texture is a complex function of material and process parameters. Parameters such as beam power, build speed and hatch distance, the type of material being processed, layer thickness and build orientation can all impact surface finish (AMSC 2017).

10.6.1 Standards for Surface Finish of AM Parts

There are many challenges to the standardization of surface finishing techniques. Uniform material removal around complex and internal structures is one and can be extremely difficult to achieve. For SLS, the total thickness of material removal that includes both surface asperities and subsurface porosity can be estimated to exceed 250 μm (AMSC 2017). Other important part feature considerations include edge retention and definition, surface roughness variation throughout the length of internal passages, extent of bell mouthing in internal passages, and achieving the required final surface roughness values for exterior and internal or blind features within a part. Standards for reliable evaluation methods, such as CT (computed tomography) scanning with high resolution for evaluation of internal passages and surface roughness, are needed for inspection. Due to its influence on final material performance, current surface finish definitions, measurements and application standards are needed for AM materials.

Surface texture is currently characterized via waviness, roughness and profile. Definitions and interpretations of surface finish specifications are included in the standards listed here.

- ISO 4287:1997 Geometrical Product Specifications (GPS)—Surface Texture: Profile Method—Terms, definitions and surface texture parameters—Amendment 1: Peak count number
- ISO 25178-1:2016 Geometrical Product Specifications (GPS)—Surface Texture: Areal—Part 1: Indication of surface texture
- ISO 25178-2:2012 Geometrical Product Specifications (GPS)—Surface Texture: Areal—Part 2: Terms, definitions and surface texture parameters
- ASME B46.1-2009 Surface Texture (Surface Roughness, Waviness and Lay)
- ISO 1302:2002 Geometrical Product Specifications (GPS)—Indication of surface texture in technical product documentation
- ISO 12085:1996 Geometrical Product Specifications (GPS)—Surface texture: Profile method—Motif parameters
- SAE AS291F—Surface Texture, Roughness, Waviness and Lay (Stabilized: Sep 2014)
- ANSI/ASME Y14.36M 1996 (R2008)—Surface Texture Symbol

10.6.2 Methods of Surface Finishing

There are several methods available for measuring the texture of a surface, including non-contact and contact approaches. These are applicable to a variety of materials, though none are specific to those produced via additive manufacturing.

Validation of surface finish may be particularly difficult on wire-like features. The following list will likely apply to planar or wide surfaces; thin wires or lattice struts do not lend themselves to stylus techniques, and other methods may be required.

Methods include:

- mechanically abrasive techniques
- epoxy filling
- electro plating
- vapour finishing

10.6.2.1 Mechanical Abrasion Techniques

Mechanical abrasion techniques such as shot peening or media blasting are likely be applied easily to AM materials, but may require investigation into their effects on fatigue life when work hardening effects and the amount of material lost per second become significant. This is an essential step in the post-processing of powder-based additive manufacturing. A highly skilled operator is required to blast surfaces consistently; there is thus variability on the surface finish of the part. This can be improved by using jigs and fixtures to control the amount of abrasion on one area of the product, but it is not guaranteed. Care must be taken for assembly parts as media blasting can wear small features used in assemblies such as parting lines. These can be taped or concealed in a jig if blasting multiple items. The part should be continuously rotated to prevent over-blasting in the same area.

Vibratory bowls are beneficial in finishing batches of parts that are essentially a hands-free process; these bowls are mainly used for improving the surface finish of SLS parts as surface finish is significantly lower in SLS polymers compared to VAT polymer methods. Vibratory finishing is used after sandblasting of SLS parts and time for the process is dependent on the surface finish required. It is important to fill a bowl with a ceramic abrasive medium of a size and shape that will fall away from the smallest spaces within 3D printed components. Alternatively, for finishing external surfaces only, larger media should be selected to ensure that none becomes embedded inside the component. Due to the abrasive nature of vibro-finishing, sharp edges, pointed features or raised features from the general surface are rounded and this must be considered for assembly parts. Parts with thin edge detail should not be finished in this way as the feature will be worn away. Similarly, large sections of mesh are unlikely to survive vibratory finishing. Typical or suggested processing time is 4 h, during which 50–100 μm is removed from the surface of the printed components—this is material dependent and can vary. Surface features should not be smaller than a 1 mm cube, while column details of 2 mm cross-section should be regarded as a minimum. Wall thicknesses down to 1 mm may be finished, although this increases to 2 mm where thin and thick sections meet. These dimensions can vary depending on the material being finished. Parts also become more flexible after vibratory finishing, as the plastic absorbs water, particularly nylons. Detergents are used with the water to remove the media taken from the part and flushing it from the system into waste. Chemicals are not dissolved in the water for these applications, as no specific surface cleaning or any other treatment is required (PDJ Vibro bowl LTD n.d.).

10.6.2.2 Electroplating

Non-abrasive methods, such as plating, may also be applicable to AM materials, as these are more dependent on material chemistry. Electroplating is a process in which an electrical current moves metal ions through an acid bath from a source metal (the anode) to the substrate (the cathode). It requires that the substrate have a conductive surface to complete the circuit and transfer metal particles onto the surface. Plating is often applied for cosmetic reasons, but also increases strength and surface durability. In tests, the tensile strength of an electroplated FDM test bar increased 10–12 times and the results of the flexural tests showed an increase of 21–24 times. In order for the metal to properly adhere to the part's surface, it must be extremely smooth; there is also a risk of warping distorting the part with the current passing through (Raja 2016).

Electroless plating is powered by a chemical reaction, as opposed to an electrical circuit. Electroless plating processes provide two advantages over electroplating. First, they can plate on non-conductive surfaces with proper pre-treatment. They do not depend on the geometry of the work piece and provide an even plating coat regardless of the complexity of the part (Raja 2016).

10.6.2.3 Solvent Vapour Smoothing

Solvent vapour smoothing may be applied to some polymeric AM materials. The process is highly dependent on the solvent and material chemistry. Vapour smoothing can address many geometries that abrasive methods cannot; however, it can cause warping and deformation in thin areas of the part being treated. After placing the product in a vapour chamber where it is exposed to the solvent, the part is immediately put in cooling chamber to stop the liquefaction. This cooling ensures that only the surface melts and the object's shape is maintained.

The process also fills any pores in the item's exterior and seals the surface, which makes it especially useful for items intended to hold liquids or gases (Sharretsplating.com n.d.).

10.6.2.4 Organic Coatings: Painting, Priming

Organic coatings (primers, paints) can be employed to improve aesthetics. There are currently no standards associated with these finishing properties. In order to achieve a good surface finish using paint, parts require sanding to a high level to achieve a smooth finish; any blemishes on the surface of the part will be significantly more visible with the application of paints or primers. This would be labour intensive and is not recommended for large batches of manufacturing, unless automated. The part will not achieve an airtight seal on the exterior and can still be susceptible to absorption etc. Organic coating should be avoided as it can be expensive and time consuming to achieve without adding any significant functional value to the product.

10.6.2.5 Dyeing of Additively Made Polymer Parts

In cases where the colour of the AM part is not of the desired colour or is unsuitable for part quality, several methods can be used to improve the parts' aesthetics. Some types of AM parts can be effectively coloured by simply dipping the part into a heated dye water bath of the appropriate colour. This method is limited to parts created from polymer powder bed systems, as the porosity in these parts leads to effective absorption of the dye (Gibson et al. 2010). Suitable additive polymers for dyeing are PA11, PA12 and PA6, and also TPU (DyeMansion GmbH n.d.).

Dyeing can be performed manually, but automated solutions are available; however, colour range is currently limited for automated solutions. Automated colouring systems can provide a repeatable and controllable dying solution that can be easily integrated to any manufacturing line. Market leaders such as Dye Mansion GmbH provide a clean and automated system called the DM60. During this process, the colour pigments and chemicals from the colour cartridges react with the surface of the raw parts and infiltrate into their pores. In this way, the dye reaches areas that otherwise would not be covered via spray painting. It is suitable for consistent dyeing of blind holes and internal channels. A stirrer at the bottom ensures that the parts do not stack, and all surfaces are accessible. Penetration depth of the dyes is around 0.2 mm (DyeMansion GmbH n.d.).

New surface treatments are currently under development for SLS technologies such as PUSh™. The results show that the PUSh™ process dramatically increases surface smoothness and elongation at break in thin specimens. It decreases the surface strength but effects are negligible in larger samples (Crane et al. 2017). This technology has been heavily researched by the Advanced Manufacturing Technologies Unit, University of Sheffield in the UK, which has spiralled into PostPro3D, an automated patent-pending post-processing solution.

10.6.2.6 Surface Textured Parts Directly from CAD (Computer-Aided Design)

The ability of adding a surface texture (Fig. 10.2) digitally to a part directly onto the surface area is achievable only in 3D printing. This is currently available in VAT polymer technologies; it can be applied to SLS and other polymer additive

Fig. 10.2 Knurl texture applied to a part produced via Carbon M2

technologies but due to resolution, the desired surface textures may not be achieved. Previously, adding surface textures to polymer products would be done by inversely texturing the mould tool of the product. This is an extremely expensive and highly skilled task and increases the cost of the mould tool. Surface texturing essentially adds no cost to an additively made part as it is added directly to the CAD of the part. Leaders in this field currently are Carbon 3D, where surface texturing can be added to the part just before printing. Other third party software is available for applying texturing to STL files which is a difficult and tedious task, but choice of finish is unlimited and can be customised by the designer.

There are several unknowns with surface texturing that may need to be explored for medical applications. The following are just a few that could affect part qualification:

- Mechanical part properties
- Tribology performance
- Consistent and even texturing on curved surfaces that may be positioned at an angle to the light/laser source.
- Solvent residue, uncured polymer residue in trapped areas of the surface texture
- Condition of the surface texture after post-processing handling
- Does the surface texturing add to or subtract from the surface of the part, if so how much material is added?

A study from Hong et al. focused on friction and wear of additively made surface textured polymer parts from a multi-jet printer. Under low load conditions, no obvious coefficient of friction behavioural tendency was observed for four types of 3D-printed surface textures (untextured, circles concave, circle convex and square concave). Surface texture does not affect the friction and wear performance of 3D-printed surfaces at lower load conditions. However, they did find that under higher load conditions, the coefficient of friction and wear loss of surfaces with a convex surface texture are higher than those of an untextured surface (Hong et al. 2017). This result can be expected but more research is required in the area and different surface texturing methods and texture need to be investigated.

The benefits of engineered functional surfaces also open the door to new applications other than aesthetic patterning, adding interesting uses and benefits in areas such as biomimicry, hydrophobicity, friction control and flow control (Holmes 2017). The technology also adds benefits for designers and design review, where textures are typically added at the very end of the product development cycle. With the benefits of 3D printing, any iteration or modification can be quickly changed without additional costs or interruption and can be tested the whole way through the product development phase unlike traditional surface texturing on injection mould tools.

10.7 Sterilisation Process Considerations for AM Products

Sterilisation is one of the fundamental processes implemented during the production of all medical devices. The basis for this process is to ensure that the products are free of any potential microbial contamination. Organisms can be removed by exposing them to deadly conditions (e.g., oxidation, intoxication) or by destroying the cell structures that are essential for life (i.e., DNA or protein structures) (Murray 2013). There are a number of published standards that give guidance on the validation of sterilization processes most commonly used for medical devices (AMSC 2017):

- ISO 11135:2014—ethylene oxide sterilization
- ISO 11137 series—radiation sterilization
- ISO 17665-1:2006 (R2013) series—moist heat sterilization
- ISO 20857:2010 (R2015)—dry heat sterilization

Each type of sterilization could potentially affect the mechanical properties of an AM produced part. With the increased rate of development of new materials for additive manufacturing, particularly that of the photopolymer resins, significant research may be required in sterilization of the material unless already stated by the material manufacturer. A thorough investigation should be carried out on material properties post-sterilization particularly for any manufacturer following ISO 13485. Depending on the type of sterilisation involved, it may be accomplished by partial, sub-lethal or repetitive processing, using a representative product and/or biological challenges. It is essential that the entire validation is documented to provide evidence for a high degree of assurance that the specified process will consistently produce a product meeting the predetermined conditions and quality attributes (Murray 2013).

Sterilisation is essential for all devices and products whose end use is expected to impose on the body's defence mechanisms, to come into contact with damaged tissue or to be implanted into the body. It has become increasingly complex due to the requirements of not exposing patients to infections caused by viable organisms present on the instruments/devices during their care (Murray 2013).

After some sterilization process such as with ethylene oxide (EtO), biocompatibility testing is required post sterilization to ensure that no EtO residue remains on the product which could be detrimental to patient health; it is also essential to test for any manufacturing chemicals and residues that may remain on the product from post-processing steps and in-line cleaning chemicals particularly for porous structure products.

10.8 Design Considerations for Post-processing of AM Manufactured Devices

When designing a part to be additively manufactured, considerations must be taken for post-processing requirements within the early design phase.

10.8.1 Design for Post-processing Operations

Most AM parts will require some post-processing. Traditional post-processing methods may not be applicable or may require tailoring to be suitable for AM parts. However, design considerations to facilitate post-processing for AM parts can reduce overall production costs. This may include a means to fixture or index the part as well as ways to reduce or eliminate the need for supports in the build. Mechanical properties may need to be validated to complete qualification and certification of the AM produced medical part. Standards in place and currently being developed for medical devices should aid in the design parameters. Some standards include:

- ISO 14971 Risk Management for Medical Devices
- FDA Technical Considerations for Additive Manufactured Medical Devices
- ASTM WK60265—New Guide for Assessing the Removal of Additive Manufacturing Residues in Medical Devices Fabricated by Powder-bed Fusion

10.8.2 Design for Surface Roughness/Fatigue

The surface roughness of parts is significantly greater when using AM. This can be of significant concern for fatigue-critical parts and gas or fluid flow in internal passages for heat transfer and pressure drop effects. There are several finishing processes that can enhance the surface finish of an additively produced product. Since material may be removed, AM products should be designed oversized to allow for material removal. There currently are no standardised design guides to assist the design team in the new product development (NPD) department in designing for this.

Test coupons can be created and tested to see the effects on the processing of the material and logged into the design history file as a DOE. The thickness of the coupons can be measured before and after the chosen surface finish method to gauge the average amount of material removed. The difference in thickness can be used as a baseline to add material to the product that is removed in finishing processes and the product can reach tolerances set out in the design specification.

10.8.3 Design for Solvent Washing

Most parts 3D printed using vat polymers must be washed or soaked in IPA or other solvents (material dependent) to remove uncured residue form the surface of the part. Depending on the material, the parts may be placed in a solvent bath, an orbital shaker or automated agitation system and exposed in the solvent for a period of time which is defined by the manufacturer of the material. Parts and part features should be designed to be strong enough to withstand the movement during agitation bath and potential impact from other parts within the soak. Good design for additive manufacturing would be the ability to wash the parts while still attached to the build plate to reduce operator handling; this is very much geometry dependent and must be tested for each design to verify and validate the washing of parts.

10.8.4 Design for Thermal and UV Curing

Designers need to understand how post-processing heat treatments can impact the material properties and the intent of the design. Thermal curing can be used to tailor and improve mechanical properties and is an essential post-process step for some polymers as it allows the final crosslinking phase to stabilise the material. Bowing and warping can occur sometimes. During this process, parts should be submerged in salt or other similar media to prevent warping or create a jig while in the oven. Some elastomeric polymers can be thermally cured and reshaped in a jig during the thermal cure phase. Submersion in salt and the making of jigs is totally dependent on part geometry. A DOE should be carried out in the design phase.

During UV curing, some UV cure systems can generate heat up to 60 °C. For thin wall sections, this can cause the thermoset polymer to warp slightly. Shrinkage may also occur, and varies from polymer to polymer.

10.8.5 Design for Inspection

Though AM may enable more complex designs, the need for inspecting critical features and dimensions, including internal surfaces, should be considered in a part and build design. For example, a poorly planned build or design may offset savings in fabrication by increasing the resources needed to verify a part's final dimensions and significantly adding to the product lead time. Including key pieces of geometry to allow for easy datum identification in the printed part can reduce inspection costs.

10.8.6 Design Parts for Safe AM Processing and Post-processing

When designing parts and building plans for fabrication by AM methods, safety must be considered for personnel operating the machines and conducting post-processing tasks. Parts should avoid designs which could trap unused liquid or powder build materials, creating potential safety hazards. Access features, such as holes, slot or vents, may be included to remove or drain out the excess material. Venting in

DLS/DLP/CLIP/SLA processes can prevent any trapped volumes within a part. Some bottom-up printers can be subjected to vacuum due to continuous draw of the build platform in some technologies and venting can be beneficial in preventing part blow outs and deformation during printing. It is best practise to split parts similar to that of injection moulding that may have a trapped volume. This can reduce the post-processing time and reduce the amount of material used within a build. Splitting parts can also reduce the risk of print failure and reduce time spent post-processing and cleaning residues and material wastage.

If parts need to be cut from a build platform, the layout should be planned to reduce the risk of breakage during the removal process and risk of injury for the operator. Avoiding large surface areas on the platform can make the removal process significantly quicker and reduce the risk of damaging parts. Large surface areas should be avoided in polymer printing as it can cause, particularly in light processing, heat build-up during the print. This should be avoided to prevent part distortion and other deformities.

Prior to printing, the build file and build parameters should be reviewed to determine the likelihood of a successful build and to assess the risk to the equipment from the build file and parameters. Measures should also be included to ensure that the design has been optimized to reduce the need for support material potentially reducing build time, and by re-orientating the part to an efficient layout.

10.9 Design Tolerances for AM Polymer Material

Design tolerances are statistically derived material properties based on a defined set of data and analysis methods. They are limits of stress, strain or stiffness that are allowed for a specific material, configuration, application, and environmental condition (Boshers 2001). For the widespread adoption of AM for the manufacturing industry, design tolerances must be developed for both materials and different technologies. The use of design for AM and build testing on production machines during the later prototyping stages during the design phase can help set achievable tolerances for the finalised design specification and speed up the production for a more streamlined product transition to manufacturing. Consideration for manufacturing should always be acknowledged in the design phase of any product development.

The development of standard test methods, specifications and best practice guides are much needed in the industry and will allow for the standardization of additively manufactured materials design data. The information obtained through these standards and specifications can be used for statistical analysis of design tolerances; there is no accepted or approved statistical analysis procedure for additively manufactured materials at this moment in time. Once these design tolerances for each material is established, the application of AM components can be accelerated. For now, as best practise, standards for traditional polymer testing should be considered before a product is produced.

The following test standards and specifications are published for use with additively manufactured polymer materials (AMSC 2017):

- ISO/ASTM 52900:2015—*Additive Manufacturing—General principles—Terminology*
- ISO/ASTM 52901—*Additive manufacturing—General principles—Requirements for purchased AM parts*
- ISO/ASTM 52921:2013—*Standard terminology for additive manufacturing—Coordinate systems and test methodologies*
- ISO 527-4 *Test conditions on isotropic and orthotropic fibre-reinforced plastics composites*
- ISO 527-5 *Test conditions for unidirectional fibre-reinforced plastics composites*
- ASTM F2971-13—*Standard practice for reporting data for test specimens prepared by additive manufacturing*

Standards in development for polymers in AM:

- SAE AMS-AM AMS7100 *Fused Filament Fabrication Process*
- SAE AMS-AM AMS7101 *Material for Fused Filament Fabrication*
- ASTM F42 WK55297 *New Guide for Additive Manufacturing—General Principles—Standard Test Artefacts for Additive Manufacturing*
- ASTM F42 WK53878 *New Specification for Additive Manufacturing—Material Extrusion Based Additive Manufacturing of Plastic Materials—Part 1: Feedstock materials*
- ASTM F42 WK53879 *Additive Manufacturing—Material Extrusion Based Additive Manufacturing of Plastic Materials—Part 2: Process Equipment*
- ASTM F42 WK53880 *Additive Manufacturing—Material Extrusion Based Additive Manufacturing of Plastic Materials: Final Part Specification*
- ASTM F42 WK56649 *New Guide for Standard Practice/Guide for Intentionally Seeding Flaws in Additively Manufactured (AM) Parts*

10.10 Summary

Post-processing should be considered from the very early stages of new product development within the medical device industry. The ability to apply design for manufacturing rules and apply it to the post-processing line can effectively reduce the time in removing a part from the build chamber and bring it to the end user with ease. The ability to 3D print a part is relatively straightforward, but the ability to post-process a part repeatedly and reliably each time to meet qualification standards may prove to be a larger challenge than the product itself. Until standards have been produced to provide clear guidance for medical polymers in additive manufacturing, the currently existent medical device standards should be followed as a manufacturing guideline.

References

America Makes & ANSI Additive Manufacturing Standardization Collaborative (AMSC). (2017). *Standardization roadmap for additive manufacturing*. Washington: Author. (June), Public Draft.

Boshers, C. (2001). Design allowables. In *ASM handbook volume 21, Composites* (p. 1). Novelty: ASM International.

Crane, N. B., et al. (2017). Impact of chemical finishing on laser-sintered nylon 12 materials. *Additive Manufacturing, 13*, 149–155. https://doi.org/10.1016/j.addma.2016.10.001.

Dizon, J. R. C., et al. (2018). Mechanical characterization of 3D-printed polymers. *Additive Manufacturing, 20*(January), 44–67. https://doi.org/10.1016/j.addma.2017.12.002.

DyeMansion GmbH. (n.d.). *DYEMANSION DM60*. München: Author. dyemansion.com.

Evans, N. T., et al. (2016). Impact of surface porosity and topography on the mechanical behavior of high strength biomedical polymers. *Journal of the Mechanical Behavior of Biomedical Materials, 59*, 459–473. https://doi.org/10.1016/j.jmbbm.2016.02.033.

FDA. (2017). *Technical considerations for additive manufactured medical devices - Guidance for Industry and Food and Drug Administration Staff*. Rockville: Author.

Gibson, I., et al. (2010). *Additive manufacturing technologies*. Cham: Springer.

Holmes, S. (2017). *Textures for 3D printed surfaces shows intent of Carbon to bridge design and manufacture workflow*. London: DEVELOP3D.

Hong, Y., et al. (2017). Friction and wear of textured surfaces produced by 3D printing. *Science China Technological Sciences, 60*(9), 1400–1406. https://doi.org/10.1007/s11431-016-9066-0.

International Organization for Standardization. (n.d.). *ISO/TC 261 Additive manufacturing*. Geneva: Author.

Murray, K. A. (2013). *Exploiting the use of electron beam technology to investigate the effects of irradiation on commodity/medical polymers and to optimise process efficiency*. Ph.D. Thesis.

Oskui, S. M., et al. (2016). Assessing and reducing the toxicity of 3D-printed parts. *Environmental Science & Technology Letters, 3*(1), 1–6. https://doi.org/10.1021/acs.estlett.5b00249.

PDJ Vibro bowl LTD. (n.d.). *Guidelines for using vibratory bowls to finish 3D printed parts*. Milton Keynes: Author. https://www.vibratoryfinishing.co.uk.

Raja, K. (2016). A review on chemical processes for plastics substrates used in engineering industries. *International Journal of ChemTech Research, 9*(7), 354–365.

Safka, J., Ackermann, M., & Martis, D. (2016). Chemical resistance of materials used in additive manufacturing. *MM Science Journal, 2016*(6), 1573–1578. https://doi.org/10.17973/MMSJ.2016_12_2016185.

Sharretsplating.com. (n.d.). *Guide to surface finishing for 3D-printed parts*. Emigsville: Author.

The National Research Council of the National Academies. (2014). Chapter: 3 Technical challenges for the use of additive manufacturing in space. In *3D printing in space*. Washington: National Academies Press. https://doi.org/10.17226/18871.

Verhaagen, B., Zanderink, T., & Fernandez Rivas, D. (2016). Ultrasonic cleaning of 3D printed objects and cleaning challenge devices. *Applied Acoustics, 103*, 172–181. https://doi.org/10.1016/j.apacoust.2015.06.010.

Westphalen, D., Roth, K. W., & Brodrick, J. (2003). Fuzzy logic for controls. *ASHRAE Journal, 45*(6), 31. https://doi.org/10.1201/9781420035520.ch3.

Zguris, Z. (n.d.). *How mechanical properties of stereolithography 3D prints are affected by UV curing* (pp. 1–11). Somerville: Formlabs. https://doi.org/10.1126/science.aal0157.

Chapter 11
Regulatory Considerations for Devices Manufactured Using Additive Manufacturing Technologies

John McDonald

Abstract This chapter reviews the regulatory framework for medical devices manufactured using additive manufacturing technologies in various countries and regions. The medical device regulations and guidelines published by these countries stipulate the specific regulatory requirements which medical device manufacturers must fulfil in order to bring their devices into the market. This chapter reviews the current regulations concerning, in particular, CE marking and US FDA regulation, and provides an overview of the Australian, Canadian, Brazilian and Japanese regulations. The review discusses routes of conformity for medical devices, device classification requirements and considerations when making changes to mass-produced medical devices. The methods which manufacturers can use to bring devices, such as custom-made medical devices, to the market will also be outlined. The review found that a regulatory framework currently exists to bring devices manufactured using additive manufacturing technologies to market. This framework, however, is more evolved in some countries in comparison to others, in particular, in relation to testing considerations for devices manufactured using such practices. Additionally, clarity would prove beneficial for manufacturers in relation to, for example, specific regulatory guidance documents from the regulatory authorities.

Keywords Regulatory · CE marking · Custom medical devices · Regulatory guidance

11.1 Introduction

The following chapter aims to provide an overview of the regulatory framework governing medical devices manufactured using additive manufacturing technology. The discussion includes conformity assessment procedures for CE Marking, ISO standards and medical device classification and outlines the requirements imposed

J. McDonald (✉)
Teleflex Medical Euorpe Ltd, Athlone, Co. Westmeath, Ireland
e-mail: john.mcdonald@teleflex.com

D. M. Devine (ed.), *Polymer-Based Additive Manufacturing*,
https://doi.org/10.1007/978-3-030-24532-0_11

by the regulatory authorities in different jurisdictions in relation to devices manufactured using additive manufacturing technologies.

The current medical device regulatory framework was in place before the concept of additive manufacturing. In order to sell a medical device in the European market, a manufacturer must have a CE mark, which is a declaration that the product complies with the Essential Requirements of the European Medical Device Directive 93/42/EEC. By having a CE Mark, the manufacturer gains access to all European Union (EU) and international markets where CE Marking is accepted. The Medical Device Regulation 2017/745 (MDR) was published in the Official Journal of the European Union on 5 May 2017 and will supersede Directive 93/42/EEC. The MDR, which comes effective in May 2020, significantly raises the bar in terms of the regulatory requirements governing the sale of medical devices in the European market. In particular, there are very stringent requirements around clinical evidence, the rules regarding equivalency are considerably more rigorous, and there are more requirements for the manufacturer to perform Post Market Clinical Follow-up (PMCF). In addition, there are many other regulatory requirements to fulfil, including increased post-market surveillance, traceability, labelling, identification of hazardous substances and economic operator additional responsibilities. In order to introduce a medical device into the US market, the manufacturer must comply with the United States Food and Drug Administration (USFDA) Code of Federal Regulations (CFR). Furthermore, should a manufacturer wish to sell their medical devices in Canada, the devices must also comply with the Health Canada Medical Device Regulation SOR/98-282. Other countries may also have their unique national regulations with which the manufacturer needs to comply with in order to licence and sell their medical devices.

11.2 Conformity Assessment

Conformity assessment is the method by which a manufacturer demonstrates that their devices comply with the requirements of Directive 93/42/EEC. A Notified Body performs the conformity assessment process, and when a medical device is deemed to conform to CE mark requirements, the Notified Body provides the CE mark for the device. The routes of conformity are described in the annexes of Directive 93/42/EEC, for example, Annex II (full quality assurance system) or Annex V (production quality assurance). The role of the Notified Body is to provide a conformity assessment service according to Directive 93/42/EEC, taking into account all the relevant guidance documents and standards. The Notified Body is an independent certification organisation that is "notified" by a European Member State's Competent Authority (in Ireland, the Health Products Regulatory Authority (HPRA)) to determine if a medical device meets the applicable requirements for CE marking. For Class I medical devices, the manufacturer declares conformity with the legislation and CE marks the device (self-declaration of compliance). Article 14 of Directive 93/42/EEC describes the registration of persons responsible for placing devices on the market and requires that certain persons placing devices on the market must register their contact details and details of their device(s) with the Competent Authority in the Member State, where they have their registered place of business—in Ireland, this would be the HPRA and in the UK, this would be the Medicines and Healthcare products Regulatory Agency (MHRA).

Although ISO standards are not mandatory for medical devices, standards provide manufacturers with a presumption of conformance to specific regulatory requirements. Manufacturers can utilise these horizontal and vertical standards to comply with regulatory requirements. Recognised, harmonised standards for medical devices are published in the Official Journal of the European Union, which includes the titles of the standard, information regarding whether the standard has superseded another standard and the date of first publication. The European Commission provides a range of European Medical Device Vigilance System (MEDDEV) guidance documents to assist manufacturers in implementing the medical device regulations, which serve to promote a common approach to be followed by manufacturers and Notified Bodies that are involved in conformity assessment procedures. For example, guidance MEDDEV 2.7.1 Clinical Evaluation: A Guide for Manufacturers and Notified Bodies under Directives 93/42/EEC and 90/385/EEC informs the manufacturer on how to prepare a clinical evaluation that will fulfil the regulatory requirements and meet the expectations of the Notified Body.

11.3 Medical Device Classification

Medical devices classifications follow a risk-based approach based on the vulnerability of the human body taking into account the potential risks associated with the devices. Medical devices made from additive manufacturing technologies follow the same classification criteria as medical devices manufactured using traditional manufacturing methods. The medical device classification approach allows the use of a set of criteria that can be combined in various ways to determine classification, for example, the duration of contact with the body and degree of invasiveness. These criteria can then be applied to a vast range of different medical devices and technologies and are referred to as the "classification rules" which are set out in Annex IX of Directive 93/42/EEC and MEDDEV 2.4/1 REV.9. Device Classes range from Class III devices (high risk) to low-risk Class I devices. The other classifications as per Annex IX of 93/42/EEC are Class IIa and IIb devices. In the MDR, the classification criteria are described in Annex VIII, Chapter 1.

The FDA categorises medical devices into one of three classes—Class I, II or III—based on their risks and the regulatory controls necessary to provide a reasonable assurance of safety and effectiveness. Device classification depends on the intended use of the device. Class I devices generally pose the lowest risk to the patient and/or user, and Class III devices pose the highest risk. Each class of device is associated with specific regulatory requirements. At the lowest level, Class I devices are subject to "general controls," which include requirements for registration, good manufacturing practices, medical device reporting and, in some cases, pre-market notification. Since they represent the lowest risk, many Class I devices are exempt from certain requirements. Class II devices also have general controls and must contend with additional special controls, such as specific performance standards and surveillance requirements. Class III devices require general controls and additionally must have pre-market approval from the FDA.

11.4 Current Regulation Requirements Regarding Additive Manufacturing Devices

Specific regulations governing medical devices produced by 3-D printing are limited, and for the foreseeable future, the UK, EU, Canada and Australia are treating such devices as custom-made devices (Schuh and Funk 2019). However, in line with ongoing advancement in the field of additive manufactured medical products, the FDA continues to communicate with the public regarding regulatory considerations for such products (Ricles et al. 2018).

Indeed, certain countries have taken steps to provide guidance to relevant stakeholders regarding bringing medical devices to market which utilise additive manufacturing technologies. For example, the FDA has published a guidance document for Industry and FDA staff entitled "Technical considerations for additive manufactured medical devices, 2017". In Australia, the Australian Therapeutic Goods Administration (TGA) is also consulting with the relevant stakeholders to provide an opportunity for the public to contribute to the development of a regulatory framework for personalised medical devices, such as those enabled by 3-dimensional (3D) printing. Comments which emerge from this consultation process will enable the TGA to identify issues and ensure that the regulation of the devices is sufficiently robust and appropriate to the level of risk. The consultation is seeking options to (1) potentially re-define custom-made devices, (2) introduce new definitions within the personalised medical devices vocabulary, (3) clarify entities who may be liable as a manufacturer of a supplied device, (4) harmonise regulatory approaches with other jurisdictions and (5) include special provisions for certain dental and other low-risk devices to reduce unnecessary regulation burden (TGA Consultation paper 2017).

The EU, Brazil's ANVISA and Health Canada have not yet published any guidance document for additive manufacturing technologies and the Japanese regulatory agency (PMDA) has released some technical guidance around the "evaluation criteria for orthopaedic surgical implants manufactured using 3-dimensional layering technology." While this document from the PMDA (available in Japanese only), provides technical guidance for medical devices that are produced using 3D printing technologies, it does not address the regulatory pathway for such devices.

In rare cases, additive manufacturing technologies may raise questions of safety and/or effectiveness (Technical Considerations for Additive Manufactured Medical Devices, FDA Guidance 2017). The FDA has consequently developed a guidance document outlining the Agency's initial thinking on the technical considerations specific to devices using additive manufacturing, the broad category of manufacturing encompassing 3D printing. 3D printing has become a leading manufacturing technique in healthcare and medicine for a wide range of applications, including dentistry, tissue engineering and regenerative medicine, engineered tissue models, medical devices, anatomical models and drug formulation (Liaw and Guvendiren 2017).

Under the current European CE mark regulation, all objects that meet the definition of a medical device, including those that are manufactured with the use of additive manufacturing technologies, are subject to the regulation. As per Directive 93/42/EEC, a "medical device" refers to any instrument, apparatus, appliance, software, material or other article, whether used alone or in combination, including the software intended by its manufacturer to be used specifically for diagnostic and/or therapeutic purposes and necessary for its proper application, intended by the manufacturer to be used for human beings for the purposes of:

- diagnosis, prevention, monitoring, treatment or alleviation of disease,
- diagnosis, monitoring, treatment, alleviation of or compensation for an injury or handicap,
- investigation, replacement or modification of the anatomy or of a physiological process,
- control of conception and which does not achieve its principal intended action in or on the human body by pharmacological, immunological or metabolic means, but which may be assisted in its function by such means.

The basic principle of 3D printing is that the part is formed by laying down material, layer upon layer until the final design is complete. This is accomplished without a mould and is the opposite of conventional subtractive manufacturing, such as milling where an object is formed by cutting a shape from a larger piece of starting material (TGA). 3D printing is a type of additive manufacturing, and although there are several types of additive manufacturing, the terms 3D printing and additive manufacturing are often used interchangeably. 3D printing is a process that creates a three-dimensional object by building successive layers of raw material, with each new layer attached to the previous one until the object is complete. Objects are produced from a digital 3D file, such as a computer-aided design (CAD) drawing or a Magnetic Resonance Image (MRI) (3D Printing of Medical Devices, Food and Drug Administration).

Mass produced 3D printed medical devices following additive manufacturing processes are subject to the same regulatory requirements as conventionally manufactured devices. In the case of CE marking, for devices above Class I, the manufacturer must have a Quality Management System that is certified by the European notified body and the manufacturer must have evidence that each device it produces complies with the relevant Essential Requirements or Annex I of the MDD (under MDR this will be the GSPRs—General Safety and Performance Requirements). The European notified body must review this evidence (in the format of a technical file or design dossier). The manufacturer is furthermore required to undergo on-site audits by the certifying notified body.

The medical device Technical File or Design Dossier (prepared for CE marking) is a compilation of evidence generated by the manufacturer to demonstrate conformity with the Essential Requirements of Directive 93/42/EEC. The Technical file includes a description of the device in addition to supporting data to demonstrate safety and performance. Included within the technical file documentation is the Essential Requirements document itself (Annex I of Directive 93/42/EEC), which lists each regulatory requirement with which the medical device must conform. Within the Essential Requirements document, the manufacturer documents the standards or other means it may have used to demonstrate conformance to the requirement along with reference to the supporting evidence. This document connects the regulatory requirements to the supporting evidence and shows the methodology (e.g. the ISO standard) used to collate the evidence. The technical documentation will also house the supporting validations including sterilisation validation documentation, biocompatibility test reports, clinical evidence, labelling and instructions for use, packaging validations and risk management documentation.

11.5 Custom-Made Medical Devices

Directive 93/42/EEC and the MDR stipulate that while custom-made devices are not required to be CE marked when they are first placed on the market, they must meet the relevant requirements of the directives that apply to them. In addition, manufacturers of custom-made devices do not require the intervention of a notified body. Directive 93/42/EEC defines a "custom-made device" as any device specifically made in accordance with a duly qualified medical practitioner's written prescription which gives, under his responsibility, specific design characteristics and is intended for the sole use of a particular patient. The above-mentioned prescription may also be issued by any other person authorised by virtue of his professional qualifications to do so. Mass-produced devices, which are adapted to meet the specific requirements of the medical practitioner or any other professional user, are not considered custom-made devices. The MDR adds further clarity to the definition of a custom-made device and defines it as "any device specifically made in accordance with a written prescription of any person authorised by national law by virtue of that person's professional qualifications which gives, under that person's responsibility, specific design characteristics, and is intended for the sole use of a particular patient exclusively to meet their individual conditions and needs". Mass-produced devices, on the other hand, which are adapted to meet the specific requirements of any professional user and devices that are mass produced using industrial manufacturing processes in accordance with the written prescriptions of any authorised person are not considered custom-made devices. Both Directive 93/42/EEC and MDR make it clear that mass-produced devices using additive manufacturing technologies are not considered as custom-made devices.

The Health Canada Medical Devices Regulation SOR/98-282 defines a custom-made device as a medical device, other than a mass-produced medical device, that (a) is manufactured in accordance with a healthcare professional's written direction giving its design characteristics, (b) differs from medical devices generally available for sale or from a dispenser, and (c) is (1) for the sole use of a particular patient of that professional or (2) for use by that professional to meet special needs arising in the course of his or her practice.

When 3D printed medical devices were first introduced, Australia and other jurisdictions introduced exemption provisions for regulating custom-made medical devices to cover individual cases where commercially available mass-produced products were inadequate or insufficient to fulfil the needs and requirements of a particular patient. For example, the Australian provisions for custom-made devices exempt them from the necessity of being included in the Australian Register of Therapeutic Goods (ARTG). As a result, manufacturers and sponsors are exempt from many of the conventional regulatory requirements, such as manufacturing site inspections and the need for third-party certification. These provisions were based on the assumption that such devices would primarily comprise low-risk medical devices. While this assumption regarding risk classification was accurate at the time the provisions for custom-made medical device were introduced, the situation has since evolved (TGA). Currently, 3D printed implants or surgical tools made for a particular patient or physician fit the Australian definition for custom-made devices. As a result, such devices are exempt from being included in the ARTG and must comply with the requirements under the conformity assessment procedures for custom-made medical devices specified in the regulations (TGA).

Similar to conventionally manufactured mass produced medical devices, 3D printed mass-produced medical devices are not custom made and as such are regulated under the existing regulatory framework. High-risk implantable medical devices are manufactured under strictly controlled conditions and are subject to the defined pre-market testing and regulatory requirements to ensure that they conform to the essential principles for safety and performance. However, such strict regulatory oversight is not the case with a majority of similar high-risk 3D printed implants in Australia, which are currently captured under the exemption for custom-made medical devices (TGA). Till date, marketed medical products manufactured using additive manufacturing have been reviewed and regulated under the pre-market notification 510(k) process in the USA with more than 80 devices having received 510(k) clearance (Ricles et al. 2018). The cleared medical devices include orthopaedic implants, surgical guides, dental devices and devices for craniofacial applications.

11.6 Changes to Mass-Produced Medical Devices and Their Manufacturing Processes

For any mass produced medical device, manufacturers should rely on existing guidance documents for the appropriate regulatory pathway, for example, the US FDA's guidance document for Deciding When to Submit a 510(k) for a Change to an Existing Device. As per the FDA guidance document, any changes to a legally marketed device should consider the role played by the Quality System (QS) regulation, 21 CFR Part 820, regarding changes to devices. In the case of certain types of changes to devices, the Agency believes that submission of a new 510(k) is not required and that reliance on existing QS requirements is the least burdensome approach to reasonably assure the safety and effectiveness of the changed device. Regardless of whether a change requires pre-market review, the QS regulation requires manufacturers of finished medical devices to review and approve changes to device design and production (21 CFR 820.30 and 820.70) and document changes and approvals in the device master record (21 CFR 820.181). Any process whose results cannot be fully verified by subsequent inspection and testing must be validated (21 CFR 820.75), and changes to the process require review, evaluation and revalidation of the process where appropriate (21 CFR 820.75(c)). The net effect of the QS regulation is to require that when manufacturers of a finished medical device make a change to the design of a device, there is a process in place to demonstrate that the manufactured device meets the change in design specifications.

With regard to changes to CE marked devices, manufacturers may follow NBOG 2014-3 (NBOG 2014) Guidance for manufacturers and Notified Bodies on reporting of Design Changes and Changes of the Quality System, which provides guidance to the manufacturer regarding what changes fulfil the criteria for a substantial change submission. For example, a change to the manufacturing process, facility or equipment that affects the safety or performance of a device is described as a substantial change and may require submission to the Notified Body. The manufacturer should discuss this topic with their Notified Body and agree on expectations regarding substantial change submissions.

11.7 Additive Manufacturing Device Performance Testing Considerations

The type of product performance testing carried out on devices following additive manufacturing processes is generally the same as if the device was manufactured using a traditional manufacturing method (Technical Considerations for Additive Manufactured Medical Devices, FDA Guidance 2017). Moreover, the information, characterisation and testing necessary for a device made through additive manufacturing depends on a variety of factors, including, but not limited to, whether it is an implant, an instrument, single-use versus reusable, load-bearing, and/or available in pre-specified standard sizes or is patient matched (Technical Considerations for Additive Manufactured Medical Devices, FDA Guidance 2017).

The manufacturer may follow the requirements as specified in Annex I of Directive 93/42/EEC for CE marking and relevant requirements. These regulatory requirements will guide the manufacturer towards what testing requirements one should consider. The requirements of performance testing include material characterisation and biocompatibility, sterilisation (if provided sterile), packaging and transportation considerations, usability, clinical evaluation and other performance criteria. Such criteria can be met by demonstrating conformance to ISO standards. For example, compliance to ISO 10993 can be used for material chemical and physical characterisation and biocompatibility evaluation, while compliance to ISO 11607 (ISO 2006) can be used for packaging performance and compliance to EN62366 (ISO 2015) can be used to demonstrate usability requirements.

The ISO 10993 set of standards, the parent document being "Biological evaluation of medical devices—Part 1: Evaluation and testing within a risk management process" (ISO 10993-1:2009), provides a framework and set of tests from which the manufacturer can select the appropriate recognised tests to demonstrate compliance. For example, in the case of material chemical characterisation, part 18 of ISO 10993 (ISO 2005) provides several options for the manufacturer to chemically characterise the medical device and determine patient exposure. Part 13 can be used to identify and quantify degradation products from polymeric medical devices should such degradation products be expected to occur. Indeed, ISO 10993 (ISO 2018) (Part 1) outlines a set of biological evaluation test strategies that the manufacturer can select to evaluate the medical device biocompatibility properties. Testing should always be performed on the final, finished device as presented to the user. Biological evaluation tests include tests, such as sensitisation, irritation, systemic toxicity, chronic toxicity, genotoxicity, implantation and hemocompatibility. Of course, all the tests do not apply to every medical device, and consequently, the standard provides a set of criteria based on the nature of body contact and duration of medical device use that manufacturers can follow when selecting relevant biocompatibility tests. The nature of body contact would be considered if the device was in contact with skin, mucosal membrane, tissue or blood for example and duration of device refers to how long the device was in contact with the patient. For example, some devices are designed for use for a short period of time (minutes), as in the case of an intermittent urinary bladder catheter while other devices are intended for use for a more extended periods of time (number of weeks), as in the case with an indwelling urinary catheter. In general, the more invasive the device and the longer the duration of contact with the patient, the more rigorous the biocompatibility testing requirements.

The medical device must also demonstrate that it is usable by the end user, which will require the manufacturer to perform a usability study or validation. The usability engineering (human factors engineering) process permits the manufacturer to assess and mitigate risk associated with correct use and use errors, that is, normal use. This may be achieved by demonstrating compliance to EN62366 (Medical devices—Application of usability engineering to medical devices).

Note that the performance testing may follow FDA published consensus standards[1] if it is planned to market the medical device in the US.

Product performance testing may include material property testing including, for example, tensile strength, fatigue and abrasive wear testing. Testing should be performed on the final product as presented to the user. If this is not possible, a rationale should be provided as to why the test article is representative of the finished product. Worst case device configurations representing the most significant challenge for the device should be considered when selecting products for testing. As per the FDA guidance and nature of additive manufacturing, devices may have an orientation relative to the build direction and location within the build space. Indeed the orientation and build location can affect the final properties and should be considered when conducting device mechanical testing. Specifically, the build orientation (including worst-case orientation) of devices or components should be identified for each performance test and if the orientation changes with device size or design, the worst-case orientation should be identified for each configuration. Since the effect of orientation can vary depending on the manufacturing technology used, a baseline study of the machine/material combination used may help determine the degree to which the build orientation will affect mechanical properties (Technical Considerations for Additive Manufactured Medical Devices, FDA Guidance 2017).

Device dimensions can also be affected by orientation and location within the build space. The FDA guidance recommends that dimensional tolerances and dimensional measurements are specified for the worst case configurations. Variability should also be addressed and samples obtained by multiple build configurations.

11.8 Conformity Assessment Procedure for Custom-Made Device

Custom-made medical devices must also comply with the relevant essential requirements of Directive 93/42/EEC to ensure adequate safety and performance. In the case of custom-made devices for special purposes, the manufacturer may follow Annex VIII of Directive 93/42/EEC and draft a declaration of conformity before placing that custom-made device on the market. The regulatory requirements for custom-made devices in Europe as set out in Annex VIII of Directive 93/42/EEC require the manufacturer to prepare a statement which includes the following elements: name and address of the manufacturer; data allowing identification of the device in question; statement that the device is intended for exclusive use by a particular patient, in addition to the name of the patient; name of the medical practitioner or other authorized person who issued the prescription and, where applicable, the name of the clinic concerned; specific characteristics of the product as indicated by the prescription; statement that the device in question conforms to the essential

[1] https://www.accessdata.fda.gov/scripts/cdrh/cfdocs/cfStandards/search.cfm

requirements set out in Annex I and, where applicable, indicating which essential requirements have not been fully met, together with the grounds.

In addition, as per Annex VIII of Directive 93/42/EEC, the manufacturer must also undertake to make the following available for the competent national authorities: documentation, indicating manufacturing site(s) and allowing an understanding of the design and manufacture and performances of the product, including the expected performances, so as to enable the assessment of conformity with the requirements of the Directive. For custom-made devices, the manufacturer must undertake to review and document experience gained in the post-production phase, including the provisions referred to in Annex X (clinical evaluation), and to implement appropriate means to apply any necessary corrective action. This undertaking includes an obligation on the part of the manufacturer to notify the competent authorities of the following incidents immediately on learning of them and the relevant corrective actions: (1) any malfunction or deterioration in the characteristics and/or performance of a device, as well as any inadequacy in the labelling or the instructions for use which might lead to or might have led to the death of a patient or user or to a serious deterioration in his state of health and (2) any technical or medical reason connected with the characteristics or performance of a device for the reasons referred to in subparagraph (1) leading to systematic recall of devices of the same type by the manufacturer.

11.9 Summary

Mass produced medical devices manufactured using additive manufacturing technology follow the same regulatory requirements and submission expectations as the classification and/or regulation followed by devices manufactured using conventional practices. In Australia, the TGA has taken some action regarding the regulatory approaches currently available for devices manufactured utilising additive manufacturing technologies and have taken such action considering the lack of third party or regulatory oversight of custom-made devices and their manufacture and the increasing supply of high-risk devices under the custom-made device exemption. Some of the concerns are that the current regulatory framework does not adequately mitigate risks to patients receiving implantable devices that fit the definition of custom made (TGA Consultation paper 2017). In the US, more than 80 devices utilising additive manufacturing technologies have been cleared using the 510(k) process.

The medical device regulation is evolving and will continue to evolve to embrace the new opportunities and challenges presented over time including additive manufacturing technologies and how to bring devices following such manufacturing processes to market. Although, a regulatory framework exists that enables manufacturers to bring medical devices manufactured using additive manufacturing technologies to market; more regulatory guidance would be helpful to guide manufacturers through the regulatory requirements, highlighting the specific areas of focus in particular for these technologies. Indeed, Liaw and Guvendiren (2017) suggest that the regulations for conventional medical products could be tailored for 3D-printed products to offer a much clearer framework for this highly customisable manufacturing process, to ensure safety and effectiveness of the products.

References

FDA. (2017). *Technical considerations for additive manufactured medical devices guidance for industry and Food and Drug Administration Staff Document issued on December 5, 2017.* Rockville: Author.

ISO. (2005). *ISO 10993-18:2005 Biological evaluation of medical devices -- Part 18: Chemical characterization of materials*. Geneva: Author.

ISO. (2006). *ISO 11607-1:2006 Packaging for terminally sterilized medical devices -- Part 1: Requirements for materials, sterile barrier systems and packaging systems*. Geneva: Author.

ISO. (2015). *EN 62366 Medical devices - Application of usability engineering to medical devices*. Geneva: Author.

ISO. (2018). *ISO 10993-1:2018 Biological evaluation of medical devices – Part 1: Evaluation and testing within a risk management process*. Geneva: Author.

Liaw, C. Y., & Guvendiren, M. (2017). Current and emerging applications of 3D printing in medicine. *Biofabrication, 9*(2), 024102.

NBOG. (2014). *NBOG's best practice guide-2014-3. Guidance for manufacturers and notified bodies on reporting of design changes and changes of the quality system.*

Ricles, L. M., Coburn, J. C., Di Prima, M., & Oh, S. S. (2018). Regulating 3D-printed medical products. *Science Translational Medicine, 10*(461), eaan6521.

Schuh, J. C. L., & Funk, K. A. (2019). Compilation of international standards and regulatory guidance documents for evaluation of biomaterials, medical devices, and 3-D printed and regenerative medicine products. *Toxicologic Pathology, 47*, 344.

TGA. (2017). *TGA Consultation paper. Proposed regulatory changes related to personalised and 3D printed medical devices. Version 1, November 2017.* Woden: Author.

Chapter 12
Additive Manufacturing: Future Challenges

John G. Lyons and Declan M. Devine

Abstract Since additive manufacturing processes, as we know them today, were invented in the mid-1980s, there has being an upsurge in its use as a bona fide manufacturing technique. Additive manufacturing has entered the space of biomedical applications where teeth retainers and hearing aids are commonly produced by this technique with the latter almost exclusively made so. This appears to be only the tip of the iceberg but there are a number of challenges, in terms of the process and materials, which must be overcome before the additive technique becomes dominant in mainstream manufacturing. As these are overcome, other challenges will arise such as standardisation, inspection, business models and, of course, unexpected consequences. Nevertheless, the advantages of additive manufacturing will ensure that it has a key role to play in biomedical applications.

Keywords Additive manufacturing · Patient specific · Future challenges

12.1 Introduction

Since the dawn of the new millennium, an explosion in computing power, innovative design methodologies, industrialised additive manufacturing and hybridised manufacturing have given us a peek at the potential impact of new technologies on manufacturing industries, global consumers and the environment. Additive manufacturing has the capability of revolutionising entire sectors, redrawing the boundaries of traditional industries and truly globalising and personalising the supply chain, but the technology currently lies at a nexus, and must begin to transition from a much-hyped potentially revolutionary new technology, to a dependable, repeatable manufacturing tool with widespread adoption. Since its inception in the mid-1980s, additive manufacturing has spawned a plethora of thematically related but

J. G. Lyons (✉) · D. M. Devine
Athlone Institute of Technology, Athlone, Co. Westmeath, Ireland
e-mail: slyons@ait.ie

D. M. Devine (ed.), *Polymer-Based Additive Manufacturing*,
https://doi.org/10.1007/978-3-030-24532-0_12

technologically differentiated processes including stereolithography (SLA), fused deposition modelling/fused filament fabrication (FDM/FFF), laminated object manufacturing (LOM), direct light processing (DLP), selective laser sintering (SLS), direct metal laser sintering (DMLS), electron beam melting (EBM), material jetting (MJ), drop on demand (DOD), binder jetting (BJ) and many others.

The proliferation of additive manufacturing and its assimilation into the everyday lives of consumers, professionals and industries will undoubtedly change the way we live, the way we work and the way we experience goods and services, and it will do so in ways both subtle and seismic over the next several decades. The full exploitation of additive manufacturing is levered by freedom of design compared to traditional manufacturing approaches—complexity is possible; however, to reach its potential, additive manufacturing technology must mature in key areas (Fig. 12.1).

Market forecasts, financial forecasts, even weather forecasts are something we encounter every day; technological forecasting is more complex as technology development is influenced by a myriad of direct and indirect drivers—in fact, the only thing we can predict about the future with certainty is that in 20 years' time, there will still be people somewhere, trying to predict the future! With that caveat in mind, this chapter will explore the potential future avenues and challenges for the development of additive manufacturing.

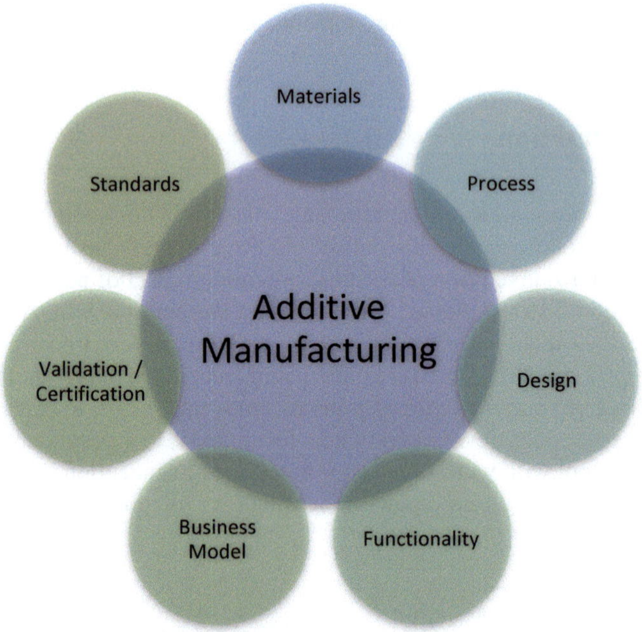

Fig. 12.1 Key areas for development in additive manufacturing

12.2 Process

The wide variety of different approaches to additive manufacturing have resulted in specific processes which each have an extensive range of benefits but which also have inherent weaknesses affecting widespread implementation, for example; heat is widely utilised in most additive manufacturing processes. The mechanism and intensity of the heat applied varies from high in 3D printing of poly etherether ketone (PEEK) to the ambient temperatures utilised for some binder jetting applications. It is this fragmentation of processes that may slow the development of the additive manufacturing sector, as resources both in academia and industry continue working in multi-disciplinary fields utilising varying technologies under the catch-all term of 'additive manufacturing'. However, it has been predicted that the industry will move from this fragmented approach to a partnership model through mergers and acquisitions (3D Printing for Healthcare Applications 2016). Despite this diversity, it is possible to describe (with some generalisations) the challenges affecting future process development. One of the main competitive advantages traditional manufacturing has over additive manufacturing is the speed of production. Low-volume production using additive manufacturing is faster today than conventional manufacturing with lower (and often non-existent) tool costs, quicker lead times and few resource constraints. However, higher volumes are significantly slower, and 3D printers simply cannot compete with injection moulding and casting machines when it comes to the speed of high-volume manufacturing. Upscaling the productivity of additive manufacturing systems is a necessary step to overcome cost hurdles in series production. New systems and new approaches to engineering, chemistry and design are quickly transforming additive manufacturing: although Carbon's SpeedCell™, Hewlitt Packard's MultiJet Fusion, Arburg Freeformer and Stratasys multi-cell platform all approach the additive manufacturing process differently, they have a common goal—to increase production speed. While these systems are based on layer-by-layer printing, researchers at the University of California, Berkeley have developed 'the replicator' which builds a 3D image by scanning an object from multiple angles and then projecting it into a tube of resin which solidifies under exposure to light. While this technology is limited to small parts, it has enormous potential due its relative speed of production (Kelly et al. 2019). Nevertheless, there are some key areas that have been highly optimised in traditional manufacturing but are only beginning to be explored in relation to additive manufacturing processes. These include:

- Loading and unloading procedures
- Robotic automation
- Statistical process control

All of which may offer gains when applied to emerging technologies.

More widespread adoption of additive manufacturing technologies in the aerospace and automotive industries, as well as in the medical sector will continue to fuel advancements in technology that increase the speed of available technologies. However, the speed that a printer can manufacture a part is intrinsically linked to several factors—all of which will continue to evolve over the next number of years.

12.3 Design

There are few constraints in terms of medical device design when it comes to additive manufacturing—you can literally make anything you want based on routine clinical imaging. However, the functionality of the said part is a completely different matter! Design for a successful additive manufacturing process must take into account both the additive manufacturing technology to be utilised, its limitations and tolerances as well as the materials available. Gao et al. (2015) notes that the new areas open for such explorations include topologies and geometries that take advantage of additive manufacturing, for example, envisioning new types of mechanisms, materials and structures with multi-scale and multi-resolution, for programmable matters (Tibbits 2014), cellular materials (Ajdari et al. 2012), deployable structures (Maheshwaraa et al. 2007) and biomimetic materials (Rim et al. 2011). These can help dictate the reaction of surrounding tissue, promote favourable conditions for implant encapsulation by fibrous tissue where tendons need to glide over the implant and facilitate bone ongrowth for implant anchoring in orthopaedics.

With the renewed interest in colonisation of the Moon and even Mars, additive manufacturing will play a key role in the treatment of these pioneers. With scarce supplies and resources, the ability to print customised implants or devices with little waste will be essential. It is conceivable that many of these implants will be produced from biobased polymers grown locally. NASA has been experimenting with additive manufacturing in microgravity and in 2014, in collaboration with the company Made in Space Inc, printed a ratchet wrench on the International Space Station. This illustrated the utility of 3D printing in these environments. The design freedom associated with additive manufacturing in space exploration range from simple CubeSats printed with PEEK, manufacturing and assembling complex structures in orbit to construction of domiciles on moons and planets on future manned missions. In the short term, Allevi's ZeroG extruder and similar innovations will enable printing with biomaterials such as hydrogels or hyperelastic bone in space.

Major software companies like Siemens, Autodesk and Dassault Systems are racing to create in-depth 3D design and engineering tools to unlock the full potential of 3D printing. Software design and simulation tools like Netfabb and Materialise will allow engineers to gain a clearer understanding of the technology and maximise its capabilities for series part production. A challenge for designers will be the development of scalable designs which can be used for in situ printing of regularity-approved implants in the hospital setting. The advent of Industry 4.0 and the Internet of Things will play an important role in this as it will allow the additive manufacturing equipment developed to, not only be connected to the internet at all times but leverage advanced computing power in the cloud as well as have the capacity to access much larger datasets to aid in final design selection. This will in turn improve artifical intelligence to perform more and more complex processing operations, drive defect detection prior to or during the print and speed up time to market utilising generative design methodologies. Designers must continue to work to exploit the geometric and material complexity enabled by additive (and hybrid) manufacturing but in the future, machine learning may allow for additive manufacturing technologies that can learn from their own mistakes and iterate designs for efficiency and performance without significant human guidance.

12.4 Metrology

Additive manufacturing uptake will continue to grow as long as the parts manufactured are capable of meeting the requirements they were designed for. The first step in this process is to confirm their properties utilising metrology. As additive manufacturing enables a degree of part complexity difficult to achieve utilising traditional manufacturing methods, conventional metrology can struggle to accurately characterise the parts, which can lead to quality control and quality assssurance challenges in the highly regulated medical technology sector. As the additive manufacturing tools coalesce in the future into industrially viable approaches, the opportunities for in-line metrology will grow; this will increase the speed of the process and its efficiency, where parts noted to be flawed during production will be dispositioned prior to build completion, thereby freeing up production assets, reducing opportunity cost losses and saving materials.

Inspection and quality assurance has been an important driver of all manufacturing methods for all of recorded history. However, standardisation is not yet well established. The additive manufacturing industry is starting to respond to the need for standardisation at a global level. The uptake and development of these standards is slow due to commercial pressures and the fragmentation of the technologies under the 'additive manufacturing' umbrella. Metrology and the development of adequate and universally applied standards will improve speed and reproducibility of parts. These concerns are much more in the fore of the manufacturing industry than with the 'maker movement' which is based on people creating products for their own use or small-scale sale from their own homes. Nevertheless, this group of people have helped to drive the public understanding of the technology which ultimately has lead to its widespread acceptance. Physical measurement of the parts is important but so too is the modelling of the structures being created, as the parts often contain complex internal structures; and due to the layer by layer nature of the additive manufacturing process, the parts often contain complex microstructures. The prediction of the composition/microstructure and the formation of residual stresses during the process are regarded as two major difficulties in additive manufacturing processes. Having good models of the process is the base for improving the process and having better quality parts. Metrology tools, models and validation procedures for additive manufacturing will help to drive market uptake of the technology, drive approval by regulatory agencies of parts produced and enable consumer confidence in a technology dogged by the term 'prototype'.

12.5 Materials

Functional parts manufactured utilising additive manufacturing require a suitable design, process and crucially, appropriate feedstock materials. A necessary major step for the future evolution of additive manufacturing is to increase the functionality of fabricated products. Again, the fragmentation of the manufacturing technologies available results in material development efforts being spread across multiple areas. But additive manufacturing is now pushing the frontier of new breeds of design approaches and tools.

When it comes to feedstock materials, one major improvement to drive future development is common—standardisation. Batch to batch variations in shape and particle size (powders), viscosity and shelf life (liquids), and ovality and melting point (filaments) are present in many commercially available materials. The supply chain for many of these materials will mature in the short term leading to more repeatable and reproducible qualified materials being available for research and manufacturing. For additive manufacturing to continue to flourish, the maturing material supply chains will need to leverage economies of scale to drive higher production rates of materials which should lower the cost of the materials, incentivising the adoption of the technology.

There is a range of material related challenges to be addressed to continue the development of additive manufacturing. For polymer filament materials, shrinkage after printing which results in part distortion and layer to layer bonding are areas for increased focus. Photopolymerisation-based polymer systems continue to require increased shelf lives, and physical ageing will remain a challenge without further development. Photopolymerised polymer systems also pose a sustainability challenge as they will prove difficult to recycle. The microstructure of the material is hugely influenced by the printing process and the feedstock materials. Material development will need to occur in tandem with high-resolution material-specific analysis materials and mechanical property characterisation to properly develop materials that can repeatedly meet the requirements of commercial manufacture. Material developments most likely to impact the healthcare sector in the medium terms include polymers which are reinforced with graphene and carbon fibre-reinforced and conductive resin-based polymers for use in 3D printing of entire devices with electro or wireless connectiveity which will aid remote in vivo monitoring (3D Printing for Healthcare Applications 2016). One of the promising areas for additive manufacturing is designing and synthesising functionally gradient materials (FGM). The concept of FGM is to change the microstructure and composition gradually such that the performance of the built part is optimal. Such FGM can impart local properties as needed so as to custom-tailor the mechanical, thermal and electrical properties within the same part. FGM material can be designed in a discrete or continuously varying manner. For the former, each layer will have a different material or composition when a part is built such that its properties are gradually changed. For the latter, such a gradual change will be achieved by continuously varying the constituent material composition (Gao et al. 2015).

Hybridised manufacturing and the melding of materials and production methods will help additive manufacturing attain its potential. One way to achieve this is an approach known as functional grading, in which gradients of single or multiple materials are utilised intelligently to meet several criteria simultaneously. Combined approaches will likely feature widely as the technology matures as few applications can be satisfied with materials that cannot achieve a complex set of requirements.

The emergence of additive manufacturing for pharmaceutical and bioprinting applications will blur the boundaries between controlled substances, engineering materials and living tissue. This will lead to challenges for supply chains, storage and handling, and give rise to important ethical debates. While 'bioficial', lab-produced, 3D bioprinted, complex human organs are closer today than ever before, the technology has some way to go. This need for bioprinting is supported by the several emerging application fields within tissue engineering and regenerative medicine (Knowlton and Tasoglu 2016), including:

- organ-on-a-chip for medical research (Knowlton et al. 2016; Lee and Cho 2016);
- replicating diseased tissues, such as tumour-on-a-chip, for investigating the process of diseases (Dai et al. 2016; Mandrycky et al. 2016);
- regenerating human tissues, such as skin, cartilage and bone, for transplantation (Bajaj et al. 2014; Gao and Cui 2016);
- conducting stem-cell research (Tasoglu and Demirci 2013).

Lower cost printers will drive research into materials development. Key focus on the enabling of vascularisation of tissue will help drive the development of technology. It is likely that in the medium term, due to their lower level of intricacy, cartilage and skin will be more easily replicated with more complex organs that require a degree of sophistication not being yet available. Availability of suitable cell lines and matrix materials and a robust mechanism for ensuring the integrity of the research carried out in this realm is vital for the technology to evolve to its potential. Materials for pharmaceutical dosage forms and bone implants manufactured by additive manufacturing are already widely available and have been increasing in their adoption and commercial availability. It is likely that this trend will continue; however, the high cost of appropriate biomaterials continue to stymy research into new approaches with the technology and will likely hamper the propagation of the approach until material supply chains have matured.

12.6 Business Models

With the advent of the internet, consumers have rapidly become involved in the production process—blurring the line between production and consumer, creating prosumers. The 'maker movement' together with the falling prices of consumer 3D printers have leveraged the technology away from 'hobbyist' approaches into more mainstream. The ability of companies to create and monetise their products are governed by their business model and while prosumers will never compete with large companies such as BMW who recently celebrated the millionth part fabricated utilising additive manufacturing, they dream of a world with 'Star Trek' style replicators able to create parts in their homes and thus have the power to subvert traditional business models of supply for medical devices. There is a current trend where surgeons are producing surgical models and jigs within the hospital settings. It is very likely that this will grow in the future with eventually, full implants being produced within the hospitals. These advances are also likely to enter the pharmacy setting where 3D printing of polypills which release drugs in controlled manners can reduce pill burden on patients.

The future of additive manufacturing will have a very direct effect on companies' business models. Increased customisation of parts coupled with patient-specific implants, will drive the use of additive manufacturing in medical devices. This will build on uses such as Align Technology's Invisalign Orthodontics, patient-specific PEEK implants from companies like Xiiloc Medical and the quiet revolution that additive manufacturing has brought to the hearing-aid industry. Changes in production through improved design and rapid tooling will allow for faster time to market and more frequent design changes. Additive manufacturing will allow for direct manufacturing by the consumer (thereby, allowing the company to sell the designs etc to leverage its intellectual property). The supply chain will be revolutionised as parts can be made quicker closer to use; storerooms will be a thing of the past as parts are printed to order and tried and tested optimised production processes such as Kaizen and Just In Time, will need to be re-evaluated. It is likely that a proliferation of online design marketplaces, additive manufacturing bureaus and online resources will emerge with crowdsourced business models to disrupt traditional manufacturing of low-cost items. Rapid tooling will lower the cost and speed up the turnaround time of manufacturing tool creation, enabling limited product customisation for traditional manufacturing and help to drive the bespoken part availability. Indeed, even in the educational sector, traditional models of education will be re-envisioned as additive manufacturing makes the production of models for surgical practice more commonplace, allowing students to see, touch and feel their designs and allows educators the freedom to create instructional tools previously unavailable.

A 2014 Deloitte study pointed to two major areas of influence for 3D printing in automotive applications: as a source of product innovation and as a driver of supply chain transformation. Over the past nearly half-decade, these predictions have shown to be spot-on as new vehicle models come out faster and sleeker, with digital supply chains reshaping logistics. Movements like those embodied by the e-NABLE community, utilising basic additive manufacturing to 3D print open source prosthetics worldwide is an example of how technology can quickly alter how patients and consumers interact with traditionally mass-manufactured advanced goods. Currently, additive manufacturing is best suited to high-value, low-volume parts, but as the technologies continue to develop, this is likely to change and as with all business model shifts, cost versus benefit will drive change.

12.7 Dangers of Technology Development

It would be unwise to not consider the myriad of possible dangers caused by the additive manufacturing revolution. As with all new and emerging technologies, the development of additive manufacturing is vulnerable to the law of unintended consequences (intervention or development in any complex system tends to create unanticipated and often undesirable outcomes). For example:

- Assembling chemical compounds on a molecular level using a chemical 3D printer is possible. In 2012, Professor Lee Cronin, the University of Glasgow's Regius Chair of Chemistry pioneered the development of a prototype 'Chemputer' that allowed the universal assembly of complex molecules on demand, democratising the ability to discover and make new molecules. Of course, the dark side of this technology is the potential for future DIY chemists to create anything from cocaine to sarin.
- The controversy and debate surrounding 3D printed guns reached an inflection point in 2018 after the U.S. State Department reached a settlement that allowed Cody Wilson (a self-proclaimed crypto-anarchist) and his non-profit organization, Defense Distributed, to upload CAD files of functional DIY firearms online, claiming that the initial ban was in violation of the U.S. First Amendment. This ruling was subsequently overturned and Defense Distributed was forced to take down the 3D files. But by then, the organization's website had already supplied thousands of people across the world access to 3D firearm blueprints, and copies of the digital files remain available on the internet even today. In early 2019, the French 3D printer manufacturer DAGOMA launched an anti-3D printed gun campaign called Operation Harmless Guns. Collaborating with the advertising agency TBWA\Paris, the company hopes to thwart the proliferation of 3D printed guns by flooding the internet with hundreds of non-functional models. But the advent of the maker movement and the continuing development of additive manufacturing technology may soon put weapons and other restricted devices within reach of the everyday consumer with little in the way of legislative or technological safeguards.
- Conversations about the moral, ethical, and legal issues surrounding bioprinting have started, but they will inevitably stir a lot more controversy as the technology becomes more advanced and commonplace. Printing human organs on demand, no longer necessitating the need for living or deceased human donation, creating alternatives to animal testing and avoiding species boundary crossing are all undeniably positive applications for the technology, but will the law of unintended consequences lead to a tiered system of therapeutic organ replacement? With those who can afford to pay for self-generated organs living longer; perhaps enjoying a higher quality of life, while others wait until a human organ donor becomes available and then undertake a punishing drug regime for the rest of their lives to prevent episodes of rejection of the transplanted organ. Who owns tissue printed using the technology—the donor of the cells or the owner of the printer? As the technology develops, it is vitally important that ethical implications of technology leaps are understood and legislated for and that robust standards and usage guidelines are implemented.

12.8 Summary

Additive manufacturing has already infiltrated the everyday lives of countless people. However, this is likely to be the tip of the iceberg for additive manufacturing. While Star Trek style replicators may be a long way into the future, versions of this futuristic machine are already available. Many of these advances are aimed at the medical device sector due to the complexity and need for individualisation of medical devices. However, there remains a number of challenges which must be overcome before additive manufacturing can replace traditional manufacturing processes. While overcoming these challenges may increase the reach of additive manufacturing, it is essential that standardisation and legislation keep pace with developments to prevent unintended consequences.

References

Ajdari, A., Jahromi, B., Papadopoulos, J., Nayeb-Hashemi, H., & Vaziri, A. (2012). Hierarchical honeycombs with tailorable properties. *International Journal of Solids and Structures, 49*(11–12), 1413–1419.

Bajaj, P., Schweller, R. M., Khademhosseini, A., West, J. L., & Bashir, R. (2014). 3D biofabrication strategies for tissue engineering and regenerative medicine. *Annual Review of Biomedical Engineering, 16*, 247–276.

Dai, X., Ma, C., Lan, Q., & Xu, T. (2016). 3D bioprinted glioma stem cells for brain tumor model and applications of drug susceptibility. *Biofabrication, 8*, 045005.

Frost and Sullivan. (2016). *3D printing for healthcare applications*. Mountain View: Author.

Gao, G., & Cui, X. (2016). Three-dimensional bioprinting in tissue engineering and regenerative medicine. *Biotechnology Letters, 38*, 203–211.

Gao, W., Zhang, Y., Ramanujan, D., Ramanib, K., Chen, Y., Williams, C. B., Wang, C., Shin, Y. C., Zhang, S., & Zavattieri, P. D. (2015). The status, challenges, and future of additive manufacturing in engineering. *Computer-Aided Design, 69*, 65–89.

Giffi, C., Ganggula, B., & Illanda, P. (2014). *3d Opportunity in the automotive industry*. New York: Deloitte University Press.

Kelly, B. E., Bhattacharya, I., Heidari, H., Shusteff, M., Spadaccini, C. M., & Taylor, H. K. (2019). Volumetric additive manufacturing via tomographic reconstruction. *Science (New York, N.Y.), 363*, eaau7114. https://doi.org/10.1126/science.aau7114.

Knowlton, S., & Tasoglu, S. (2016). A bioprinted liver-on-a-chip for drug screening applications. *Trends in Biotechnology, 34*, 681–682.

Knowlton, S., Yenilmez, B., & Tasoglu, S. (2016). Towards single-step biofabrication of organs on a chip via 3D printing. *Trends in Biotechnology, 34*, 685–688.

Lee, H., & Cho, D.-W. (2016). One-step fabrication of an organ-on-a-chip with spatial heterogeneity using a 3D bioprinting technology. *Lab on a Chip, 16*, 2618–2625.

Maheshwaraa, U., Bourell, D., & Seepersad, C. (2007). Design and freeform fabrication of deployable structures with lattice skins. *Rapid Prototyping Journal, 13*(4), 213–225.

Mandrycky, C., Wang, Z., Kim, K., & Kim, D. H. (2016). 3D bioprinting for engineering complex tissues. *Biotechnology Advances, 34*, 422–434.

Rim, J., Zavattieri, P., Juster, A., & Espinosa, H. (2011). Dimensional analysis and parametric studies for designing artificial nacre. *Journal of the Mechanical Behavior of Biomedical Materials, 4*(2), 190–211.

Tasoglu, S., & Demirci, U. (2013). Bioprinting for stem cell research. *Trends in Biotechnology, 31*, 10–19.

Tibbits, S. (2014). 4d printing: Multi-material shape change. *Architectural Design, 84*(1), 116–121.

Index

D. M. Devine (ed.), *Polymer-Based Additive Manufacturing*,
https://doi.org/10.1007/978-3-030-24532-0

D

MIX
Papier aus verantwortungsvollen Quellen
Paper from responsible sources
FSC® C105338

If you have any concerns about our products,
you can contact us on
ProductSafety@springernature.com

In case Publisher is established outside the EU,
the EU authorized representative is:
Springer Nature Customer Service Center GmbH
Europaplatz 3, 69115 Heidelberg, Germany

Printed by Libri Plureos GmbH
in Hamburg, Germany